KINEMATICS, DYNAMICS AND STRUCTURE OF THE MILKY WAY

ASTROPHYSICS AND SPACE SCIENCE LIBRARY

A SERIES OF BOOKS ON THE RECENT DEVELOPMENTS
OF SPACE SCIENCE AND OF GENERAL GEOPHYSICS AND ASTROPHYSICS
PUBLISHED IN CONNECTION WITH THE JOURNAL
SPACE SCIENCE REVIEWS

VOLUME 100
PROCEEDINGS

KINEMATICS, DYNAMICS AND STRUCTURE OF THE MILKY WAY

PROCEEDINGS OF A WORKSHOP ON "THE MILKY WAY"
HELD IN VANCOUVER, CANADA, MAY 17–19, 1982

Edited by

W. L. H. SHUTER

Department of Physics, University of British Columbia, Vancouver, B.C., Canada

D. REIDEL PUBLISHING COMPANY

DORDRECHT : HOLLAND / BOSTON : U.S.A.
LONDON : ENGLAND

Library of Congress Cataloging in Publication Data

Main entry under title:

Kinematics, dynamics, and structure of the Milky Way.

(Astrophysics and space science library ; v. 100)
"Workshop was co-sponsored by the University of Maryland and the
University of British Columbia"–T.p. verso.
Includes indexes.
1. Milky Way–Congresses. I. Shuter, W. L. H. (William Leslie
Hazlewood), 1936–   II. University of Maryland, College Park.
III. University of British Columbia. IV. Series.
QB857.7.K56   1983   523.1'13   82–24050
ISBN-13:978-94-009-7062-5   e-ISBN-13:978-94-009-7060-1
DOI: 10.1007/978-94-009-7060-1

Published by D. Reidel Publishing Company,
P.O. Box 17, 3300 AA Dordrecht, Holland.

Sold and distributed in the U.S.A. and Canada
by Kluwer Boston Inc.,
190 Old Derby Street, Hingham, MA 02043, U.S.A.

In all other countries, sold and distributed
by Kluwer Academic Publishers Group,
P.O. Box 322, 3300 AH Dordrecht, Holland.

D. Reidel Publishing Company is a member of the Kluwer Group.

This Workshop was co-sponsored by the University of Maryland
and the University of British Columbia

TABLE OF CONTENTS

PREFACE

The idea of holding this workshop on "The Milky Way" arose at the conference dinner of a meeting on "Regions of Recent Star Formation" held at Penticton in June 1981. Leo Blitz (now at the University of Maryland) and I decided that there was a need, and agreed that we would organize one in Vancouver in the Spring of 1982.

The purpose of the workshop was to have an intensive exchange of ideas between some of the most active workers in the field regarding the recent work which has been significantly changing our concepts of the Milky Way. To achieve this we limited the number of participants, and planned the program so that there was ample time for discussion.

The meeting appeared to work very well, both scientifically and socially, and this volume contains 50 of the 55 papers that were presented. The discussion was very lengthy, but since the papers were written up after the meeting many of the points raised have been incorporated in the publications, and it seems pointless to reproduce it here.

Leo and I would like to thank the many people who helped to make the meeting a success:

C.V. Finnegan (Dean of Science at UBC) and Frank J. Kerr (Provost of MPSE at the University of Maryland) who welcomed the participants on behalf of the sponsoring Universities.
Bart Bok who opened the scientific proceedings, and Maarten Schmidt who gave the closing summary.
The Scientific Organizing Committee: Leo Blitz (Chairman), Frank Kerr, Ivan King, Jill Knapp, Jerry Ostriker, Bill Shuter.
The Local Organizing Committee: Leo Blitz, Chris Chan, Bill McCutcheon, Bill Shuter (Chairman) assisted by Arnold Gill, Brian Glendenny, Jean McCutcheon, Kochu Menon, Maureen Ponton, Carmen de Silva, Bev Shuter.
The Invited Reviewers: John Bahcall, Bob Brown, Ken Freeman, Frank Kerr, Jill Knapp, Donald Lynden-Bell, Jerry Ostriker, Pat Thaddeus, Roland Wielen.
The Session Chairmen: Sidney van den Bergh, Butler Burton, Carl Heiles, Ivan King, C.C. Lin, Per Olof Lindblad, Kochu Menon, Vera Rubin, Allan Sandage.

We wish to thank our financial sponsors:
NSERC who provided a conference grant, the UBC Emergency Grants Committee, the Dean of Science at UBC, the Department of Physics at UBC, the National Science Foundation through Frank Kerr's grant AST-8021283 and the Department of Astronomy at the University of Maryland.
Finally, we wish to thank the authors for the prompt preparation of their papers during a very busy Summer.

W.L.H. Shuter
Department of Physics
University of B.C.
Vancouver
Canada V6T 1W5

W. L. H. Shuter (ed.), Kinematics, Dynamics and Structure of the Milky Way, ix.
Copyright © 1983 by D. Reidel Publishing Company.

Dr. John Bahcall, Institute for Advanced Study
Dr. Tom Bania, Boston University
Dr. Frank N. Bash, University of Texas, Austin
Dr. Sidney van den Bergh, Dominion Astrophysical Observatory
Dr. Leo Blitz, University of Maryland
Dr. Hans Bloemen, Sterrewacht, University of Leiden
Dr. Bart J. Bok, Steward Observatory
Dr. Robert L. Brown, NRAO
Dr. W. Butler Burton, Sterrewacht, University of Leiden
Chris Chan, U.B.C.
Dr. Richard S. Cohen, Goddard Institute for Space Studies
Dr. David Crampton, Dominion Astrophysical Observatory
Dr. John M. Dickey, NRAO
Michel Fich, University of California, Berkeley
Dr. M.P. Fitzgerald, University of Waterloo
Dr. Doug Forbes, University of Nebraska, Lincoln
Dr. Ken C. Freeman, Mt. Stromlo Observatory
Dr. Carlos Frenk, University of California, Berkeley
Dr. Phil C. Gregory, U.B.C.
Jon Haass, M.I.T.
Dr. Hugh C. Harris, Dominion Astrophysical Observatory
Dr. Michael G. Hauser, NASA Goddard Space Flight Center
Dr. Carl E. Heiles, University of California, Berkeley
Dr. Lloyd Higgs, Dominion Radio Astrophysical Observatory
Dr. Kimmo A. Innanen, York University
Dr. Ed Jenkins, Princeton University Observatory
Dr. Frank J. Kerr, University of Maryland
Dr. Ivan R. King, University of California, Berkeley
Dr. Gillian R. Knapp, Princeton University
Shrivinas Kulkarni, University of California, Berkeley
Dr. C.C. Lin, M.I.T.
Dr. Per Olof Lindblad, Stockholm Observatory
Dr. Harvey Liszt, NRAO
Dr. Jack L. Locke, Herzberg Institute of Astrophysics
Dr. Felix J. Lockman, NRAO
Dr. Donald Lynden-Bell, Cambridge
Dr. William H. McCutcheon, U.B.C.
Dr. James W-K Mark, University of California, Livermore
Dr. Hans Mayer-Hasselwander, M.P.I. fuer Extraterrestische Forschung
Dr. T.K. Menon, U.B.C.
Dr. Felix Mirabel, University of Puerto Rico
Dr. Jeremiah Ostriker, Princeton University Observatory

Dr. Michael W. Ovenden, U.B.C.
Dr. William L. Peters, University of Texas, Austin
Dr. Maurice H.L. Pryce, U.B.C.
B. Cameron Reed, University of Waterloo
Dr. Harvey B. Richer, U.B.C.
Dr. William Roberts, Jr., University of Virginia
Dr. Vera C. Rubin, Carnegie Inst. of Washington
Dr. Allan Sandage, Mt. Wilson Observatory
Dr. David Sanders, University of Massachusetts
Dr. Maarten Schmidt, California Institute of Technology
Dr. Philip E. Seiden, IBM Research Center
Dr. William Shuter, U.B.C.
Dr. Luis de Sobrino, U.B.C.
Dr. Philip Solomon, SUNY
Dr. Antony A. Stark, Bell Laboratories
Dr. Linda Stryker, Dominion Astrophysical Observatory
Dr. Patrick Thaddeus, Goddard Institute for Space Studies
Dr. Barry E. Turner, NRAO
Dr. Arthur R. Upgren, Van Vleck Observatory
Dr. Simon White, University of California, Berkeley
Dr. Roland Wielen, Technische Universitaet, Berlin
Dr. James Wright, National Science Foundation
Dr. Judy Young, University of Massachusetts
Dr. Chi Yuan, SUNY

NEW TRENDS IN MILKY WAY RESEARCH

Bart J. Bok
Steward Observatory, University of Arizona

I feel honored indeed to have been asked to present the opening remarks at this beautifully planned Workshop on problems relating to the structure, kinematics and dynamics of the Milky Way System. These are exciting days for we have in recent years witnessed remarkable leaps forward in at least five areas of research relating to our Galaxy:

(1) Evidence that our familiar Galaxy, with its central bulge, disk with spiral structure, and long-recognized modest halo is embedded in a huge outer halo or corona with a mass at least five times that estimated for our traditional Galaxy and a radius three or more times the traditional value.

(2) Thanks mostly to advances in instrumentation in the near and far infrared, in radio astronomy, and in X-ray and Gamma-ray astronomy, we are constantly learning new facts about the central regions of our Galaxy, including the very center itself.

(3) The recent CO radio-molecular studies are blowing new life into the study of the spiral structure of our Galaxy. This applies to spiral features in the inner parts (the beautiful results by Cohen et al. deserves special mention) and especially to features that are traceable in H I and in CO in the nearer parts of the outer halo or Corona.

(4) The study of globular star clusters shows great promise for the future, but we must admit that the situation has become rather confusing. A few years ago it seemed as though real advances were being made in the study and interpretation of abundances of the elements from the spectra of stars in globular clusters, but today the whole story seems confused. We have made great advances in research on the dynamics of globular clusters, but much remains to be done on the dynamics of the globular cluster system, especially on the interpretation of the observed motions of the remote globular clusters that may define the outer rim

1

W. L. H. Shuter (ed.), Kinematics, Dynamics and Structure of the Milky Way, 1–7.
Copyright © 1983 by D. Reidel Publishing Company.

of the outer halo or corona of our Galaxy.

(5) One of the healthiest and most active fields for research on
our Galaxy relates to star birth and early evolution. Here radio-
molecular, near and far infrared work, and research based on
observations in the old-fashioned optical region, are continuing to
reveal new vistas. We seem to have succeeded pretty well in pin-
pointing the places in our Galaxy where the action lies.

It is well for me to mention briefly the books and survey
articles that describe the work now in progress plus future Milky Way
research. I find it easy to do so, for in 1981 I published the Fifth
Edition of the book by Priscilla and myself, THE MILKY WAY. You may
remember that we published the Fourth Edition in 1973 (two years before
her death) and, when it was done, she and I expected that the need for
the fifth would not occur until the later 1980's. Far from it. By
1978 it was clear that a new edition was urgently wanted, for the fourth
was out of date five years after it was published. A major overhauling
was clearly in order.

My friend, Dimitri Mihalas, felt apparently the same way about his
book GALACTIC ASTRONOMY, for which a fine second edition appeared in
1981, this one written jointly with James Binney. There will be two
volumes: The first one deals with structure and kinematics, the second
(by Binney and Tremaine) with dynamics.

There are other books, papers and reports that help the worker in
the field find his way about. I like to refer often to the recent
volume on THE STRUCTURE AND EVOLUTION OF NORMAL GALAXIES, a series of
lectures edited (for NATO) by Fall and Lynden-Bell.

Very soon we shall all receive the IAU Reports for Commissions
which will be discussed at the Patras General Assembly. Watch
especially for the Reports of Commissions 24, 25, 28, 30, 33, 34, 37
and 40; over the past few years I have found many uses for the special
Long Report for Commission 33 prepared in 1979 by Frank Kerr for
Montreal and I look forward to receiving Kuzmin's Report for Patras.

To complete this section, I should refer briefly to the series of
articles in our area published in Scientific American for 1981 and 1982.
The most recent one in the group is in the July, 1982 issue by Charles
Lada and the one before it is by Leo Blitz in the April, 1982 issue.
Lada's article deals with outflows of gas from young stars and Blitz's
with Giant Molecular Clouds. The year before, March, 1981, Scientific
American published my article on our Galaxy.

We are now faced with the reality of a 4-component Galaxy. The
central component, the bulge, reaches to about 5 kpc from the galactic
center. The second component is the central disk, in which we have
spiral structure and in which our sun is located. It stretches to about
15 kpc from the center. The two components are enveloped in a

presumably womewhat flattened halo (or spheroid) with a maximum radius
of 20 kpc. The masses are:

1. disk: 8×10^{10} $M_\odot$

2. bulge: 2×10^{10} $M_\odot$

3. halo (or spheroid): 0.5×10^{10} $M_\odot$

We add to this the outer halo (or corona), with a minimum mass and
radius of 6×10^{11} $M_\odot$ and 60 kpc.

Who started the work that led to the general adoption gradually of
the enlarged and very massive Galaxy? The concept arose when the
results of several, rather unrelated, researches were combined. I shall
list them briefly and in no particular order:

(1) Einasto was one of the first to suggest the larger mass and
radius on the basis of his analysis of the motions of galaxies of the
Local Group (see Kerr's 1979 Report).

(2) Ostriker, Peebles and their associates, influenced by
Lynden-Bell (as I see it), gave good theoretical arguments in favour of
a massive outer halo (or corona).

(3) Morton Roberts and Bosma gave radio astronomical evidence for
flat rotation curves in various galaxies, thus suggesting that there
was much matter in the outer parts of galaxies, to well beyond the
observable limits of spiral structure in most spiral galaxies. The
work of Vera Rubin, Ken Ford and Thonnard resulted in the discovery
optically of the prevalence of flat rotation curves in most spiral
galaxies. One asks, naturally, if our Galaxy possesses possibly a flat
rotation curve between 12 and 25 kpc from the center? The answer --
Knapp, Gunn and Tremaine -- Shuter -- Blitz -- Bahcall, Soneira and
Schmidt -- seems to be a resounding YES. As a matter of fact, I fear
that it will take a lot of courage for someone to present at this
Workshop a rotation curve that rises or falls between 15 and 25 kpc
from the center! Vera might object.

(4) Hartwick and Sargent gave in a way the most convincing
evidence for our Galaxy to have a radius of at least 60 kpc and a
minimum mass of 6×10^{11} $M_\odot$, where they showed that the radial velocities of 11 or so outlying globular clusters were high enough to
indicate the validity of their estimated high mass for the corona.

To learn more about the properties and dynamics of the halo and at
least the nearer parts of the outer halo and corona, we may well wish
to follow the approaches that were first outlined by Oort and Hill, by
Vashekidse and by MacRae and myself in the nineteen thirties and early
forties. They were attempts to obtain the density distributions
perpendicular to the galactic plane and -- through the study of the

radial velocities for a fair sample of the stars -- the velocity
distributions perpendicular to the galactic plane. The two are very
directly related as long as there exists a fair approach to dynamical
equilibrium for the stars under consideration. Fields in the directions
of the galactic poles and in high to intermediate galactic latitudes are
suited for this type of analysis. The basic approaches are still those
presented in my small book (1937) THE DISTRIBUTION OF THE STARS IN SPACE,
where I stressed the use of the (m, log π) Table method. A modern
version, based on far more comprehensive standards than were available
45 years ago, has been developed by Bahcall and Soneira, joined more
recently in their work by Maarten Schmidt. In this type of work the
basic precision color-magnitude data of Weistrop represent the right
sort of basic material for this type of analysis. I understand that,
during this Workshop, Sandage will present an outline of the Selected
Area project that he now has underway at Mount Wilson Observatory. I
hope that in these researches northern and southern galactic latitudes
may receive equal emphasis.

What are the objects that populate, however thinly, the parts of
the outer halo or corona that are beyond the distance of rather few
kiloparsecs that we reach in our color-magnitude count approach?
Frankly, I do not know and I suspect that no one in this dignified
Workshop has the answer. Old red dwarfs have been suggested, but the
work by Upgren and his colleagues does not indicate that they are
present in sufficient numbers to provide the missing mass. Dead dwarf
stars, and possibly tiny black holes have also been suggested, but
neither of these hold much attraction. Another possible source for the
missing mass are neutrinos left over from the Big Bang. They are a
group of particles that physicists like to bring up, but as long as no
one knows if neutrinos have mass and -- if they do -- what their mass
is, this is not as yet an attractive proposition.

The best we can probably do is to set up reasonable structural
and dynamical models for the mass and velocity distribution of stars
and clusters, and a few dwarf galaxies in the outer halo and corona,
and see what velocity distributions and rotation curves result. The
dynamical astronomer with access to a fair-sized computer can attempt
to construct and check different models against observed radial
velocities.

Whatever approaches we decide on in the end, there lies a lot of
obvious research ahead of us. Precision photometry, color-magnitude
arrays and star counts will supply the basic information that is needed
for a start. The Space Telescope promises to give really faint limits
of magnitude for which precision data seem to be within reach. The
papers by Bahcall et al. show the way and follow the star count
approach we must.

I shall have to be brief in my presentation of the backgrounds of
other topics that will have our attention during the next three days.
While we cannot and should not take action with regard to the

fundamental constants A, B, R_o and Θ_o, we must consider to which extent
these local constants may have to be revised.   There is little doubt
that R_o, the distance from the sun to the galactic center, must be
revised downward from the "official" value R_o = 10 kpc.   As of now,
R_o = 8.5 kpc seems a reasonable compromise, but we may have to come down
to R_o = 7 kpc, or thereabouts. Θ_o may also have to come down a bit from
its present value Θ_o = 250 km s^{-1}, perhaps to Θ_o = 220 km s^{-1}.   Even
A -- which has the official value at the sum of +15 km s^{-1} kpc^{-1} may go
down to A = +12 km s^{-1} kpc^{-1}.   B at -10 km s^{-1} kpc^{-1} may go down to
-12 km s^{-1} kpc^{-1}.

    Because my time for this presentation is limited, I shall not speak
to length now about the globular clusters of our Galaxy, except to say
that the study of their physics and dynamics is most important if we
wish to understand the origin and evolution of our Galaxy.   We now have
the possibility of obtaining a good spectrum for almost any globular
cluster star we wish to observe.   All of us in the field realize that
even greater opportunities will arise when the New Generation Telescopes
and the Space Telescope get into action.   The study of globular clusters
as individuals should flourish in the decades to come.

    Astronomers interested in our Galaxy will increasingly have to pay
attention to the Central Bulge and the very clearly marked Center itself.
Two and a half years ago I drafted the new version of Chapter 8 of
THE MILKY WAY, the chapter that deals with the Central Bulge and the
Galactic Center.   I wish I could update it right now.   Thanks to
Charles Townes' skillful use of the 12.8μ line of ionized neon, the
velocities for knots close to the galactic center have been measured and
they indicate that there are several million solar masses of matter very
near the Center controlling these motions.   It looks as though there is
a central mass possibly as high as 50 million solar masses within a
parsec of the center.   What is it?   Maybe a giant black hole caused by
the falling on top of each other of lots of old dead stars?   With the
aid of the VLBI, Balick and Brown have found a strong radio-emitting
object within 10 astronomical units of the center.

    The Central Bulge has many features about which we shall hear at
this Workshop.   In the outer parts, there is a ring of Giant Molecular
Clouds.   Then -- at 3 kpc from the Center -- there is the Oort-Rougoor
spiral feature, or smoke ring, which is rotating and expanding with
velocities -- the range 50 to 135 km s^{-1}, a feature most marked in H I.
At 1.5 kpc follows the Burton-Liszt tilted feature, mostly a ring of
H I and H_2.   Another smoke ring follows at 0.3 kpc from the Center, a
mixture of H I, H_2 and H II -- probably a region of star formation.   At
10 pc there is a warmish cloud (5000° K) and inside it -- within 3 pc --
are the Rodriguez and Chaisson compact clouds.   Along with the powerful
Balick-Brown radio source comes the long-known Becklin et al. infrared
source.   All of this is still highly preliminary, for much work on the
center is underway, most notably the Heiligman-Bell Telephone Survey
of ^{13}CO of the surroundings of the Center.

Infrared work from Mauna Kea shows that there is a thin haze of cosmic (silicate) dust overlying the center, which appears to have a ring-like concentration. The total luminosity of this source is very large. Is the source in some way related to the suspected giant black hole? X-ray and Gamma Ray observations yield further clues. The custering of X-ray sources around the Center is remarkable and the strength of the Center in the first Gamma Ray maps of the heavens is striking.

I am sorry that there are no specific papers on the Galactic Center on our program. The Center bears watching and should be talked about.

There are several excellent papers on spiral structure on our program. I am looking forward to hearing them, especially the papers of such possible structure in the inner parts of the outer halo and corona. Many years ago first Weaver and then Verschuur suggested the presence of such features, which seem closer to reality now than they appeared ten years or so ago. And -- to top it all -- Blitz, Kulkarni and Heiles will present their four-armed spiral pattern emerging from the Giant Molecular Clouds and the H II regions of the inner parts of our Galaxy.

My principal concern about galactic spiral structure is that the distances used in the current work are still largely kinematic. As Burton and his associates have demonstrated, kinematic distances are always to be suspect, even when the rotation curve is well-established. And they are doubly uncertain when the rotation curve itself is not based on solid spectral-magnitude-distance measurements. Fortunately Fitzgerald, Jackson and Moffat have extended the open cluster and association system to a distance of 17,000 pc in the direction of the anti-center, but most distances beyound are quite uncertain. Optical distances of high caliber are needed for reliable rotation curves.

I shall be brief about the problems of interstellar matter and star formation in our Galaxy, for I have written about these topics at length in my April 1981 article in Sky and Telescope. I wish to make two major points:

(1) Giant Molecular Clouds. The discovery of the GMC's represents one of the greatest advances in the study of the interstellar medium. I first learned about them from Phil Solomon and associates. Leo Blitz tells me there are 4,000 of them in our Galaxy, so that is the official figure. Their masses run as high per GMC as 100,000 to one million solar masses and their radii are of the order of 50 pc, more or less. These giant blobs, each of them more massive than an average globular cluster, move about in or close to the central galactic plane. Not only are they the obvious breeding grounds for young stars and associations, or clusters, but they are also our Galaxy's basic massive units that affect the motions **and** **stability** of loosely bound star groupings, open and moving clusters for example. Since there are plenty of good radio

astronomers present at our Workshop, the GMC's will not be overlooked in our deliberations. However, I must report that I have been shocked on three or four occasions by the ignorance of them shown by optical astronomers and especially by workers in galactic dynamics. This situation must be remedied.

(2) The most startling development of the past two years has been the realization that very peculiar forces are at work in connection with processes of star formation. Lada's Scientific American article gives several examples of unexpected movements of gases, near to or ejected by a protostar into space -- often in the form of Herbig/Haro objects -- with velocities of 70 or more km s^{-1}. Modern infrared techniques permit us to locate single protostars or associations of protostars deep inside dark clouds. CO radio radial velocities show that the velocities of the gases inside the cloud are generally of the expected size of one to several km s^{-1}. But then, unexpectedly high CO velocities are found very close to the protostar, where bi-polar ejections occur with speeds of the order of 70 or more km s^{-1}. What is going on? One of the finest examples of this sort of behavior is in and near the globule ESO 210-6A. Schwartz discovered that there are two Herbig/Haro objects (now numbered 46 and 47) associated with the globule. The spectra were studied by Dopita and I obtained a fine photograph (IV N hypersensitized emulsion) of the globule and HH46 and 47 (connected by an umbilical cord) on Valentine's Night 1979, at the prime focus of the CTIO 4-meter reflector in Chile. This combination of objects has been studied recently by John Graham who reports that there are a couple of HH objects on the other side of the globule. One group of HH objects has a negative radial velocity of a little more than 100 km s^{-1}, whereas the ones on the other side have comparable positive radial velocities -- all from Graham's measurements with the spectrograph on the Yale one-meter reflector at CTIO. A CCD photograph (near 1μ) with the CTIO 4-meter reflector, prime focus, shows that deep inside the globule there is an infrared protostar directly on the line connecting the Herbig/Haro objects. The simplest explanation is that the HH Objects are the results of bi-polar ejection by the protostar.

An this is a good note on which to end my Introductory Remarks.

MAGNETIC FIELD STRENGTH MEASUREMENTS IN TWO TYPES OF REGION

Carl Heiles
Astronomy Department, University of California, Berkeley

Thomas H. Troland
Physics and Astronomy Department, University of Kentucky,
Lexington

We report the first measurements of interstellar magnetic fields using Zeeman splitting of the 21-cm line seen in emission, made with the Hat Creek 85-foot telescope. We have five measurements in two different types of region.

One type of region is the shock front behind an expanding HI shell. At one position in each of two shells, the measured line-of-sight field components are both about 7 microGauss. We discuss in some detail the physical nature of one of these shells located in Eridanus. The magnetic field severely limits the density enhancement in the cool post-shock gas, and it probably has some effect on the overall dynamics of the shell.

The other type of region is clouds associated with a star-forming region--specifically the Orion region. In the large HI cloud that envelopes the ionized and molecular clouds, the line-of-sight field strength is 10 microGauss. In the large CO cloud just south of the Orion nebula, the field strength is also about 10 microGauss. These results are discussed using the virial theorem. In the HI cloud, gravity appears to be weaker than the kinetic and magnetic forces, making this cloud unstable to expansion. In the CO cloud, gravity and kinetic forces balance, and the magnetic field may be important in disrupting the cloud.

These results are more fully discussed in the 1 September issue of <u>Astrophysical Journal Letters</u>.

W. L. H. Shuter (ed.), Kinematics, Dynamics and Structure of the Milky Way, 9.
Copyright © 1983 by D. Reidel Publishing Company.

A PHOTOMETRIC STUDY OF THE LOWER MAIN SEQUENCES OF THE HYADES
AND THE FIELD STARS

A. R. Upgren and E. W. Weis
Van Vleck Observatory, Wesleyan University
Middletown, CT, U.S.A.

In recent years, the Van Vleck Observatory has undertaken a program
of broad-band photoelectric photometry of the stars on its parallax program and of similar stars which are possible members of nearby clusters.
Most of the cluster and field stars are K and M dwarfs inhabiting the
lower main sequence. The photometry includes the R and I bands of the
Kron system as well as the normal B and V magnitudes. The data has been
obtained with the 1.3m and no. 2 0.9m telescopes of the Kitt Peak National Observatory and the 0.6m Perkin telescope of the Van Vleck Observatory.

In addition to providing apparent and absolute magnitudes and color
indices as a regular component of the Van Vleck parallax lists, this
program seeks to compare the loci of the lower main sequences of the
nearby clusters to that defined by the nearby field stars in order to
detect and examine any systematic differences which may be found. The
program has the further result of providing valuable additional information on the probability of membership of stars in cluster regions whose
proper motions have indicated an association with the cluster.

Photometry has been obtained of stars in the region of the Hyades,
Praesepe and Coma clusters whose proper motions are indicative of
membership, with special emphasis on the first and nearest of these
clusters. To date, four lists of photometry of possible Hyades members
have been published (see the most recent paper by Weis and Upgren
(1982a) and references cited therein) and a fifth is in preparation.
Together, the four published lists extend photometry to more than 600
stars in the cluster region, or about twice the number with previous
photometry. They include almost all of the stars brighter than visual
magnitude 16 in any of the proper motion surveys and all stars in van
Bueren's (1952) list of bright members. Only about 80 stars with V < 16
remain to be observed because no identification charts are presently
available for them, but the Lick Observatory is in the process of making
positive identifications of these remaining stars, and photometric data
is being obtained for them as well. The total number of stars in the
entire range +3 < V < +16 which appear to be probable members of the

W. L. H. Shuter (ed.), Kinematics, Dynamics and Structure of the Milky Way, 11–13.
Copyright © 1983 by D. Reidel Publishing Company.

Hyades from their photometry as well as their proper motions is 315, of which 213 also have radial velocities confirming membership. The additional unobserved stars are expected to increase this number by less than ten percent. The luminosity function based on these data is in agreement with that of Oort (1979). Although the proper motion survey of Pels et al. (1975) upon which Oort's function is based missed some confirmed members brighter than V = 12, and contained very few fainter than this magnitude, the assumptions which he made lead to a realistic luminosity function for the Hyades stars identified by the proper motion surveys. However, considerable evidence exists which reveals that the proper motion studies are incomplete at magnitudes fainter than about 12. The distribution of these faint probable members is much more confined to the central region of the cluster which has been more intensively searched for faint members than the outlying areas, and none of them lies to the east of R.A. 4^h35^m (1950).

Proper motion members are more difficult to detect in the eastern part of the Hyades region because it lies closer to the convergent point and the proper motions are smaller, and also because it lies at a lower galactic latitude and the density of background nonmembers is higher. Probably for these reasons, the proper motion surveys are incomplete especially in the eastern regions and the true luminosity function of the cluster should be somewhat higher than the observed function found by Oort (1979) and in the present study for the fainter stars.

The Hyades main sequence bluer than 1.5 in B-V or 1.0 in R-I (Kron) is known to closely resemble that of the nearby field stars in color-magnitude diagrams after application of the Lutz-Kelker correction to the latter and a cluster distance modulus of 3.25 (Upgren 1978). Only recently, however, has photometry become available for the dwarf M stars below these color limits. Photoelectric observations have now been made of a large number of faint stars most of which have trigonometric parallaxes more than 20 times as large as their published errors. Preliminary comparison shows that this reddest portion of the Hyades main sequence containing about 87 stars lies above that of the field stars throughout the range 1.0 < R-I (Kron) < 1.4. These results and the photometric data for the field stars have been submitted for publication (Weis and Upgren 1982b) and are in general agreement with a similar photometric investigation made by Stauffer (1982) of the Hyades main sequence brighter than M_V = 13.5.

The discrepancy between the positions of these faint dM stars in the Hyades and their counterparts among the nearby stars is unlikely to be explained by either the presence of unsuspected binaries among the cluster stars or a depth effect within the cluster. It also does not appear to be evidence for continuing pre-main sequence contraction, since the time scale required for star formation in the cluster would be much too long.

References

Bueren, H. G. van 1952, Bull. Astron. Inst. Neth., 11, p. 385.

Oort, J. H. 1979, Astron. Astrophys. 78, p. 312.

Pels, G., Oort, J. H. and Pels-Kluyver, H. A. 1975, Astron. Astrophys. 43, p. 423.

Stauffer, J. 1982, Astron. J. 87, p. 899.

Upgren, A. R. 1978, in I.A.U. Symp. 80, The HR Diagram, p. 39.

Weis, E. W. and Upgren, A. R. 1982a, Publ. Astron. Soc. Pacific 94, p. 475.

Weis, E.W. and Upgren, A. R. 1982b, Publ. Astron. Soc. Pacific, submitted.

THE COORDINATION OF SPACE AND GROUND-BASED PARALLAX PROGRAMS FOR
IMPROVEMENT OF THE STELLAR LUMINOSITY FUNCTION

A. R. Upgren
Van Vleck Observatory, Wesleyan University
Middletown, CT, U.S.A.

Recent studies have shown that the general stellar luminosity
function does not increase monotonically with decreasing luminosity (or
mass) but reaches a maximum at an absolute magnitude of about +6 or
nearly one solar mass, and a minimum at slightly fainter values, near
one-half solar mass. It has been customary for investigators to adopt
the smoothed function of Luyten which increases steadily throughout this
range of absolute magnitude. However, the dip in the function between
absolute magnitudes of about +6 to +9, found long ago by van Rhijn, has
recently been reconfirmed by Wielen (1974) from an analysis of the Catalogue of Nearby Stars of Gliese (1969). Lately, Upgren and Armandroff
(1981) have investigated the nearby stars listed in Gliese's catalogue
and its recent supplement (Gliese and Jahreiss 1979). They found that
all main sequence stars brighter than M_V = +9 (spectral class dM0) as
well as the giants and subgiants are uniformly distributed throughout
the volume of space within the limiting distance of 22 pc from the Sun.
They were able to conclude for several reasons that no statistically
significant number of these stars within 22 pc remain to be discovered
or identified; thus, the nearby stellar sample is complete and the
Wielen function is correct. However, incompleteness becomes very significant for the dwarf M stars, even those just fainter than M_V = +9.
Since most of the details of this study have been published (Upgren and
Armandroff 1981), no further discussion will be given here.

We can use this information to derive an experimental design maxi-
mizing the information received from a combination of parallax programs
using space satellites now in the planning stage along with existing
ground-based telescopes. In this paper, two major goals are discussed.
The first of these is the improvement of the HR diagram or more specifically, the mean and dispersion of the main sequence and of the sequences
of the giant and subgiant stars. The second is an improvement in the
knowledge of the space densities of each kind of star; that is, an improvement in the luminosity function itself. This would be accomplished
by extending the completeness distance limit for the brighter stars well
beyond the limit of Gliese's catalogue. It is our intent here to maxi-
mize the information necessary to determine these data by recognizing

15

W. L. H. Shuter (ed.), Kinematics, Dynamics and Structure of the Milky Way, 15–19.

the constraints inherent in each program and utilizing their advantages.

The principal instrument is the HIPPARCOS Space Astrometry Satellite. The instrumentation and goals of this telescope have been discussed previously in several places (see e.g. the contributions by
Gliese, Schmidt-Kaler and Pagel in Barbieri and Bernacca, 1979). Its
coverage is planned to include the measurement for parallax and proper
motion of all 40,000 stars brighter than the eighth visual apparent magnitude with an error in each parallax of about ±0."002. Between magnitudes 8 and 11 (the faintest limiting visual magnitude), the mean parallax error is predicted to rise from ±0."002 to about ±0."005 and is somewhat dependent upon galactic latitude. The percentage of stars included
would decrease rapidly from 100% at V = 8 to zero at V = 11.

In contrast to this, the two largest ground-based parallax programs
at the present time are located at the U.S. Naval and Van Vleck Observatories and determine trigonometric parallaxes with errors averaging
about ±0."005 per star. The U.S. Naval Observatory program is limited to
M dwarf and white dwarf stars fainter than about M_V = +9 (Harrington and
Dahn 1980) and the Van Vleck Observatory program centers on dwarfs and
subdwarfs of spectral classes G, K and early M, mostly in the range in
M_V from +6 to +10 (Upgren and Breakiron 1981). The Greenwich, Lick,
McCormick, Mt. Stromlo and Sproul Observatories are also actively measuring parallaxes of similar stars in smaller numbers with about the same
error. (It should be noted here that all of these programs are of much
higher precision than their predecessors whose results are listed in the
General Catalogue of Trigonometric Stellar Parallaxes and its supplement
of 1963, and which had a much larger mean error in parallax amounting to
±0."016.)

In order to determine the optimal contributions of HIPPARCOS and
other space and ground-based telescopes to the two objectives mentioned
above, we shall assume a uniform and isotropic density of all kinds of
stars in transparent space. Although fine tuning may be necessary to
adjust this assumption more closely to reality, it is likely that it is
valid since the scale heights in distance perpendicular to the galactic
plane of the kinds of stars actually observed is very large compared to
the practical limit in distance of all foreseeable parallax observations,
and the total interstellar absorption within this distance limit is negligible. The value of a parallax is proportional to the ratio of the
error in parallax to the parallax itself, or σ_π/π. Although the size of
this ratio necessary for a "good" parallax is entirely arbitrary, we
shall adopt a ratio of one-fifth as the upper limit for parallaxes of
acceptable quality; thus, $\sigma_\pi/\pi < 0.2$. One reason for accepting this
limit is that the systematic effect found by Lutz and Kelker (1973) is
evaluated to about this upper limit. A second reason is that the uncertainty in M_V for a ratio of 0.2 is about half a magnitude, or about
equal to its uncertainty from other secondary luminosity calibrations,
and the trigonometric parallax loses its ability to assign luminosities
with higher precision than these other methods.

The total weight of a group of parallaxes can be assumed here to be represented by the sum of their inverse variances; thus, $W = \Sigma (\pi/\sigma_\pi)^2$. We can see from Figure 1 of Upgren and Armandroff (1981) that a little more than 100 stars are found among the nearby stars tabulated by Gliese for each whole magnitude interval in M_V from +5 to +9. We shall thus define a weight, W_o, representing the total weight of the parallaxes of 100 stars in a one-magnitude interval uniformly distributed in space to the limiting distance of Gliese's catalogue with standard errors of ±0".002 on average. Throughout the interval $+5 \leq M_V \leq +9$, the parallax weight of all of Gliese's stars in a one-magnitude interval will slightly exceed W_o but brighter than +5 it will not, and fainter than +9, a high-velocity bias will be present due to incompleteness among even the nearest stars. For brighter portions of the main sequence and for the evolved stars, it is necessary to extend the distance limit in order to obtain a total parallax weight equal to W_o, thus assuring a precision in the location (mean and dispersion) of the upper main sequence equal to that of the lower main sequence. The total weight of the parallaxes of a group of stars (assuming uniform space distribution) varies directly with the volume of space, or with the cube of the distance covered, but it also varies inversely with the square of the distance if the mean parallax error of a star does not vary with distance. Thus, the total weight to distance d or W_d is equal to $dn_o W_o/100 d_o$ where d_o is the distance limit of Gliese's catalogue and n_o is the number of stars within $M_V \pm 1/2$ and within d_o (n_o is assumed to be 100 for $5 \leq M_V \leq 9$). The distances necessary for weight W_o to be achieved for each whole magnitude interval are given in Table I. The information appearing in all but the last two rows applies to main sequence stars only. Data for the subgiants and giants appear in these last two rows. Mean absolute visual magnitudes of +3±1 and +1±1 were adopted for the subgiants and giants, respectively.

The first three columns of the table list, in order, the absolute magnitude interval, the approximate number of stars, n_o, within the distance limit d_o, and the expected number with $\sigma_\pi/\pi < 0.2$ in the entire sky, labelled $n_{0.2}$. The distance limit, d, necessary for a total parallax weight amounting to W_o, and the approximate total number of stars to be expected within that distance, are given in the next two columns. The final column shows the apparent visual magnitude of the stars, V_d, which corresponds to the absolute magnitude shown in the first column at the distance limit, d. We can make several comments based upon the data shown in Table I. The number $n_{0.2}$ for stars with $M_V \leq +8$ is actually the total expected number of stars in the whole sky within 100 pc of the Sun since their parallaxes must be larger than +0".010 in order to be more than five times as large as their expected errors. For the fainter stars, $n_{0.2}$ drops sharply because the parallax error increases towards fainter magnitudes as mentioned above. The limiting distance is reduced and the total volume is reduced greatly, hence the much smaller number of stars ideally accessible to HIPPARCOS. We can also see that a sufficient number of main sequence stars between absolute magnitudes +1 and +6, as well as giants and subgiants necessary for a total parallax weight of W_o, would be included among the totality of stars brighter than

TABLE I.   PARALLAX WEIGHT BY MAGNITUDE

$M_V \pm 1/2$	n_o	$n_{0.2}$	$d(pc)$	n_d	V_d
-1	1	100	2000	10^6	11
0	3	300	600	10^5	10
+1	10	1000	200	10^4	8
2	20	1900	100	2500	7
3	35	3300	60	900	7
4	60	5600	33	300	7
5	110	10000	20	100	7
6	120	11000	20	100	8
7	110	10000	20	100	9
8	120	12000	20	100	10
9	150	5000	<20	100	12
10	240	4000	<20	100	13
11	360	3000	<20	100	14
subgiants	45	4200	45	500	7
giants	20	1900	100	2500	7

apparent magnitude +8 which the HIPPARCOS program plans to observe in
any event, hence no additional stars of these absolute magnitudes need
be added to its program.   For stars outside this range in M_V, more stars
are needed.   At the bright end, a sharp limit occurs at magnitude +1.
For yet brighter stars, no number will yield sufficient information for
the main sequence to be well positioned in the HR diagram and secondary
distance methods must continue to provide their luminosities.   For magnitudes +6 to +8, the stars in Gliese's catalogue will suffice, and
since many of them are fainter than apparent magnitude +8, they must be
added to the HIPPARCOS program.   For magnitudes +9 and fainter, incompleteness is known to exist even within the Gliese sample and thus, a
high-velocity bias is unavoidable.   We should also keep in mind that
many of the stars fainter than $M_V = 6$ are also fainter than V = 8 and
the parallax error of HIPPARCOS approaches that of the presently active
ground-based parallax programs.   It is thus best in planning a program
list of the fainter G and K dwarfs for HIPPARCOS, or for any space program, to coordinate it with the Van Vleck and other less active existing
programs which observe these stars, especially as these programs can
work to fainter apparent magnitudes than can the space telescopes whose
apparent visual magnitude limit is near +11.   It is questionable whether
the M dwarfs should be observed at all from space (other than for orbital motion) since the U.S. Naval Observatory program currently observes
hundreds of these stars at apparent magnitudes as faint as 17.

For the second major goal discussed here, the mean space densities
of stars, the requirements are less stringent.   The stellar density
needs only to be determined from counts of stars likely to be within the
100 pc limit if it is known that the sample is complete.   One must be
aware that the properties of any stellar sample limited in distance or
apparent magnitude may be subject to a systematic error as a result

(e.g. Lutz 1979). The most severe problem is that beyond the distance limit of Gliese's catalogue, our knowledge of stellar types is far from complete. We are forced to rely, for the most part, on objective-prism surveys for prior knowledge of the stellar properties of interest (e.g. spectral type and apparent magnitude) and these cover only perhaps ten percent of the sky. Star counts to a spectroscopic distance well in excess of 100 pc can be made only in these regions. From Table I, we can see that stellar densities are not likely to be significant for stars brighter than about absolute magnitude +1 (spectral class A0 on the main sequence).

In sum, the HIPPARCOS satellite appears to be most effective for the common stars between M_v of +1 to +7 or perhaps +8. Its program does not replace but rather complements the existing ground-based parallax programs, and maximum coordination between them is likely to prove fruitful and avoid needless duplication. The American Space Telescope is not planned to provide parallaxes for more than a few tens of stars, although these will be of very high precision. Its contribution to astrometry is of most value only if the stars chosen for observation include the parallax standard stars, now being selected for observation for all active parallax programs. In order to be most successful, a combined program to improve space densities and the HR diagram must include photometry of all program stars. Photoelectric observations, at least in the broad bands, is a regular feature of the Van Vleck, U.S. Naval and several of the other currently active parallax programs. Some plan for photometry of stars observed with the space telescopes would make their parallaxes of much greater use to astronomy.

References

Gliese, W. 1969, Veroeffentl. Astron. Rechen-Inst. Heidelberg No. 22.
Gliese, W. 1979, in Barbieri, C. and Bernacca, P. L. (eds.), European Satellite Astronomy, Tip. Antoniana, Padova, Italy, p. 195.
Gliese, W. and Jahreiss, H. 1979, Astron. Astrophys. Suppl. 38, 423.
Harrington, R. S. and Dahn, C. C. 1980, Astron. J. 85, 454.
Lutz, T. E. 1979, Monthly Notices Roy. Astron. Soc. 189, 273.
Lutz, T. E. and Kelker, D. H. 1973, Publ. Astron. Soc. Pacific 85, 573.
Pagel, B. E. J. 1979, in Barbieri, C. and Bernacca, P. L. (eds.), European Satellite Astronomy, Tip. Antoniana, Padova, Italy, p. 211.
Schmidt-Kaler, Th. 1979, in Barbieri, C. and Bernacca, P. L. (eds.) European Satellite Astronomy, Tip. Antoniana, Padova, Italy, p. 203.
Upgren, A. R. and Armandroff, T. E. 1981, Astron. J. 86, 1898.
Upgren, A. R. and Breakiron, L. A. 1981, Astron. J. 86, 776.
Wielen, R. 1974, I.A.U. Highlights of Astronomy 3, 395.

CHANGES IN INTERSTELLAR ATOMIC ABUNDANCES
FROM THE GALACTIC PLANE TO THE HALO

Edward B. Jenkins
Princeton University Observatory

ABSTRACT

A few, specially selected interstellar absorption lines have been
measured in the high resolution, far-ultraviolet spectra of some 200
O and B type stars observed by the International Ultraviolet Explorer
(IUE). This study indicates that for lines of sight extending beyond
about 500 pc from the galactic plane, the abundance of singly-ionized
iron atoms increases relative to singly-ionized sulfur. However, the
relative abundances of singly-ionized sulfur, silicon and aluminum do
not seem to change appreciably. A likely explanation for the apparent
increase of iron is the partial sputtering of material off the
surfaces of dust grains by interstellar shocks. Another possibility
might be that the ejecta from type I supernovae enrich the low density
medium in the halo with iron.

I. INTRODUCTION

Many absorption lines which appear in ultraviolet stellar spectra
reveal the gas-phase abundances of elements in the interstellar medium
(Spitzer and Jenkins 1975). In past years, the spectrometer on the
Copernicus satellite provided much information in this regard, but the
limited sensitivity of this instrument has restricted surveys of the
interstellar gas to within a few kpc of the sun. The International
Ultraviolet Explorer (IUE) can, in its high resolution mode, observe
the spectra of early-type stars as faint as 10th magnitude, thus
greatly extending our distance range. Early in the life of this
satellite, Savage and de Boer (1979) capitalized on IUE's high
sensitivity by observing lines of sight through our galactic halo
toward stars in the Large Magellanic Cloud. This paper will report on
the material in front of a large selection of stars widely scattered
about the galactic halo. The result for these stars will be compared
with those from other stars in the plane of the galaxy. For reasons
discussed below, however, the analysis of each spectrum will be
restricted to a few, specially selected absorption lines.

W. L. H. Shuter (ed.), Kinematics, Dynamics and Structure of the Milky Way, 21–30.
Copyright © 1983 by D. Reidel Publishing Company.

Under most circumstances, interstellar lines recorded by IUE are not suitable for deriving column densities. As a rule, a line which is strong enough to measure reliably in a single exposure is, with the usually found velocity dispersion b < 10 km s⁻¹, badly saturated and on the flat portion of the curve of growth. Only when many exposures are co-added or special, noise-suppressing reduction procedures are invoked can one hope to obtain a believable measurement (see, e.g. York and Jura (1982)). The survey being considered here makes use of single exposures which were analyzed only with the standard spectrum reduction procedure (IUESIPS). As a consequence, we must adopt a conservative viewpoint in the data interpretation and concentrate on a restricted set of conclusions which have the least sensitivity to the unknown behavior of the radial velocity profiles.

If absorption lines from two atomic species have the same equivalent width divided by wavelength W_λ/λ, we may infer that the ratio of their column densities is inversely proportional to the ratio of the respective products of transition f-values and wavelengths, provided the two velocity profiles are identical. The requirement for profile similarity becomes more critical if the lines are highly saturated. In general, the profiles <u>can</u> be somewhat different, but we may still cope with saturated lines if we add a qualification to the conclusion on column density ratios: that the results apply only to gases at velocities displaced from the line core, at approximately unit optical depth in each case. Since we will be considering moderately strong and probably badly saturated lines in this study, the column density comparisons will apply only to <u>disturbed</u> gases, at a velocity displacement on either side of the line core of roughly one-half of the equivalent width (see Figure 1). Such restricted conclusions are of use in qualitative comparisons between disk and halo gas, and they should be valid as long as the profile differences are not as extreme as the hypothetical, pathological situation illustrated in Figure 2.

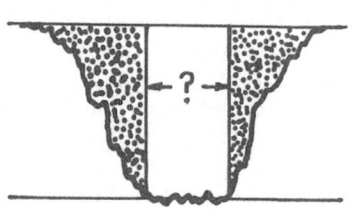

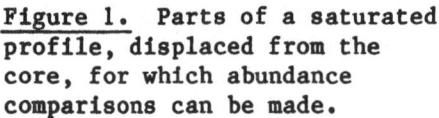

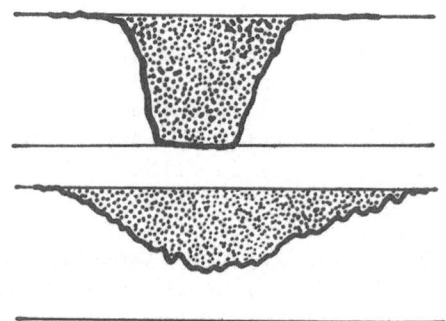

Figure 1. Parts of a saturated profile, displaced from the core, for which abundance comparisons can be made.

Figure 2. Two profiles with identical equivalent widths which have such a large difference in velocity behavior that the comparisons discussed here lose their meaning.

If the two lines have unequal equivalent widths, we can, strictly speaking, only state the column density ratio in terms of an inequality, i.e., the ratio is greater or less than the amount corresponding to lines of equal strength. However, information on the slope of the curve of growth may be available from two or more lines of differing strength from a given atom or ion. In this case, one can deduce an approximate column density ratio for unequal lines, provided the slope is not very small.

II. DATA

For this survey we make use of IUE high-resolution spectra of approximately 200 stars, 41 of which are more than 500 pc from the galactic plane. About half of the stars at large z distances and a few of the low-z stars were observed by the author in 1978 and 1979. The remaining spectra, and by far the largest portion, were furnished by the National Space Science Data Center (i.e., results from other observing programs now released to the public domain).

Distances to the stars are inferred from the MK classifications, their corresponding absolute magnitudes and the stars' apparent magnitudes. Since the scale height of O and B type stars is less than a hundred pc, stars at z distances greater than a kpc or so are very rare. Hence, we must consider the danger that in the necessarily selective process of finding suitable target stars at large z, our sample may be badly contaminated by underluminous, nearby stars whose distances are greatly overestimated. Our concern is aggravated by a nagging theoretical question: if such stars are normal and were born in the galactic plane, how do we reconcile their long time in transit away from the plane with their relatively short main-sequence lifetimes[1] (Greenstein and Sargent 1974)? Pettini and West (1982) have given a rather complete overview of the observational evidence and the controversies in the literature bearing on the stars' distances and present a fairly optimistic viewpoint that the inferred distances are not significantly overestimated in a systematic fashion. Of course, if our investigation of interstellar lines shows that the apparently high-z stars seem to differ from ones at low z, that evidence alone would confirm independently that the high latitude stars are not nearby interlopers.

After a perusal of the many absorption lines within the coverage of IUE's short-wavelength camera, the following three pairs of lines were found, empirically, to have nearly similar equivalent widths.

ion	$\lambda(\text{Å})$	f-value[2]		ion	$\lambda(\text{Å})$	f-value[2]
FeII	1608	0.062	....	SII	1253	0.0107
SiII	1808	0.0055	....	SII	1250	0.0053
AlII	1670	1.88	....	SiII	1304	0.147

All of these ions are in the dominant stage of ionization expected
within HI regions; i.e., each is at the lowest stage which has an
ionization potential greater than that of hydrogen. Moreover, the
ionization potentials of FeII, SiII and AlII are very similar to each
other (~ 17 eV), but that of SII is somewhat larger (~ 23 eV).

    Figures 3, 4 and 5 show how the ratios of the equivalent widths
of the three pairs of lines vary with the z distances of the target
stars. A significant increasing trend can be seen for the ratio of
FeII and SII as lines of sight go beyond |z| > 500 pc. This

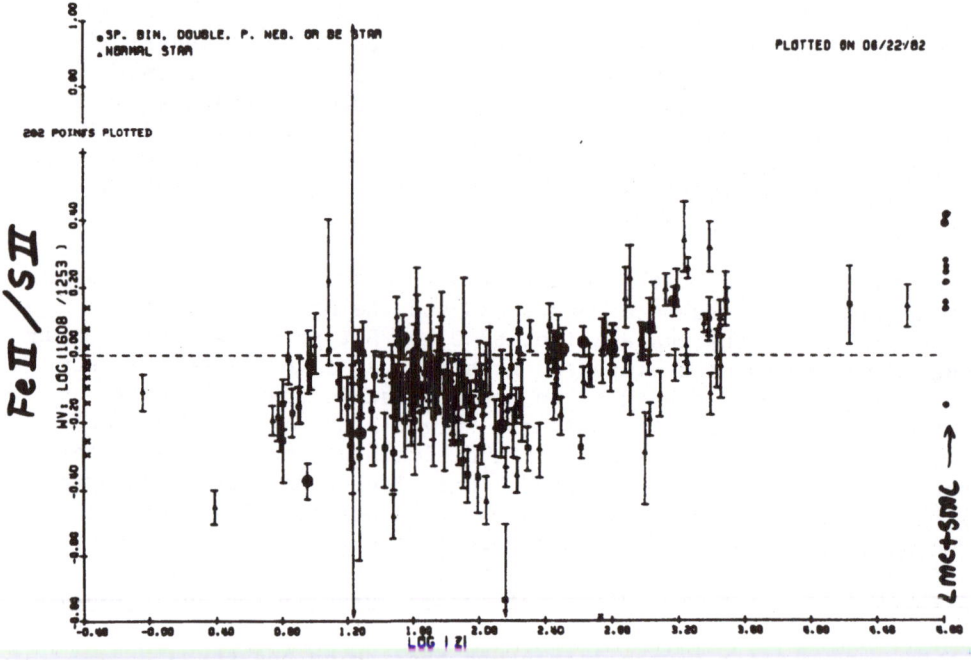

(9) Z DEPENDENCE OF EQ. WIDTH RATIO

Figure 3. Logarithm of the ratio of the W_λ/λ of FeII (λ1608) and SII
(λ1253) absorptions vs. the logarithm of the star's distance away from
the galactic plane (in pc). The horizontal dashed line representing
equal strengths corresponds to log $\left[N(FeII)/N(SII)\right]$ = -0.87.
Measurements toward the Large and Small Magellanic Clouds taken by
Savage and de Boer (1981) are plotted at the far right (without error
bars). Errors in the equivalent widths due to noise were estimated
from the amplitudes of fluctuations in the continuum levels near the
respective lines; the resulting ±1σ errors in the ratios are shown.
In practice, comparisons of multiple measurements of some lines show
that the true errors are about 1.5 times the calculated errors,
indicating other sources of uncertainty of undetermined origin.
Hence, error bars should be interpreted only as an indication of
relative reliability from one measurement to another. Circled points
are for absorption lines having $|v_{LSR}|$ > 20 km s^{-1}.

enhancement of the Fe feature over that of S is also seen in the
Magellanic Cloud stars. The SiII/SII and AlII/SiII plots seem to show
virtually no change for the high-z stars, although the LMC + SMC data
show some increase in the ratios. Typically, the strengths of the two
SII lines (at 1250 and 1253 Å) differ by less than $2^{1/2}$, which means
that the changes in relative abundances are somewhat more than twice
the variations (in the logarithm) of the plotted line strength
ratios.

The AlII and SiII lines shown in Figure 5 are generally stronger
than those which are compared in Figures 3 and 4. If, in general, the
wings of the velocity profiles converge to zero rapidly (i.e., as does
a gaussian distribution, for instance), the sides of the strong
absorption lines will be steep, making the curves of growth flat and
the points in Figure 5 less sensitive to changes in relative
abundance.

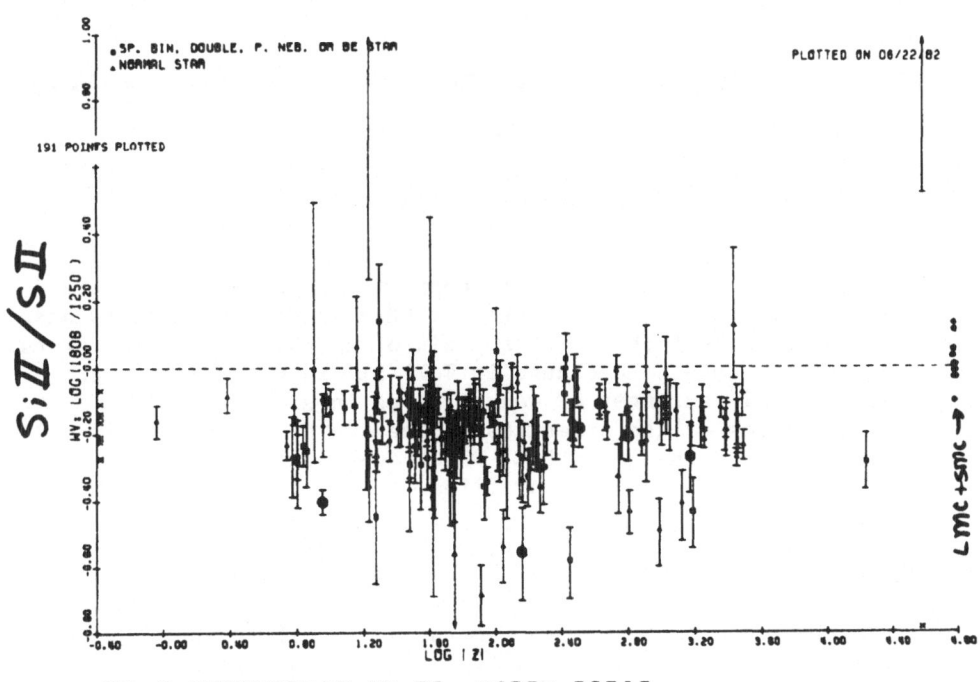

(9) Z DEPENDENCE OF EQ. WIDTH RATIO

Figure 4. Same as Figure 3, but for the ratio of SiII (λ1808) to SII
(λ1250). The dashed line of equality corresponds to log
$\left[N(SiII)/N(SII) \right]$ = -0.18.

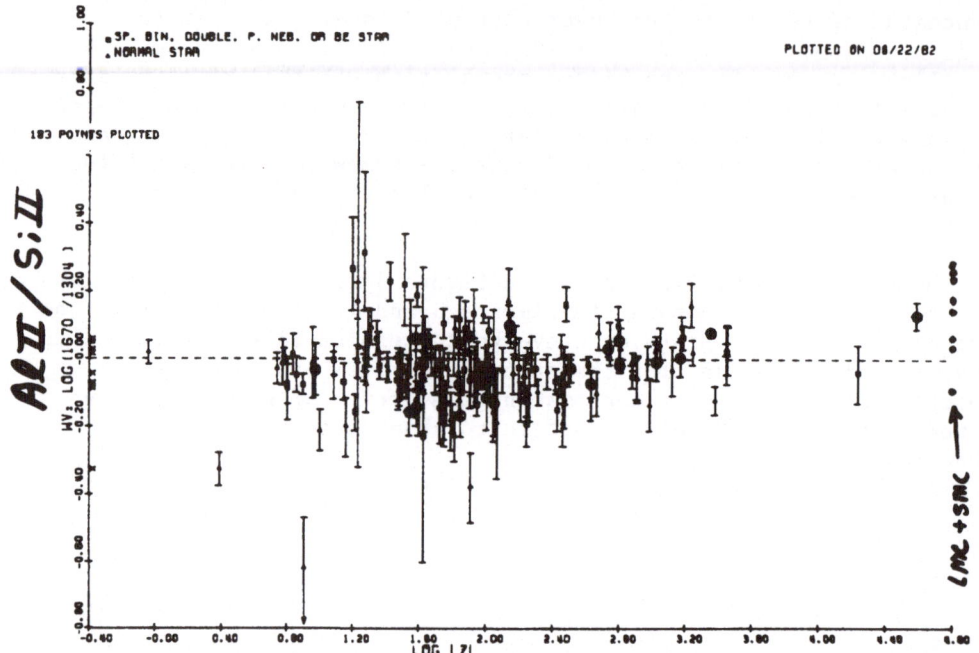

(9) Z DEPENDENCE OF EQ. WIDTH RATIO

Figure 5. Same as Figure 3, but for the ratio of AlII (λ1670) to SiII (λ1304). The dashed line of equality corresponds to log $[N(\text{AlII})/N(\text{SiII})]$ = −1.21.

As a rule, the absorption lines toward the high latitude stars are not systematically weaker or stronger than those toward stars near the plane. This is because there is a mix of nearby and distant stars in the low−z set. Thus, the change in FeII/SII is not due to there being systematically stronger or weaker lines in either set. Moreover, the ratios of the strengths of the two SII lines (λ1250 and λ1253) do not seem to change with z, which suggests that there is no strange behavior in the velocity profiles.

III.   DISCUSSION

From earlier studies with the Copernicus satellite, we know that interstellar abundances of most free atoms and ions, relative to hydrogen, are below the respective amounts corresponding to the cosmic abundance ratios. It is generally accepted that these depletions are caused by condensation onto grains, and the amounts vary strongly from one element to the next. Those atoms which easily form refractory compounds are highly depleted (some by a factor of at least 1000), while others which are left over to form volatile substances are only slightly below their cosmic abundances (to within a factor of less than 3). Dense interstellar clouds exhibit the strongest depletions (Morton 1974; Snow 1976, 1977; Snow and Jenkins 1980). Stars which

have relatively little reddening, i.e. lines of sight dominated by
intercloud material, show smaller depletions (Morton 1978, York and
Kinahan 1979, Morton and Bhavsar 1979). It is reasonable to presume
that the intercloud material is frequently exposed to shocks which
partially destroy the grains by sputtering and grain-grain collisions
(Shull 1978, Draine and Salpeter 1979, Spitzer 1976) and return atoms
to the gas phase.

Generally, the previous studies have shown that depletions of S
are very much less than those of Fe, Si or Al. For the purpose of
discussion here, we may adopt the SII lines as indicators of virtually
undepleted material, against which we compare the abundances of the
more highly depleted species. The three plots indicate that the
logarithms of the Fe, Si and Al abundances relative to that of S seem
to be midway between the solar values +0.19, +0.34 and -0.81,
respectively, and the numbers -1.70, -1.17 and -3.80 observed toward
ζ Oph (Morton 1974; Shull, Snow and York 1981; Snow and Meyers 1979),
with its classic line of sight through a dense cloud. This statement
is only approximate, however, for the reasons given earlier. The
lower depletions indicated for gas even in the galactic plane is
probably a consequence of our preferentially sampling the intercloud
material, which dominates over the cloud gas in the wings of the
strong absorption lines.

The decline in the depletion of Fe away from the plane mimics the
effect seen in high-z stars for Ca by Cohen and Meloy (1975) and Ti by
Albert (1982). It is surprising that a similar behavior is not seen
for Al and Si. A uniform, partial destruction of grains having a
homogenious composition seems to be ruled out here. Otherwise, we
would expect Al, the most highly depleted element, to show the
strongest change, which is not the case. Barlow and Silk (1977) have
proposed that thin monolayers of metals may physically adsorb to the
surfaces of silicates or the ice mantles of grains. These layers
would be easily disrupted by low velocity shocks, which could explain
the variability of depletion. In contrast to the situation for Fe,
Barlow and Silk point out that silicon is unlikely to coat grain
surfaces because the energy liberated in forming a hydride is
sufficient to overcome the binding.

In their survey of weak interstellar iron lines recorded by
Copernicus, Savage and Bohlin (1979) found that the Fe depletion was
highly correlated with the average density of gas along the various
lines of sight. One could argue that the dependence with z seen here
is a manifestation of the same effect, inasmuch as the high latitude
lines of sight have less average density than those in the plane.
Indeed, one might question whether or not there is any fundamental
difference between intercloud material at z = 0 and the gases
geographically removed from the layer of clouds centered on the
plane.

While not too common in this survey, there are a few lines of

sight in the plane which have color excesses per unit distance which
are as low as many of the high latitude cases. Twenty-five stars with
$|z| < 500$ pc and $E(B-V)/r < 0.1$ mag/kpc were found to have an average
value of -0.10 for the logarithms of the ratios of FeII to SII W_λ/λ.
Thus, we may conclude that low average densities alone do not explain
the trend that is seen for stars at high z.

If, for some reason, the moderately stirred up gas in each case
is dominated by circumstellar material, it would be conceivable that
the FeII we measure could, to varying extents, be further ionized by
stellar radiation from the target stars. Such an effect could alter
the apparent ratio of Fe to S, since SII is harder to ionize. One
might then propose that we are misled into thinking the variation in
FeII/SII is a z effect, when indeed it results only from the fact that
low-z stars are generally hotter and more able to doubly ionize the
nearby Fe. Figure 6 addresses this possibility by showing the
relationship between the target stars' effective temperatures and the
ratios of the line widths. No correlation is evident; hence the
prospects of circumstellar ionization influencing the results seems
remote.

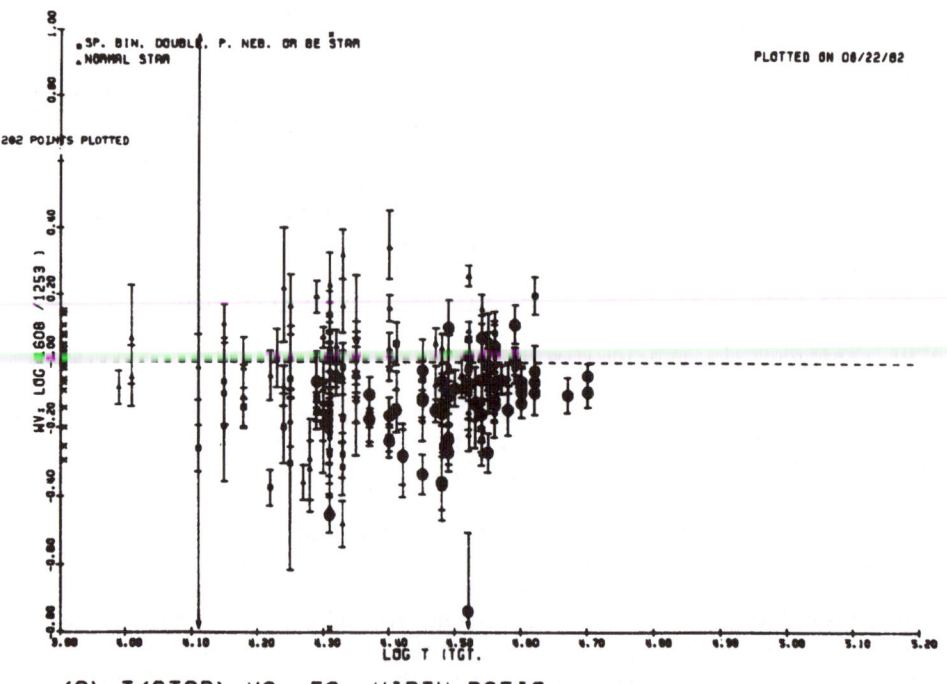

(8) T(STAR) VS. EQ. WIDTH RATIO

Figure 6. Logarithm of the FeII to SII W_λ/λ ratios (as in Fig. 3) vs.
the logarithm of the effective temperature of the respective target
stars. Circled points are for stars identified to exist within
prominent O-B associations (Humphries 1978).

So far, the discussion here has emphasized the possibility that the extra Fe in the halo is caused by a more efficient destruction of grains. Perhaps it is worthwhile to consider an alternate origin, however, such as enrichment of the low-density gas away from the plane by the ejecta of type I supernovae. Such ejecta are believed to be especially rich in iron, and the importance of such injection could be far greater in the halo than in the disk of our galaxy. From a theoretical standpoint, it is probably very difficult to estimate quantitatively the contributions from direct element enrichment by supernovae in the halo, because we can draw only upon rough estimates on the composition and amount of the ejecta, along with a very crude picture on the rate (and even the overall nature) of the cycling of high-z gaseous material.

An insight on whether the extra Fe comes from grains or supernovae may evolve from further observational studies. For instance, the methods used here could be applied to shocked, high-velocity gas components seen in the galactic plane. For these absorptions, which occur near known supernovae or active stellar associations (Jenkins, Silk and Wallerstein 1976, Cowie, Songaila and York 1979), grains are probably being destroyed. If the differential changes in depletion from one element to another are the same as in the halo gas, one would favor the grain destruction as the proper explanation for the enrichment in the halo.

This research was supported by grant NSG-5248 from the U.S. National Aeronautics and Space Administration. A vital ingredient to the success of this investigation has been the IUE project policy of providing old observations, in easily processable form, to outside researchers. The author is indebted to Dr. W. Warren and C. Perry of the National Space Science Data Center for their cheerful and effective response to a gigantic request for data.

REFERENCES

Albert, C.E.: 1982, Ap.J. (Letters), 256, pp. L9-L12.
Barlow, M.J. and Silk, J.: 1977, Ap.J. (Letters), 211, pp. L83-L87.
Cohen, J.G. and Meloy, D.A.: 1975, Ap.J., 198, pp. 545-549.
Cowie, L.L., Songaila, A. and York, D.G.: 1979, Ap.J., 230, pp. 469-484.
Draine, B.T. and Salpeter, E.E.: 1979, Ap.J., 231, pp. 438-455.
Greenstein, J.L. and Sargent, A.I.: 1974, Ap.J. (Suppl.), 28, pp. 157-209.
Humphries, R.M.: 1978, Ap.J. (Suppl.), 38, pp. 309-350.
Jenkins, E.B., Silk, J. and Wallerstein, G.: 1976, Ap.J. (Suppl.), 32, pp. 681-714.
Morton, D.C.: 1974, Ap.J. (Letters), 193, pp. L35-L40.
Morton, D.C.: 1978, Ap.J., 222, pp. 863-880.
Morton, D.C. and Bhavsar, S.P.: 1979, Ap.J., 228, pp. 127-146.
Morton, D.C. and Smith, W.H.: 1973, Ap.J., (Suppl.), 26, pp. 333-364.

Pettini, M. and West, K.:   1982, preprint.

Savage, B.D. and Bohlin, R.C.:   1979, Ap.J., 229, pp. 136-146.

Savage, B.D. and de Boer, K.S.:   1979, Ap.J. (Letters), 230,
        pp. L77-L82.

Savage, B.D. and de Boer, K.S.:   1981, Ap.J., 243, pp. 460-484.

Shull, J.M.:   1978, Ap.J., 226, pp. 858-862.

Shull, J.M., Snow, T.P. and York, D.G.:   1981, Ap.J., 246,
        pp. 549-553.

Shull, J.M., Van Steenberg, M. and Seab, C.G.:   1982, preprint.

Snow, T.P.:   1976, Ap.J., 204, pp. 759-774.

Snow, T.P.:   1977, Ap.J., 216, pp. 724-737.

Snow, T.P. and Jenkins, E.B.:   1980, Ap.J., 241, pp. 161-172.

Snow, T.P. and Meyers, K.A.:   1979, Ap.J., 229, pp. 545-552.

Spitzer, L. and Jenkins, E.B.:   1975, Ann. Rev. Astron. Astrophysics,
        13, pp. 133-164.

Spitzer, L.:   1976, Comments Ap., 6, pp. 177-187.

York, D.G. and Kinahan, B.F.:   1979, Ap.J., 228, pp. 127-146.

York, D.G. and Jura, M.:   1982, Ap.J., 254, pp. 88-93.

NOTES:

[1]   One way out of this problem, suggested by R. Polidan (private communication), is to propose that these stars have been rejuvenated recently by mass deposition from (as yet unrecognized) close companion stars which are relatively long lived but which have just recently evolved off the main sequence.

[2]   Sources for transition of f-values:
            FeII from Shull, Van Steenberg and Seab (1982)
            SiII from Shull, Snow and York (1981)
            AlII and SII from Morton and Smith (1973)

GAMMA RADIATION AS A TRACER OF THE LOCAL INTERSTELLAR GAS

J.B.G.M. Bloemen
Cosmic Ray Working Group and Sterrewacht, Leiden,
The Netherlands
On behalf of the COS-B Caravane Collaboration

1. INTRODUCTION

During a successful mission of more than 6½ years, from August 1975 until
April 1982, the ESA COS-B satellite (Scarsi et al. 1977) has observed
the sky in gamma rays of energy greater than 50 MeV. The Milky Way is
completely covered (Mayer-Hasselwander et al. 1982) and at all longitudes
the survey is extended to medium latitude regions at $|b| \lesssim 20°$ or at some
selected regions even further.
The major part of the galactic gamma-ray emission is concentrated *along
the galactic plane* and has been mapped by the COS-B experiment in
unprecedented detail. Although the physical nature is still uncertain,
it most probably consists partly of emission diffuse in nature, i.e.
originating from the interaction of energetic cosmic rays with the interstellar matter and radiation fields, and partly of emission from discrete
sources (Swanenburg et al. 1981). In spite of much work (Bignami, Caraveo
and Maraschi 1978, Protheroe et al. 1979, Hermsen 1980, Rothenflug and
Caraveo 1980, Riley and Wolfendale 1980, Salvati and Massaro 1982) the
question of the relative contribution of these two components remains
still largely open. However, the unresolved sources of gamma rays
compiled in the 2nd COS-B catalogue (Swanenburg et al. 1981) have a
very narrow latitude distribution ($< |b| > \cong 1°.5$), despite their greater
detectability away from the intense galactic-disc emission, so that it
is very likely that most of the *medium latitude* emission is diffuse in
nature. The contributions of extragalactic gamma radiation ((1.3 ± 0.5)
$\times 10^{-5}$ph $cm^{-2}s^{-1}sr^{-1}$ above 100 MeV, Thompson and Fichtel 1982) and of
gamma rays originating from the interaction of energetic cosmic rays
with radiation fields (e.g. Kniffen and Fichtel 1981) are small. Therefore gamma rays at these latitudes are mainly produced by the interaction
of energetic cosmic rays with the interstellar gas. In addition gamma
radiation doesn't suffer absorption, so that it is expected to be an
excellent tracer of the local interstellar gas distribution at medium
latitudes.
Despite possible inhomogenities in the cosmic-ray-energy density, gamma-
ray astronomy adds a most valuable new dimension to such existing and
diverse tracers as:

31

W. L. H. Shuter (ed.), Kinematics, Dynamics and Structure of the Milky Way, 31–41.
Copyright © 1983 by D. Reidel Publishing Company.

(i)   stellar reddening
(ii)   the line and continuum emission of interstellar atoms and
        molecules at radio and millimeter wavelengths.
(iii)   the distribution of the soft X-rays of diffuse thermal origin
(iv)   star counts and galaxy counts
(v)   UV absorption of H_2 Lyman lines towards bright stars
The distribution of the local interstellar gas is not well known. While
the column density of atomic hydrogen (N_{HI}) has been thouroughly mapped
by the 21 cm line emission, direct observations of molecular hydrogen
using tracer (v) are sofar restricted to a limited number ($\sim$100) of
directions (Savage et al. 1977, Bohlin et al. 1978). Large scale mapping
of the various CO (e.g. Blitz 1980) and OH (Wouterloot 1981) lines at
millimeter and radio wavelengths will give adequate coverage eventually,
but the quantitative derivation of the H_2 density is too uncertain to
allow straight forward estimates to be made without possible serious
error (see e.g. Federman et al. 1980, Frerking et al. 1981, Lequeux 1981,
Wouterloot 1981) even disregarding the high optical depth effects.
Although surveys of the reddening of stars at distances up to 2 kpc
(Fitzgerald 1968, Lucke 1978, Neckel and Klare 1980) give a good overview
of the total absorption distribution, the density of useful stars at
intermediate latitudes is not adequate to show structure on scales $\leq$10°.
Also selection biases and the patchiness of the absorption are difficult
to correct for. Star-count data are not available with wide and uniform
coverage and are restricted to regions of high contrast in absorption.
     The use of galaxy-count data (Heiles 1976, Burstein and Heiles 1978,
Strong and Lebrun 1982) allows more uniform and unbiased sampling of the
total absorption for $|b|$>10° but is limited to regions with A_{pg}<2
magnitude and is affected by galaxy clustering. This most reliable gas-
tracer known sofar, will be shown to correlate well with the gamma-ray
intensities (Section 2). The close correlation enables the independent
use of gamma-ray data to indicate the total gas distribution where no
galaxy-count data are available. Using the good statistics after the
long COS-B mission, column densities of hydrogen in molecular form can
now be predicted (Section 4).
The cosmic-ray distribution has a non-negligible impact on the gamma-ray
tracer. It is not only the (un)homogenity of the cosmic-ray-energy
density that should be taken into account, but also the fact that the
much awaited π° decay bump at 70 MeV does not show up in the measurements
implies that we have to consider both protons and electrons (dominating
the gamma-ray production at respectively high and low energies).

It is the purpose of the present paper to give a status report on our
knowledge of the local interstellar medium at medium latitudes as seen
in gamma rays detected by the COS-B experiment during the first five
years of operation time and on the correlations with the other gas
tracers (especially the galaxy counts). After an overall comparison we
will go down to molecular-cloud scales and briefly indicate the prospects
of gamma-ray astronomy.

2. GAMMA-RAY INTENSITY AND GAS COLUMN DENSITIES

In the past several authors have discussed the correlations between
gamma rays measured by the SAS-2 satellite and the interstellar gas at
medium and high galactic latitudes giving information about the local
gamma-ray emissivity per interstellar gas atom (Fichtel et al. 1978,
Lebrun and Paul 1979, Strong and Wolfendale 1981).
Lebrun et al. (1982) studied the correlation of gamma-ray intensity I_γ
measured by COS-B with N_{HI} and N_{HT} (the total gas column density as
estimated from galaxy counts) from a statistical viewpoint. They showed
that a convincing correlation between N_{HT} and I_γ exists in the latitude
range $10° < |b| < 20°$ and the correlation with N_{HI} only is considerably
worse, similar to the conclusions of the analysis of the SAS-2 data. At
the moment more gamma-ray data from later COS-B observation periods are
available, which enables us to make the spatial comparison with the
total gas column density as estimated from galaxy counts in a more
explicit form (see also Strong et al. 1982a). Also the more extensive
work that has been done to determine corrections for the variations in
the instrumental background due to solar modulation and for the variations
in the instrument response over the five years of observation time
included here (August 1975 until October 1980) makes a more detailed
analysis possible.

The comparison of I_γ with N_{HT} and N_{HI} shown here, has been performed for
latitudes $11° < |b| < 19°$ and over the whole longitude range. Both N_{HT} and
N_{HI} have been convolved with the COS-B instrumental point-spread function
in the corresponding energy range.
     The gamma-ray data have been divided into three energy intervals:
70-150 MeV, 150-300 MeV and 300-5000 MeV. The field of view for each
observation period was restricted to regions within a radius of 20°
from the pointing direction. An E^{-2} input spectrum was assumed to
include the energy response of the instrument.
     The value of N_{HT} was estimated from the Lick galaxy counts (Shane
and Wirtanen 1967, Seldner et al. 1977) assuming

$$N_{HT} = 2.0 \times 10^{21} \; Log \; (50/N_g) \; atom \; cm^{-2} \qquad (2.1)$$

as given by Strong and Lebrun (1982). The values of N_g have been averaged
over $3° \times 3°$ bins to reduce the effects of galaxy clustering and statistical
fluctuations and to obtain a reasonable estimate of $<N_g>$ for regions with
many $1° \times 1°$ bins having $N_g = 0$ as is common for $|b| < 10°$. The latter
regions with $N_g < 1$ (corresponding by equation 2.1 to $N_{HT} > 3.5 \times 10^{21}$ atom
cm^{-2}) are only used for their contribution to the distribution at $|b| > 11°$
after convolution with the COS-B instrumental point-spread function.
     Values of N_{HI} for the northern celestial hemisphere were derived
from the surveys of Weaver and Williams (1973) and Heiles and Habing
(1974). For the southern hemisphere the data of Heiles and Cleary (1979)
for $|b| > 10°$ and Strong et al. (1982b) for $|b| < 10°$ were used. Also here,
data for $|b| < 11°$ were only required for the convolution.

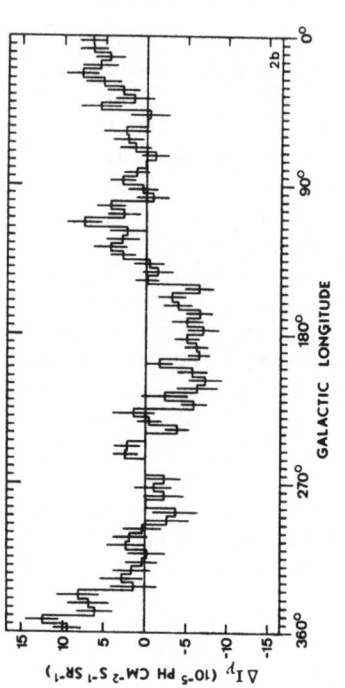

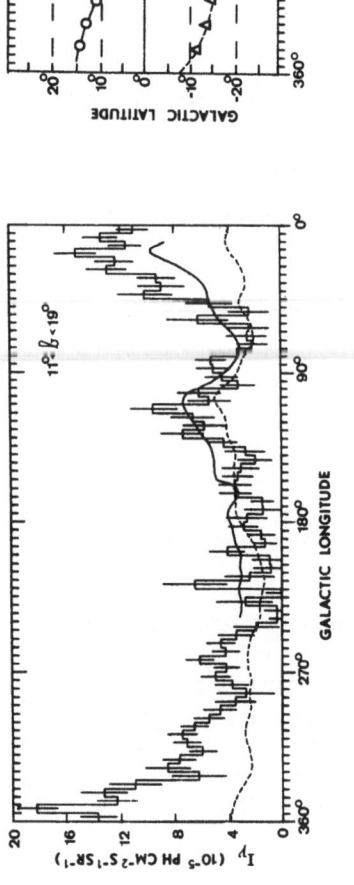

Figure 2: a. Schematic view of two local systems of Dolidze (1980) (△) and Gould (Stothers and Frogel 1974) (o).
b. Difference between average gamma-ray intensities (70–5000 MeV) in the regions 11°<b <19° and −19°<b< −11°.

Figure 1: Comparison of measured and predicted gamma-ray intensities in the energy range 70–5000 MeV. The bars show the average measured intensity with statistical errors. The predicted intensity from N_{HT} as traced by galaxy-count data is indicated by the continuous line, using the emissivities of table 1. The dashed line shows the predicted intensity for atomic hydrogen alone, using the same parameters.

Table 1: Gamma-ray emissivities determined by maximum-likelihood fitting of the total-gas-column density, as estimated from galaxy-count data, to the gamma-ray intensities in three energy ranges.

Energy range (MeV)	$q_\gamma/4\pi$ (10^{-26} atom^{-1}s^{-1}sr^{-1})
70 – 150	1.40
150 – 300	0.53
300 – 5000	0.59

For the comparison of the gamma-ray intensity and the total-gas-column density in the latitude range mentioned, it is tested that the gamma-ray emission results from the interactions between a uniform cosmic-ray flux and the interstellar gas:

$$I_\gamma = (q_\gamma/4\pi)\,\tilde{N}_{HT} + I_B \qquad (2.2)$$

(see Lebrun et al. 1982, Strong et al. 1982a), where $\tilde{N}_{HT}$ is the convolved total gas column density, q_γ the gamma-ray emissivity per hydrogen atom and I_B an isotropic, mainly residual instrumental background. Since the galaxy counts are available only down to $\delta = -25°$, this analysis is restricted to the longitude range $10° \leq l \leq 240°$. For each energy range q_γ and I_B were determinded using a maximum-likelihood method on bins of $1° \times 1°$ (see also Lebrun et al. 1982 and Strong et al. 1982a). Regions with obvious galaxy clustering were excluded. Also the region $11° < b < 19°$ and $l < 40°$ was excluded from the fitting procedure. It seems that this is the only large region where clear deviations from our testing hypothesis are present and inclusion would considerably increase the derived emissivity values. Possible reasons for these discrepancies are given in the last paragraph of this section. The maximum likelihood estimates of q_γ are given in table 1. No statistical errors are indicated, since for the moment the systematic uncertainties of $\sim 25\%$ are significantly larger. The ratio of the emissivities in the three energy ranges 0.56: 0.2: 0.23 is in good agreement with the E^{-2}-input-spectrum assumption (0.54: 0.24: 0.22). The observed intensities, averaged over regions for which COS-B exposure exists, and those predicted from N_{HT} and N_{HI} are shown in Figure 1 for the total energy range 70 – 5000 MeV. In this energy range the prediction is the sum of the predictions for the three individual energy ranges.

The agreement between the predicted emission from the total gas and the observed gamma-ray intensity is good. The discrepancies can be ascribed to instrumental uncertainties, such as the sensitivity of the instrument and the instrumental background, and/or to real deviations from the assumptions contained in equations (2.1) and (2.2). The instrumental uncertainties are not very serious, because the relevant corrections have been determined by comparison of different observations of the same regions. The real discrepancies include effects of galaxy clustering (e.g. the large cluster in Perseus affects the prediction

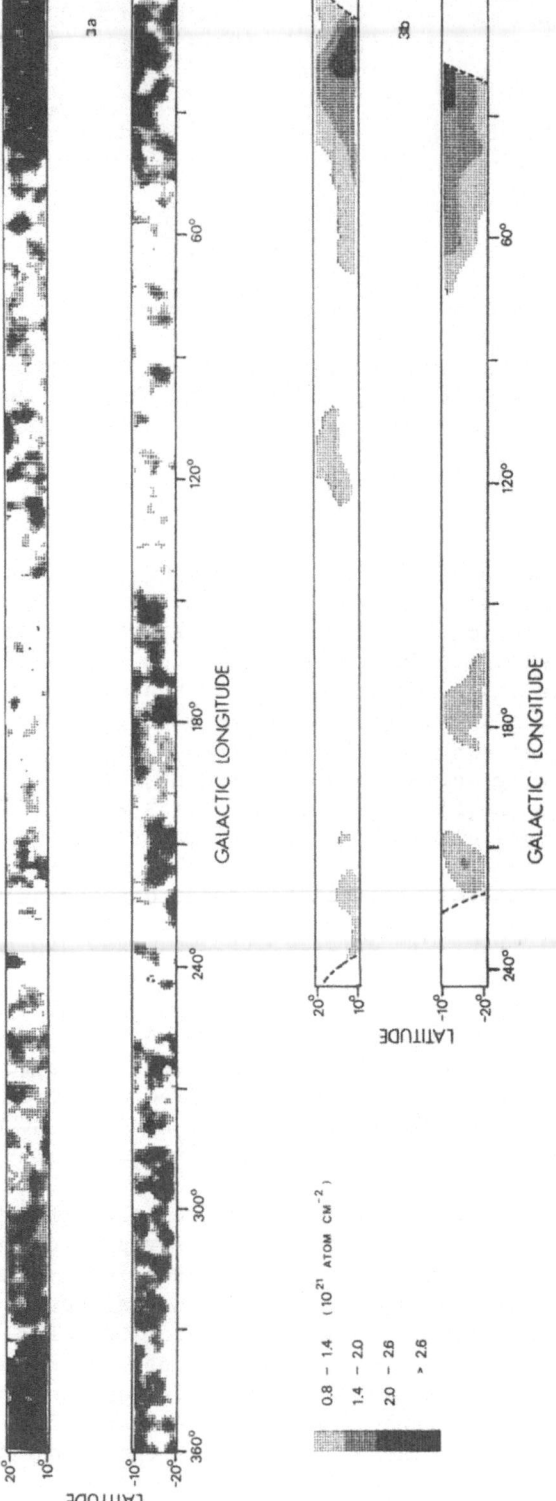

Figure 3: a. Map of molecular hydrogen column densities derived from gamma-ray excesses relative to the expectation from atomic hydrogen.
b. Map of molecular hydrogen column densities derived from total-gas-column density excesses (as derived from galaxy count data) relative to atomic hydrogen column densities. Regions outside the dashed lines correspond to areas where no Lick galaxy counts are available. The convolution is the same as for Figure 3a.

at 1=150° for negative latitudes), fluctuations in the gas-to-dust ratio, the failure of equation (2.1) as N_g tends to zero, spatial variations in the cosmic-ray intensity and the presence of genuine point sources not sufficiently intense to be independently recognized as such. Due to the non-linearity of the galaxy counts, the total-gas-column densities are underestimated at some places, reducing the dynamic range compared to the gamma-ray intensities.

The agreement appears equally good in each energy range (Strong et al. 1982a) giving no indication of an energy dependent exclusion or enhancement of cosmic-ray electrons or protons in clouds, or of a different spatial distribution of protons and electrons. The overall agreement indicates that on scales ≥10° the *product* of the gas-to-dust ratio and the cosmic-ray density does not vary by large factors.

3. THE LOCAL BELT SYSTEMS

An overall view of Figure 1 clearly indicates a strong gamma-ray excess in the longitude range 320°-40° and one in the anti centre from 1=150° to 1=220°. These excesses are associated with the well known Gould's Belt. This local disc-shaped structure of OB-, T- and R-associations and various types of diffuse nebulae and matter (Stothers and Frogel 1974, Frogel and Stothers 1979), was for the first time qualitatively seen in the SAS-2 gamma-ray data (Fichtel et al. 1975, Hartman et al. 1979). The discovery of a second structural feature similar to the Gould's Belt, but quite independent of it, has been reported by Dolidze (1980 a,b) and has been shown to be visible in the COS-B data base by Bignami (1981) and Bignami et al. (1981). A schematic view of both systems is shown in Figure 2a. Especially in the Cepheus region (1=100°-130°) the 'Dolidze Belt' excess is clearly visible in Figure 1. All these features are also outstanding excesses in the stellar reddening map of Lucke (1979). An overview of the differences between the average intensities in the relevant latitude ranges above and below the galactic plane is shown in Figure 2b as a function of longitude. The correlation with Figure 2a is striking.

4. MOLECULAR HYDROGEN

Figure 1 shows that the gamma-ray data can be much better fitted by the total-gas estimates than those from HI alone as pointed out by Lebrun and Paul (1979) using the SAS-2 data and by Lebrun et al. (1982) using COS-B data. The difference between the gamma-ray distribution and the HI estimate is an indication of the presence of H_2 . The most wide spread region where this occurs is in the longitude range 320°-40°. This supports the conclusion of Strong and Lebrun (1982) that the large absorption indicated by the galaxy-count data in these regions is a result of large N_{HT} and not of a lower gas-to dust ratio. Three other outstanding excesses of N_{HT} over N_{HI} and measured gamma-ray intensity over that estimated from HI can be seen in Figure 1: in Cepheus (1=100°-130°, b>11°), in Taurus en Perseus (1=150°-190°, b<11°) and in Orion

(1=200°-220°, b<11°). Last two complexes correspond to known nearby
molecular cloud complexes mapped in CO (Blitz 1980) and OH (Wouterloot
1981). So, for 240°≤l≤360°, where there are no galaxy-count data, the
gamma-ray excesses reveal, apparantly, large column densities of
molecular hydrogen in the longitude range 330°<l<360°, both above and
below the galactic plane. The peak at l≅355° in Figure 1a corresponds
to the ρ Oph molecular cloud, already known as a gamma-ray source
(Bignami and Morfill 1980, Swanenburg et al. 1981). The good one-dimensional correlation between N_{HT} and gamma-ray intensity enables us to
map H_2 in two dimensions using the residual gamma-ray flux (see Strong
et al. 1982a):

$$\hat{N}_{H_2} = (4\pi/q_\gamma) \ (I_\gamma - I_B) - \hat{N}_{HI}.\eqno(4.1)$$

The resulting map is shown in Figure 3a. Figure 3b shows a similar
presentation of N_{H_2} deduced from galaxy counts using equation (2.1) and

$$\hat{N}_{H_2} = \hat{N}_{HT} - \hat{N}_{HI}.\eqno(4.2)$$

Because both methods of 'H₂-mapping' have their own uncertainties
discussed before, features present in both Figures 3a and 3b have a
higher probability of being genuine H_2 features, than the ones appearing
only in one map.

The exposure of the gamma-ray measurements is now that good, that
correlations with other gastracers on smaller scales are possible. The
excellent correlation of gamma-ray measurements with CO observations of
the Orion complex is already shown by Caraveo et al. (1980, 1981), using
only a part of the COS-B exposure available. The total mass of the OriA
and OriB clouds, estimated from the gamma-ray flux and assuming the local
emissivity value, is consistent with the various estimates from CO data
($(1.5-2) \times 10^5 \ M_\odot$). Just to point out the excellent *positional*
coincidence, Figure 4 shows the main part of the Orion complex as seen
in CO millemeter emission and gamma-rays. A detailed analysis of the
whole complex and of other near molecular-cloud complexes at medium
latitudes such as Ophiuchus, Taurus, Perseus and Cepheus, is in progress.

5. CONCLUSIONS

It is shown that a good correlation between high-energy gamma rays and
the total absorption along the line-of-sight exists at intermediate
latitudes. The gamma-ray intensity at these latitudes can be interpreted
by the interaction of a uniform cosmic-ray flux with the local interstellar gas assuming a constant gas-to-dust ratio. Local galactic
features such as the Gould's Belt and Dolidze Belt are clearly revealed
by the gamma-ray data.

On smaller scales molecular cloud complexes can be seen and H_2 column
densities have been derived. The local molecular hydrogen on the
southern hemisphere is explicitly revealed for the first time by gamma-

rays. The H_2 maps provide a completely sampled coverage for all galactic longitudes and should therefore be an useful basis for future studies of local molecular-cloud complexes. With the long COS-B mission, gamma-ray astronomy provides now one of the most homogeneous and complete indicators of local interstellar matter.

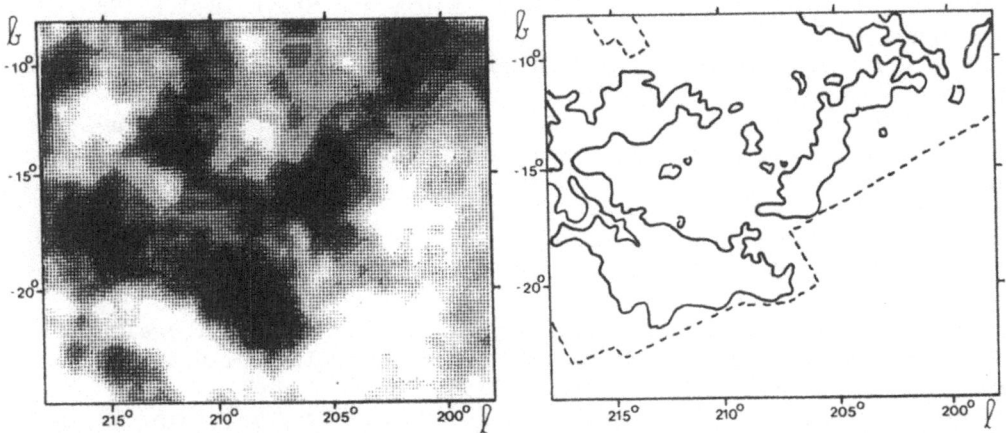

Figure 4: Half-tone plot of the gamma-ray intensities (left) and CO contour plot of 1°K peak antenne temperature (right, Columbia CO group private communication, using the data collected by Kutner et al. 1977 and Morris, Montani and Thaddeus 1977) of the Orion region. The dashed line indicates the boundaries of the CO survey.

REFERENCES

Bignami, G.F.:1981, Phil.Trans.Roc.Soc.Lond. A301, P.555
Bignami, G.F., and Morfill, G.E.:1980, Astron.Astrophys. 87, p.85
Bignami, G.F., Caraveo, P.A., Maraschi, L.:1978, Astron.Astrophys. 67, p.149
Bignami, G.F., Barbareschi, L., Bloemen, J.B.G.M., Buccheri, R., Caraveo, P.A., Hermsen, W., Kanbach, G., Lebrun, F., Mayer-Hasselwander, H.A., Paul, J.A., Strong, A.W., and Wills, R.D.: 1981, Proc. 17th Int. Cosmic Ray Conference (Paris) 1, p.182
Blitz, L.: 1980 in 'Giant Molecular Clouds in the Galaxy', eds. P.M. Solomon and E.G. Edmunds, Pergamon Press, p.1
Bohlin, R.C., Savage, B.D., and Drake, J.F.: 1978, Astrophys.J. 224, p.132
Burstein, D., and Heiles, C.: 1978, Astrophys.J. 225, 40
Caraveo, P.A., Bennett, K., Bignami, G.F., Hermsen, W., Lebrun, F., Masnou, J.L., Mayer-Hasselwander, H.A., Paul, J.A., Sacco, B., Scarsi, L., Strong, A.W., Swanenburg, B.N., and Wills, R.D.: 1980, Astron. Astrophys. 91, p.L3

Caraveo, P.A., Barbareschi, L., Bennett, K., Bignami, G.F., Hermsen, W., Kanbach, G., Lebrun, F., Manou, J.L., Mayer-Hasselwander, H.A., Sacco, B., Strong, A.W., and Wills, R.D.: 1981, Proc. 17th Int.Cosmic Ray Conference (Paris) 1, p.139

Dolidze, M.V.: 1980a, Soviet Astron. 6, p.51

Dolidze, M.V.: 1980b, Soviet Astron. 6, p.394

Federman, S.R. Glassgold, A.E., Jenkins, E.D., and Shaya, E.J.: 1980, Astrophys.J. 242, p.545

Fichtel, C.E., Simpson, G.A., and Thompson, D.J.: 1978, Astrophys.J. 222, p.833

Fichtel, C.E., Hartman, R.C., Kniffen, D.A., Thompson, D.J., Bignami, G.F., Ogelman, H., Ozel, M.E., and Tümer, T.: 1975, Astrophys.J. 198, p.163

Fitzgerald, M.P.: 1968, Astron.J. 73, p.983

Frerking, M.A., Langer, W.D., and Wilson, R.W.: 1981, preprint Bell Telephone Labs, Holmdel, New Jersey

Frogel and Stothers, R. : 1977, Astron.J. 82, p.890

Hartman, R.C., Kniffen, D.A., Thompson, D.J., Fichtel, C.E., Ogelman, H.B., Tümer, T., and Ozel, M.E.: 1979, Astrophys.J. 230, p.597

Heiles, C.: 1976, Astrophys.J. 204, p.379

Heiles, C. and Clearly, M.N.: 1979, Aust.J.Phys.Astrophys. Supp. 47, p.1

Heiles, C. and Habing, H.J.: 1974, Astron.Astrophys. Supp. 14, p.1

Hermsen, W.: 1980, Ph.D. thesis, University of Leiden

Kniffen, D.A., and Fichtel, C.E.: 1981, Astrophys.J. 250, p.389

Kutner, M.L., Tucker, K.D., Chin, G., and Thaddeus, P.: 1977, Astrophys. J. 215, p.521

Lebrun, F., Bignami, G.F., Buccheri, R., Caraveo, P.A., Hermsen, W., Kanbach, G., Mayer-Hasselwander, H.A., Paul, J.A., Strong, A.W., and Wills, R.D.: 1982, Astron.Astrophys. 107, p.390

Lebrun, F., and Paul, J.A.: 1979, Proc. 16th Int.Cosmic Ray Conference, Kyoto 12, p.13

Lequeux, J.: 1981, Comments on Astrophysics 9, p.117

Lucke, P.B.: 1978, Astron.Astrophys. 64, p.367

Mayer-Hasselwander, H.A., Bennett, K., Bignami, G.F., Buccheri, R., Caraveo, P.A., Hermsen, W., Kanbach, G., Lebrun, F., Lichti, G.G., Masnou, J.L., Paul, J.A., Pinkau, K., Sacco, B., Scarsi, L., Swanenburg, B.N. and Wills, R.D.: 1982, Astron.Astrophys. 105, p.164

Morris, M., Montani, J., and Thaddeus, P.: 1980, IAU symposium No.87, Interstellar Molecules, ed. B.H. Andrew, p.197

Neckel, Th., and Klare, G.: 1980, Astron.Astrophys. Suppl. 42, p.251

Protheroe, R.J., Strong, A.W., Wolfendale, A.W., and Kiraly, P.: 1979, Nature 277, p.542

Riley, P.A., and Wolfendale, A.W.: 1981, Rivista Del Nuova Cimento Vol.4, no.4: 1

Rothenflug, R., and Caraveo, P.A.: 1980, Astron.Astrophys. 81, p.218

Salvati, M., and Massaro, E.: 1982, Mon.Not.R.astr.Soc. 198, p.11

Savage, B.D., Bohlin, R.C., Drake, J.F., and Budich, W.: 1977, Astrophys. J. 216, p.291

Scarsi, L., Bennett, K., Bignami, G.F., Boella, G., Buccheri, R., Hermsen, W., Koch, L., Mayer-Hasselwander, H.A., Paul, J.A., Pfeffermann, E., Stiglitz, R., Swanenburg, B.N., Taylor, B.G., and Wills, R.D.: 1977, Proc. 12th ESLAB Symp., ESA SP124, p.3

Seldner, M., Siebers, B., Groth, E.J. and Peebles, P.J.E.: 1977, Astron. J. 82, p.249

Shane, C.D., and Wirtanen, C.A.: 1967, Pub.Lick.Obs., 22, p.1

Stothers,R., and Frogel, J.A. : 1974, Astron.J. 79, p.456

Strong, A.W., and Lebrun, F.: 1982, Astron.Astrophys. 105, p.159

Strong, A.W., and Wolfendale, A.W.: 1981, Phil.Trans.Roy.Soc. Lond. A301, p.541

Strong, A.W., Bignami, G.F., Bloemen, J.B.G.M., Buccheri, R., Caraveo, P.A., Hermsen, W., Kanbach, G., Lebrun, F., Mayer-Hasselwander, H.A., Paul, J.A., and Wills, R.D.: 1982a, Astron.Astrophys. (in press)

Strong, A.W., Murray, J.D., Riley, P.A., and Osborne, J.L.: 1982b, Mon. Not.R.astr.Soc. (in press)

Swanenburg, B.N., Bennett, K., Bignami, G.F., Buccheri, R., Caraveo, P. A., Hermsen, W., Kanbach, G., Lichti, G.G., Masnou, J.L., Mayer-Hasselwander, H.A., Paul, J.A., Sacco, B., Scarsi, L., and Wills, R.D.: 1981, Astrophys.J. 243, p.L69

Thompson, D.J., and Fichtel, C.E.: 1982, Astron.Astrophys. 109, p.352

Weaver, H., and Williams, R.W.: 1973, Astron.Astrophys. Suppl. 8, p.1

Wouterloot, J.G.A.: 1981, Ph.D.Thesis, University of Leiden

THE OORT COMET HALO AND GIANT MOLECULAR CLOUDS

Sidney van den Bergh
Dominion Astrophysical Observatory
Herzberg Institute of Astrophysics
Victoria, B.C.

It was first realised by Biermann (1978) that Giant Molecular Clouds (GMC's) passing close to the sun might eject significant numbers of comets from the Oort (1950) halo. This conclusion was strengthened and confirmed by Napier and Staniucha (1982) who performed numerical integrations of 33000 comet orbits subjected to perturbations by GMC's. The loss of comets may also be described by the same formalism that has been developed to discuss loss of stars from clusters moving in the galactic force field. Early work by King (1962) modified by Keenan (1981) shows that

$$r_t \sim 0.44 \ d \ (M_\odot/M_c)^{1/3} \qquad (1)$$

in which r_t is the tidal limiting radius of the Oort cloud imposed by a GMC of mass M_c passing the Sun at a minimum distance d. Substituting d = 30 pc (which corresponds to the radius of a typical GMC) yields $r_t = 6 \times 10^4$ au for $M_c = 1 \times 10^5 \ M_\odot$ and $r_t = 2.5 \times 10^4$ au for $M_c = 1 \times 10^6 \ M_\odot$. For comets with elongated orbits that pass close to the sun the corresponding semi-major axes are a = 3×10^4 au and a = 1.2×10^4 au, respectively, i.e. values that bracket the a $\sim 2 \times 10^4$ au inner boundary of the Oort Cloud. Calculations by van den Bergh (1982) show that the solar system will have suffered ~ 1 close (d < 30 pc) encounter with a GMC of $1 \times 10^6 \ M_\odot$ and ~ 10 encounters with clouds of $\sim 1 \times 10^5 \ M_\odot$ during its lifetime. Clearly such encounters will decimate the Oort Comet halo, which extends from $2 \times 10^4 < a < 1.5 \times 10^5$ au. This result implies that the Oort Cloud has to be repopulated from more tightly bound regions of the solar system or by capture of interstellar comets.

It has been shown by many authors (e.g. Noerdlinger 1977, Valtonen and Innanen 1982) that such capture of interstellar comets is a highly inefficient process. Clube and Napier (1982) have, however, suggested that the capture hypothesis might be resurected by considering the capture of comets as the sun passes through GMC's. This process is more efficient than capture from empty interstellar space because it involves a three-body interaction between a comet, the sun and a GMC. Clube and Napier show that this process could replenish the Oort comet halo if GMC's contain 0.1 cometesimals au^{-3}, i.e. 10^{21} cometesimals in

W. L. H. Shuter (ed.), Kinematics, Dynamics and Structure of the Milky Way, 43–45.

a typical cloud. In practice it is, however, impossible to build up such a high density of cometesimals because GMC's are "leaky". Cometesimals with random motions of $\sim$ 10 km s^{-1} escape freely from GMC's on a time-scale of a few million years. Objects with velocities of only a few km s^{-1} are gravitationally bound to a GMC but will escape when the GMC itself dissolves over periods of some tens of millions of years (Blitz and Shu 1980). On average a GMC therefore can only hold on to cometesimals for $\sim$ 1 x 10^7 yr. Subsequently they will escape on essentially star-like orbits. As a result the density of cometesimals can, at best, be only one or two per cent higher in GMC's than it is in the disk of the Galaxy (van den Bergh 1982).

From the fact that no strongly hyperbolic comet has ever been observed near the sun Sekanina (1976) has shown that the density of interstellar comets near the sun is < 1.4 x 10^{-4} au^{-3}. Since the density of such objects, which travel on star-like orbits, is more or less uniform throughout the disk of the Galaxy this same upper limit also applies to the density of cometesimals in GMC's. Such a value is $\sim$ 700 times lower than the density which Clube and Napier (1982) require to replenish the Oort Cloud by capture of interstellar comets. It follows that the Oort Cloud must be replenished from a source within the solar system.

Possibly the Oort Cloud, which presently contains $\sim$ 10^{11} comets, is repopulated during very close encounters between the sun and stars. During such events comets might be ejected into the Oort halo from an inner reservoir (Hills 1981) containing $\sim$ 10^{13} objects in the range 1 x 10^3 < a < 2 x 10^4 au. Since comets in this inner reservoir are quite tightly bound they will not, in general, be perturbed into orbits passing close to the sun (Tinsley and Cameron 1974, Hills 1981) by stars passing at distances $\sim$ 1 pc.

Computations by Fernández and Ip (1981) shows that this inner reservoir of comets might have been built up by (proto)-Neptune as this object started to increase in mass thus becoming able to throw cometesimals, that initially had solar distances r $\sim$ 30 au, into long-period orbits.

REFERENCES

Biermann, L.: 1978, in Astronomical Papers Dedicated to Bengt Strömgren Eds. A. Reiz and T. Andersen (Copenhagen Observatory:Copenhagen) p.327.

Blitz, L. and Shu, F.H.: 1980, Ap.J. 238, 148.

Clube, S.V.M. and Napier, W.M.: 1982, Q.J.Roy.Astr.Soc. in press.

Fernández, J.A. and Ip, W.H.: 1981, Icarus 47, 470.

Hills, J.G.: 1981, Astron.J. 86, 1730.

Keenan, D.W.: 1981, Astr.Ap. 95, 340.

King, I.: 1962, Astron.J. 67, 471.

Napier, W.M. and Staniucha, M.: 1982, M.N.R.A.S. 198, 723.

Noerdlinger, P.D.: 1977, Icarus 30, 566.

Oort, J.H.: 1950, Bull.Astr.Inst.Neth. 11, 91.

Sekanina, Z.: 1976, Icarus 27, 173.

Tinsley, B.M. and Cameron, A.G.W.: 1974, Astr.Space Sci. 31, 31.

Valtonen, M.J. and Innanen, K.A.: 1982, Ap.J. 255, 307.

van den Bergh, S.: 1982, J.Roy.Astron.Soc.Canada in press.

ON THE LOCAL STANDARD OF REST

C. YUAN
Department of Physics
The City College of the City University of New York

I. Introduction

Under the influence of a spiral gravitational field, there should be differences among the mean motions of different types of objects with different dispersion velocities in a spiral galaxy. The old stars with high dispersion velocity should have essentially no mean motion normal to the galactic rotation. On the other hand, young objects and interstellar gas may be moving relative to the old stars at a velocity of a few kilometer per second in both the radial (galacto-centric), and circular directions, depending on the spiral model adopted. Such a velocity is usually referred as the systematic motion or the streaming motion. In this paper, we shall show that the conventionally adopted local standard of rest is indeed co-moving with the young objects of the solar vicinity (also see Lin and Yuan 1974). Therefore, it has a net systematic motion with respect to the circular motion of an equilibrium galactic model, defined by the old stars.

Using the spiral pattern consistently adopted by Lin, Yuan and Shu (1969), the radial component of this motion is found to be about 7 km/sec toward the anticenter direction and the tangential component is about 5 km/sec in the direction of rotation. The outward motion of the adopted local standard of rest can be shown in excellent agreement with the observations. In particular, it resolves the long-standing problem of the differences between the northern and southern rotation curves as indicated in HI observations (Kerr 1964). The tangential component, however, is not well supported by the observations. It may suggest that the local arm, the Orion spur, is associated with a mass density which is comparable with a typical spiral arm, and therefore disturbs the systematic motion near the sun.

The observations are summarized in section 2. The results on the systematic motion based on the density-wave theory are presented in section 3. We compare and discuss the results and observations in the concluding section.

W. L. H. Shuter (ed.), Kinematics, Dynamics and Structure of the Milky Way, 47–52.
Copyright © 1983 by D. Reidel Publishing Company.

II. Observations.

The local standard of rest is determined by measuring the solar motion with respect to the local stars and interstellar gas. Since no differentiation was expected between the old stars and the young objects in their mean motions at the time the determinations were made. The adopted local standard of rest relies heavily on the data of young stars, because of their superior observational quality. In fact, B0 stars are given a weight of 6 compared to 0.3 for dM5 stars (Delhaye 1965). Thus the adopted local standard of rest practically coincides with the mean motion of the B0 stars in solar vicinity (See Table 1).

Table 1 summarizes the solar motions determined wih respect to different types of objects in the solar vicinity. The notation (U,V,W) denotes the three velocity components in the anti-center, rotational and vertical directions respectively. The table is by no means complete. However, it gives a fair representation of the existing observations.

Table 1:   Solar Motion

	U km/sec	V km/sec	W km/sec	Ref km/sec
Standard	-10.4	14.8	7.3	1
Peculiar	- 9.0	12.0	7.0	1
B0	- 9.6	14.5	6.7	1
0-B5	- 9.8	17.0	7.5	2
dK0	-10.8	14.9	7.4	1
dK5	- 9.5	22.4	5.8	1
dM0	- 6.1	14.6	6.9	1
dM5	- 9.8	19.3	8.6	1
H_2CO	-10.1	18.2	10.2	2
CO:Dark Cloud	-11.1	19.9	11 6	2
High latitude HI	-11.9	15.7	7.7	3
Molecular Clouds	- 6.0	-	-	4

1 Delhaye (1965)   3. Mast and Goldstein (1970)
2. Frogel and Stothers(1977)   4. Blitz (1982)

It is clear from table 1 that the U component of the solar motion changes very little among different types of objects, from very young to moderately old, or to be more precise, from those with low velocity dispersion to those with high velocity dispersion. The V component does display some significant change. But, the situation is complicated by the effect of the asymmetric drift. Extrapolating the data of the stars of higher velocity dispersion, Delhaye (1965) arrived at a conclusion that the true value V after the correction of the asymmetric drift lies between 10 km/sec to 13 km/sec. He adopted the value 12 km/sec, which is reflected in the peculiar solar motion listed in Table 1. If this value is indeed correct, the young objects such as the B0 stars, HI gas, etc are moving at a velocity lower than the mean motion of the old stars in the rotational direction (2.5 km/sec for B0, 3.7 for HI gas, etc.).

Another important piece of observations to be included in
ourdiscussion is the well known result of the difference between the
northern and southern rotation curves determined from the HI data
(Kerr 1964). This result should be given higher weight than those of
the solar motion, in the sense that it deals with the local motion
relative to the Galaxy as a whole, while the solar motion observations
are confined only to the immediate neighborhood of the Sun. If the
adopted local standard of rest has an outward motion u_ϖ, the
difference between the northern and southern observatins can be
expressed as

$$|v_{1s}|_{north} - |v_{1s}|_{south} = 2u_\varpi \cos 1$$

where v_{1s} is the line-of-sight velocity of HI with respect to the
adopted local standard of rest at the tangent points and 1 stands for
+1 for the northern data or -1 for the southern data. Comparing the
HI observational data with this formula, Kerr (1964) suggested a
general expansion of 7 km/sec near the sun. It might be noted that
the tangential component of the net motion of the adopted local
standard of rest has no contribution at all here.

III. Systematic Motion in the Solar Vicinity.

In response to the spiral gravitational field, the stars and the
gas would deviate from the pure circular motion of an equilibrium
model. The amount of deviation will depend on the spiral pattern, the
strength of the spiral gravitational field, the velocity dispersion
and finally the location relative to the spiral arms. Using the
spiral model that we have consistently adopted since 1969, we
calculate the systematic motions of the stars and the gas along the

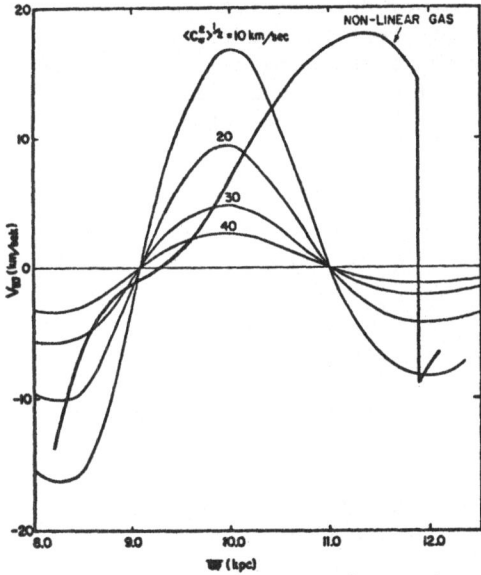

Figure 1 Radial Systematic Motions
of the Stars and Gas

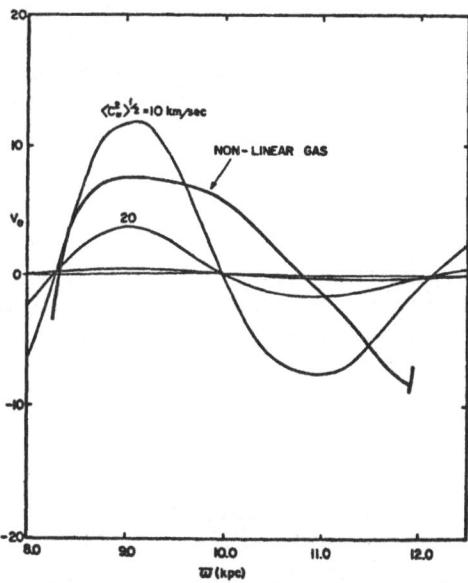

Figure 2. Tangential Systematic
Motions of the stars and the gas

line connecting the galactic center and the sun from 8 kpc to 12 kpc.
The results of the calculations are plotted in Figures 1 and 2. The
response in stars is calculated by the linear density-wave while the
response in the gas, by the non-linear theory. One of the underlying
assumptions of these calculations is that the stars and the gas are in
a quasi-stationary state. This is certainly true for the moderately
old stars, or stars with higher dispersion velocities as well as the
gas. But, it is not true for the very young stars whose have just
been formed out of gas clouds, and their initial conditions at birth
probably still play a dominant role in their present motions. Figures
1 and 2, therefore are applicable to the gas in general and to the
stars of dispersion velocity at least higher than 20 km/sec.

For stars with low dispersion velocities, their mean motion should
be determined by the method of star migration calculations (Yuan and
Grosbol 1982). Unfortunately, due to our choice of the spiral model,
the very young stars, say O-B5, whose ages are within 60×10^6 years
(Stothers 1980) would not have enough time to migrate to the solar
vicinity after their birth in the Sagittarius Arm and the Perseus
Arm. The sun is situated between these two major arms in our model
(See Figure 3). Those young stars near the sun are likely formed by
mechanisms other the shock-driven star formation proposed by Roberts
(1969) or simply delayed in formation after the galactic shock by some
reasons. Being so young, they probably still move with the same
velocity of the nearby gas clouds out of which they were formed.
Therefore, the present purpose, we shall use the results of the
non-linear gas motion for the young stars as a rule. In order to get
some feeling about the star migration calculations, the mean velocity
for B5 stars, with age ranges $55-70 \times 10^6$ (Stothers 1980) is $v_\varpi = 10$
km/sec and $v_\theta = 0$ km/sec, although the number density for those
stars formed in the Sagittarius Arm and the Perseus Arm and having
migrated to the solar vicinity is too small to be considered seriously.
The theoretical results are summarized in Table 2 in which (u_ϖ, u_θ)
is the systematic motion for the gas and (v_ϖ, v_θ), for the stars.

Table 2   Systematic Motions at the position of
the Sun (in km/sec).

Non-linear gas	$u_\varpi = 6.5$	$u_\theta = 5.0$
Stars with $\langle c_\varpi^2 \rangle^{1/2} = 20$ km/sec	$v_\varpi = 9.5$	$v_\theta = 0$
Stars with " 30 "	$v_\varpi = 5.0$	$v_\theta = 0$
Stars with " 40 "	$v_\varpi = 3.0$	$v_\theta = 0$
Star Migration for B5 stars	$v_\varpi = 10.0$	$v_\theta = 0$

IV. Discussion

We first examine the radial motion of the stars and the
gas in the solar vicinity. Let us recall that the adopted local
standard of rest is practically comoving with the B0 stars, and

there seems no significant difference among various types of object in
their radial velocity components. Furthermore, as suggested by the
difference of the northern and southern rotation curves, the adopted
local standard of rest seems to have a net motion of 7 km/sec in the
anti-center direction. The theoretical results for the radial
systematic motion listed in Table 2 are in perfect agreement with
these obvservations. The young objects, such as BO stars as well as
the HI gas are indeed expected to have a systematic motion towards the
anti-center direction with a magnitude equal to about 7 km/sec. For
the moderately old stars whose dispersion velocites are around 30
km/sec, the outgoing motion is only slightly reduced to 5 km/sec.
Only when the velocity dispersion reaches 40 km/sc or more, does the
motion of the star drop below 3 km/sec. The observational data for
the disk stars with velocity dispersion greater than 40 km/sec are
scarce. With u_{ϖ} = 6.5 km/sec the northern and the southern rotation
curves based on the 1965 Schmidt model are shown in Figure 4.

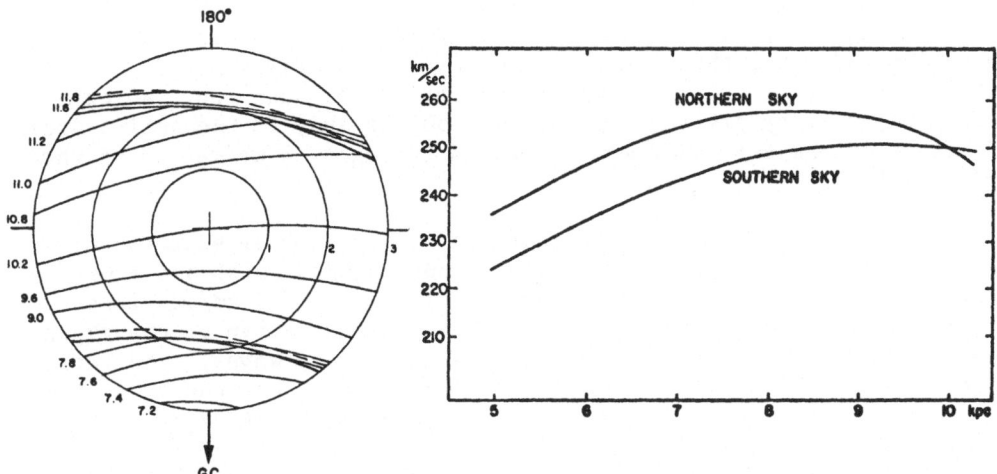

Figure 3. The adopted spiral
pattern and the gas streamlines
in the neighborhood the sun.
The darklines are the potential
minima and the thick lines are the
location of the shocks.

Figure 4. The northern and the
southern rotation curves based
on the 1965 Schmidt model and
an outgoing motion of 6.5
km/sec for the local standard
of rest.

The situation for the tangential component, however, is far from
ideal. First, there is the asymmetric drift. After its effect
carefully removed, the young objects would still move slightly faster
than the old stars in the rotational direction. The trend is
difficult to reverse even though we would like to press harder for
more correction. The result therefore is at variance with the
observation, in which the young stars and the gas lag behind the old
stars by about 3 km/sec.

A possible explanation of this discrepancy may lie on the fact
of the presence of the local arm. In our view, the local arm is not a

major arm of the basic two-arm pattern, but an inter-arm spur (Lin, Yuan and Shu 1969). The spur is comoving with the local material and it may be associated with a certain mass concentration. Using the linear analysis of the gas response, we may be able to give a order-of-magnitude estimate of the mass density required for the systematic motiin of the gas to reverse its direction . The formula is given as follows:

$$\sigma_1/\sigma_o : \hat{u}_{\varpi}/\varpi\Omega : \hat{u}_\theta/\varpi\Omega = (-k\varpi) : (1-\Omega_p/\Omega) : \kappa^2/2\Omega^2$$

where σ_1/σ_o is the ratio of the perturbed surface density to the mean density of the gas, $(\hat{u}_{\varpi}, \hat{u}_\theta)$ are the amplitudes of the velocity components, k is the radial wave number, Ω is the local angular speed, Ω_p the pattern speed, κ is the epicyclic frequency, and is the galacto-centric distance. Since the interarm spur is a material arm, $\Omega_p = \Omega$. Thus, no radial velocity component is caused by its presence, which is an excellent feature of this approach. If we take the wavelength for the Orion spur to be 2.5 kpc, or k = 2.13 kpc^{-1} and $\hat{u}_\theta$ =8 km/sec we obtain a ratio σ_1/σ_o equal to 0.8. This is somewhat greater than the mass density of a spiral arm. A concentration like this in a spiral galaxy is not uncommon. We can also check this by placing a point mass 300 pc away in the anti-center direction of the sun. The mass required for u_θ = -8 km/sec is of the order of $10^7 M_o$.

In conclusion, we believe that the conventionally adopted local standard of rest is comoving with the young objects which has a net outward motion of about 7 km/sec and a net motion of 5 km/sec in the direction of the galactic rotation. This outward motion of the young object is due to the presence of the spiral density waves. The fact that the tangential motion is not observed implies that the local arm may have a mass density like those in the regular spiral arm.

Acknowledgement. I wish to thank Professor C.C. Lin for the early collaboration on this research and for many valuable suggestions afterwards. The work is supported in part by NASA grant NGC-2348 and PSC-CUNY Research Award Program.

References

Blitz, L.: 1982, private communication
Delhaye, J. : 1965, in Galactic Structure, Stars and Stellar Systems
    Vol V, ed. Blaauw, A. and Schmidt, M., Univ. of Chicago Press,
    Chicago, pp. 61-84.
Frogel, J.A. and Stothers, R.: 1977, A.J., 82, 890.
Kerr, F.J.: 1964, I.A.U. - U.R.S.I. Symp. 20, 81
Lin, C.C. and Yuan, C.: 1975, B.A.A.S., 7, 344
Lin, C.C., Yuan, C. and Roberts, W.W.: 1978, A. A., 69, 181.
Lin, C.C., Yuan, C. and Shu, F.H.: 1969, Ap.J., 155, 721.
Roberts, W.W.: 1969, Ap.J., 158, 123
Stothers, R.: 1980, private communication
Yuan, C. and Grosbol, P.: 1981, Ap.J., 243, 432.

ON THE ANALYSIS OF MOTIONS PERPENDICULAR TO THE GALACTIC PLANE

Ivan R. King
Astronomy Department
University of California, Berkeley

Oort showed long ago in a classic paper (Oort 1932) that one can in principle derive the run of the acceleration perpendicular to the galactic plane by comparing the spatial distribution of stars in the z-direction with the Z-velocity distribution of stars of the same type. In practice, however, no one has succeeded in deriving a valid curve of the acceleration K_z; typically the curve turns over in a way that implies negative densities at some distance from the galactic plane. What I would like to do in this brief note is to suggest some possible reasons for this difficulty--specifically, that the types of stars used have been unsuitable.

Stars chosen for a K_z study must meet three criteria. (1) Their distributions of density and velocity must be in statistical equilibrium with each other. (2) Absolute magnitudes must be known for the stars whose space density is to be determined. (3) Radial velocities must be observable at 500 to 1000 parsecs from the galactic plane, in order to tie down the crucial high-velocity tail of the distribution.

This last requirement has led to the choice of stars of relatively high luminosity, lest the distant stars be too faint for radial-velocity observation. Studies of K_z have concentrated on main-sequence A stars and on K giants. Neither of these types, I shall argue, satisfies the criteria just stated.

The trouble with the A stars is that they are too young to be in equilibrium. That this is so is clearly shown by a velocity distribution given by Eggen (1965, Fig. 2). It is nothing like a smooth distribution; most of the stars in it fall in a few clumps. There is no reason to believe that the spatial distribution of such stars will be closely related to their velocity distribution.

The K giants lead us astray for a different reason: they are an inhomogeneous mixture, with main-sequence progenitors ranging from spectral class B to late F. Sandage (1957) has derived a luminosity function for such stars. About a third of them have progenitors that

53

are A5 or earlier. These stars, Sandage points out, should have a lower
velocity dispersion than the other K giants; they also have a brighter
absolute magnitude. This combination of circumstances makes the K
giants a disastrous choice for a study of K_z; because of the different
velocity dispersions their average absolute magnitude becomes
systematically fainter as one moves away from the galactic plane, and
conventional methods of deriving densities are invalidated.

The solution to the K_z problem will come, I believe, from the use
of stars lower on the main sequence. G stars would be ideal, but
considerations of limiting magnitude may put them out of reach for the
time being. Certainly an attempt should be made to use F stars for this
problem. Even though their density may turn out not to be as uniform as
we might wish (see, for example, McCuskey 1965, Figs. 10 and 11), they
are certainly a much less risky choice than A stars. I am glad to
report that such a project is under way, in the hands of a group of
Danish astronomers (Strömgren, private communication). I look forward
to seeing their results.

I am grateful to Allan Sandage for calling my attention (in the
discussion that followed the oral presentation of this paper) to his
K-giant luminosity function. This work was partially supported by NSF
Grant AST80-20606.

REFERENCES

Eggen, O.J.: 1965, in *Galactic Structure*, ed. A. Blaauw and M. Schmidt,
      p.111 (Chicago: Univ. of Chicago Press).
McCuskey, S.W.: 1965, in *Galactic Structure*, ed. A. Blaauw and M.
      Schmidt, p.1 (Chicago: Univ. of Chicago Press).
Oort, J.H.: 1932, *Bull. Astr. Inst. Neth.* 6, p.249.
Sandage, A.: 1957, *Astrophys. J.* 125, p.435.

LOCAL GALACTIC STRUCTURE AND VELOCITY FIELD

Per Olof Lindblad
Stockholm Observatory

ABSTRACT

The information contained in the local stellar velocity field is discussed. The local expansion, also reveiled by the interstellar gas, is interpreted in terms of a scenario where giant clouds are formed in galactic shocks, give birth to stars and subsequently expand.

THE LOCAL STELLAR VELOCITY FIELD

For the present analysis let us consider only components of motion parallell to the galactic plane. For a specific longitude let V_r and V_T be the radial and tangential components of motion respectively. In the case of circular motions around the galactic centre the local velocity gradients with distance are

$$\frac{dV_r}{dr} = A_c \sin 2\ell$$

$$\frac{dV_T}{dr} = B_c + A_c \cos 2\ell$$

(1)

where A_c and B_c are Oort's constants in the case of circular motion.

For a velocity field of non-circular motions we may write

$$\frac{dV_r}{dr} = K + A \sin 2\ell + C \cos 2\ell$$

$$\frac{dV_T}{dr} = B + A \cos 2\ell - C \sin 2\ell$$

(2)

W. L. H. Shuter (ed.), Kinematics, Dynamics and Structure of the Milky Way, 55–58.
Copyright © 1983 by D. Reidel Publishing Company.

where $A = A_c + f_A$

$B = B_c + f_B$ (3)

$C = f_C$

$K = f_K$

and the functions f are well known functions of the non-circular velocity components and their derivatives (Lindblad 1980).

In an analysis of the motions of O - B5 stars Tsioumis and Fricke (1979) found for stars more distant than 600 pc: $A = -B = 13$ km s^{-1} kpc^{-1}, $C = 0$, $K = 0$ i.e. circular motion and a flat rotation curve.

For stars closer than 400 pc on the other hand they found a longitude dependence of the radial velocity gradient as in Figure 1. In the analysis the stars were also divided into two groups: "Young" meaning spectral type earlier than B 2.5 and "Old" meaning later B stars. As can be seen in the figure, K and C takes values significantly different from zero in particular for the young group. This is the group of local stars that is identified with Gould's Belt. If there exists a physical model for the non-circular velocity field, then the functions f in eqs. (3) are known and a set of model parameters giving a best fit to the observations can be found.

One type of model for a non-circular velocity field is that envisaged by Blaauw (1952) for expanding associations. The only additional parameter of the model is then the expansion age of the group. Another possible model is a linear density wave. The parameters involved are then the spiral phase at the position of the sun, the inclination of the spiral pattern and the amplitude of spiral perturbation.

With the aim of performing such an analysis we are at the Stockholm Observatory carrying out a project to determine radial velocity gradients at various longitudes for different age groups of B stars on both hemispheres. A similar analysis can be made for proper motions and here the HIPPARCUS satellite may play an important role.

THE LOCAL INTERSTELLAR GAS

The very local neutral hydrogen gas which displays a wide angular distribution in galactic latitude with an inclination similar to that of Gould's Belt (Schober 1976),displays a velocity-longitude relation characteristic for that of a cloud or shell expanding in the presence of differential rotation (Lindblad et al. 1973).

The most recent modeling has been done by Olano (1982). A map of his dough-nut shaped shell swept up from the interstellar medium after the expansion had been initiated is shown in Figur 2. Olano finds an expansion age of $3 \cdot 10^7$ years and a total swept up mass of $1.2 \cdot 10^6$ M .

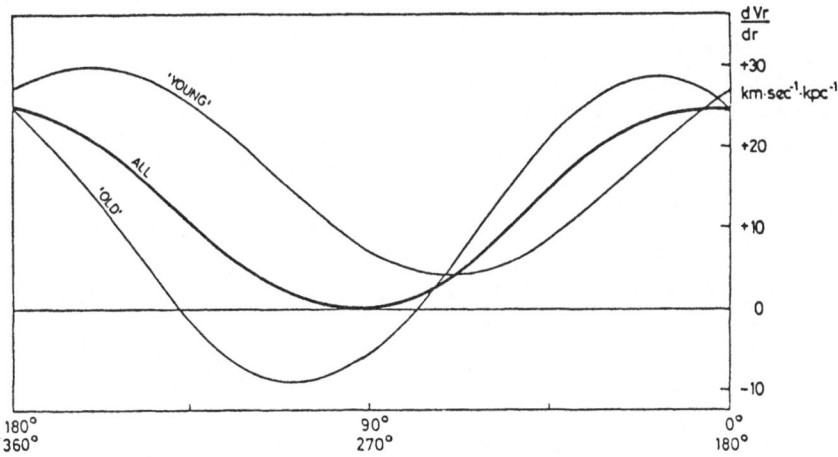

Fig. 1. Radial velocity gradients as function of galactic
        longitude for local B-stars as given by the analysis
        of Tsioumis and Fricke (1979).

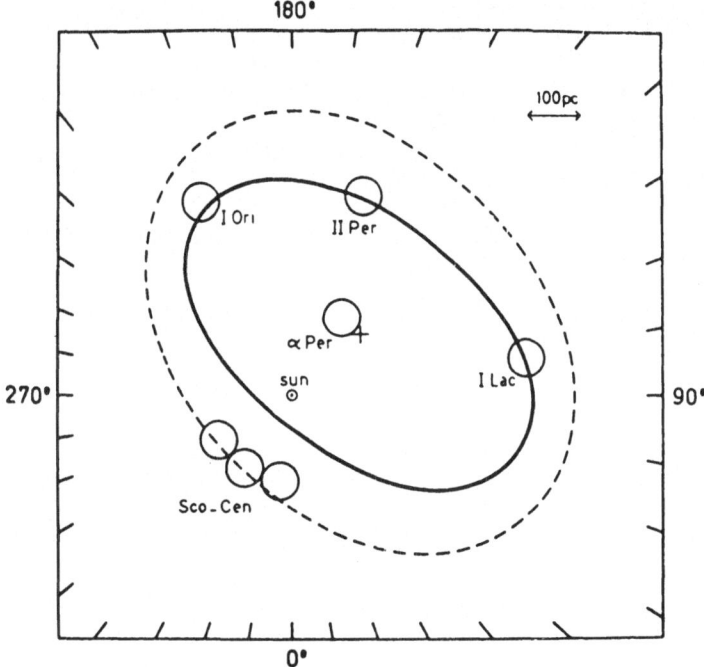

Fig. 2. Schematic representation of the positions of the
        stellar associations and the ring of gas (full-drawn
        line) related to Gould's Belt. Reproduced from Olano
        (1982).

This mass agrees fairly well with that found by Grape (1975) from HI
line profiles. The system is centered at the α Per association and
the Pleiades and displays a negligible total velocity with respect to
the Local Standard of Rest.

A POSSIBLE SCENARIO

As a possible scenario to explain these observations we may imagine
that a density wave with a large scale galactic shock compresses interstellar matter into giant clouds of the total mass of Gould's Belt.
After a first burst of star formation the stellar system expands due to
the velocity dispersion at birth. O-stars and supernovae push the
expansion of the interstellar matter into the surrounding medium. New
stellar associations may be born in the expanding front. The expansion
still continues because the surrounding matter in the inter-arm region
is by itself expanding.

REFERENCES

Blaauw, A.: 1952, Bull. Astron. Inst. Neth. 11, 414.
Grape, K.: 1975, Stockholm Obs. Rept. No. 9.
Lindblad, P.O.: 1980, Mitt. Astron. Ges. 48, 151.
Lindblad, P.O., Grape, K., Sandqvist, Aa. and Schober, J.: 1973,
      Astron. Astrophys. 24, 309.
Olano, C.A.: 1982, 'On a model of local gas related to Gould's Belt',
      Astron. Astrophys. (in press).
Schober, J.: 1976, Astron. Astrophys. Suppl. 25, 507.
Tsioumis, A. and Fricke, W.: 1979, Astron. Astrophys. 75, 1.

ON THE DETERMINATION OF R_0

M. W. Ovenden[†] and J. Byl[††]
[†]The University of British Columbia, Vancouver B.C.
[††]Trinity Western College, Langley B.C.

Abstract. The determination of R_0 by comparison of stellar radial velocities and radio tangent-point observations is shown to be unreliable because the radio and the optical observations sample different parts of the Galaxy. The most reliable method for determining R_0 seems still to be the direct analysis of the radial velocities of stars with known heliocentric distances. The value so found, $R_0 = 10.4\pm0.7$ kpc, is not sensitive to the form of the velocity field which is assumed.

1.   INTRODUCTION

In attempting to locate the kinematic centre of our Galaxy, we can use only radial velocities, since proper motions are available only for stars whose distances are small compared with the scale of the Galaxy. Optical observations of stellar radial velocities have the advantage that the heliocentric distances of the stars may also be known and the stars located within the Galaxy, but have the disadvantage that only a few radial velocities are available for stars beyond 4 kpc. Radio observations of radial velocities of interstellar features have the advantage that they cover essentially the whole of the Galaxy, but have the disadvantage that distances of individual interstellar features are (in general) not available.

This paper examines the attempt to determine the distance to the centre of the Galaxy, R_0, by combining radio and optical observations.

2.   METHOD

If at any point in the Galaxy the velocity of the centroid of stars in the neighbourhood of that point is τ, directed at right angles to the radius vector (R) to that point from the centre of the Galaxy (C), then the average radial velocity, ρ , of a star $X(r,l,b)$ with galactic co-ordinates l,b at a distance r from the position of the Sun (S) is

$$\rho = F + \tau\cos b\sin(\lambda+l)-\tau_0\cos b\sin l$$

W. L. H. Shuter (ed.), Kinematics, Dynamics and Structure of the Milky Way, 59–66.

where F is the appropriate correction for the velocity of the Sun relative to the centroid of stars in the neighbourhood of S, τ_0 is the value
of τ at S, and $\lambda \equiv \lfloor$SCX projected on to the galactic plane.   Since
$\sin(\lambda+l)=(R/R_0)\sin l$, writing $\rho' \equiv (\rho - F)$ we have

$$x(R) \equiv (\rho'R/R_0 \cos b \sin l) = R(\omega - \omega_0) = [\tau - R\tau_0/R_0] \tag{1}$$

where ω, ω_0 are the galactocentric angular velocities of the centroids
of the stars at X,S respectively. R, R_0 are the galactocentric distances
of X,S respectively; thus

$$R = [R_0^2 + r^2\cos^2 b - 2R_0 r\cos b \cos l]^{\frac{1}{2}} \tag{2}$$

Provided that everywhere $(\partial\omega/\partial R) < 0$, and in the absence of dispersion, for an object in the galactic plane the numerically maximum
corrected radial velocity ρ'_m seen from S in longitude l, corresponds to
the tangent-point for which $R=R_0\sin l$. Thus for observations restricted
to the tangent-points

$$x(l) \equiv \rho'_m(l) = R_0(\omega - \omega_0)\sin l = (\tau - \tau_0\sin l) \tag{3}$$

It should be noted that (i) from radial velocities alone, the τ-
field can be determined only with the ambiguity of a constant angular
velocity ω_c, and (ii) the τ-field does not necessarily correspond to the
field of circular velocities V.

Assuming a value for R_0, the function $x(R)$ may be found from radial
velocity observations of stars of known heliocentric distance r. The
function $x(l)$ may be found from observations of the 21-cm and other radio
lines. *Provided that the τ-field is the same for the stars and for the
interstellar matter, $x(R)$ and $x(l)$ are the same function; R_0 may be
determined as that value which gives the best match of $x(R)$ and $x(l)$.*

3.   THE FUNCTION $x(l)$

Gunn,Knapp and Tremaine(1979) showed that tangent-point observations
of the 21-cm line could be represented, over the range $0.5 < |\sin l| < 1.0$,
by the straight line $\rho'_m = +(220 \pm 3$ km s$^{-1})(1-\sin l) + (28 \pm 1$ km s$^{-1})$. More
recently Shuter (1981) has represented ρ'_m as linear in $|\sin l|$ for
$0.1 < |\sin l| < 1.0$, with slope -184 ± 9 km s^{-1}.   In both cases there are
significant systematic departures of the observations from the lines.
Gunn, Knapp and Tremaine interpret the finite intercept of their linear
regression as due to dispersion of velocities.

4.   THE FUNCTION $x(R)$

Using the same set of O- and B-stars, cepheids and clusters that
were used in Ovenden and Byl (1976) [Paper I] and Byl and Ovenden (1978)
[Paper II], we have determined the slope of the best-fit straight line

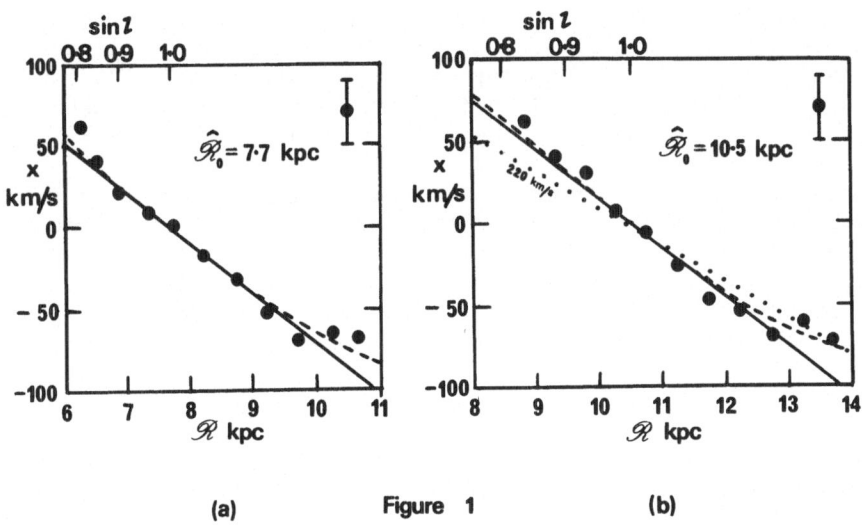

Figure 1

(a)   (b)

to the function $x(R)$ for various assumed values $\hat{R}_0$ of R_0. We find that we can represent the observations by $x = -(a\pm a')\Delta R$, $a = (33.81-0.28\hat{R}_0)$, $a' = (1.55-0.01\hat{R}_0)$, where x is in km s^{-1}; R,R_0 in kpc; $\Delta R \equiv (R-\hat{R}_0)$. The analysis was restricted to those 769 stars for which $|\sin l| > 0.5$. Figures 1(a) and 1(b) give the plots of $x(R)$ for two different values of $\hat{R}_0$. Approximately ± 8 km s^{-1} of the uncertainty of each plotted point is due to the binning. The upper scales give the appropriate values of $|\sin l|$ for tangent-point observations.

5.   COMPARISON OF $x(l)$ and $x(R)$

   Matching the slope of $x(R)$ from our analysis of the stellar radial velocities with the slope of $x(l)$ from Gunn, Knapp and Tremaine gives

   $- (33.81 - 0.28R_0)\Delta R = -220\Delta R/R_0$

$\therefore$ $R_0 = 6.90\pm 0.35$ kpc   [or 113.9 kpc]

   The same method applied to Shuter's coefficient gives

   $R_0 = 5.71\pm 0.55$ kpc   [or 115.0 kpc]

   These results assume only that the τ-fields of the stars and the interstellar matter are the same, and that x can be represented adequately as linear over the appropriate ranges of R.

6.   OPTICAL DETERMINATION OF R_o

However, only part of the optical information has so far been used. r has been used to calculate R for any assumed R_o, according to equation (2); but only the values of R appear in the definition of $x(R)$. The *explicit* use of $\rho'(r,l,b)$ can, in principle, permit an independent determination of the symmetry of the velocity field, and hence R_o .

The classic method of Trumpler and Weaver (1953) uses stars with special geometrical relationships to the Sun. The method gives inconsistent results, because (as was shown in Paper II by numerical experiments) it is sensitive to small departures from tangential motion. However, as was also shown in Paper II, a complete analysis, using all the stellar radial velocity data, yields a value

R_o = 10.4±0.7 kpc.

This is in disagreement, at the 4σ level, with the value found from the direct comparison of $x(l)$ and $x(R)$.

7.   POSSIBLE REASONS FOR DISCREPANCY

Faced with such a discrepancy, one approach is to try to find a 'best-fit' set of parameters which does least violence to the data. With this approach, a value R_o=8.1 kpc would be adopted, the discrepancies being at about the 3σ level. We believe this approach to be unwise. The discrepancy may arise because the theoretical framework within which a determination has been made may be too simplistic. In this section we discuss briefly some explanations of the discrepancy we have noted:-

(a) Systematic error in the optical analysis

To check this, we have analysed fictitious sets of radial velocities, generated by taking a defined velocity field, and adding a random radial velocity component to the motion of each fictitious star. To allow for any effects of maldistribution of the stars in our data set, the fictitious stars were given the same galactic coordinates as the real stars. We found that whatever value of R_o was used in the assumed velocity field, and whether we used a constant velocity or a velocity field defined by a power series in $(\partial\omega/\partial R),(\partial^2\omega/\partial R^2)$, we always recovered the correct value for R_o within a standard deviation that was quite comparable to that found for our actual data set. The problem is statistically well-conditioned for determining R_o.

It might be supposed that the discrepancy arises because in our previous stellar analyses we had assumed a different form of velocity-field than Gunn *et al* . Therefore we have analysed the stellar data assuming that τ is independent of R over the relevant R-range. The iteration procedure was similar to that described in Paper II. The results are shown in Table I.

TABLE I

		Solution #1		Solution #2	
V	km s^{-1}	282	± 32	289	± 32
R_0	kpc	9.88 ±	1.23	10.26 ±	1.23
u	km s^{-1}	− 12.27 ±	1.44	− 6.79 ±	0.73
v	km s^{-1}	+ 14.67 ±	0.69	+ 15.69 ±	0.59
a_R	km s^{-1}	+ 5.36 ±	1.20	0.00	
a_λ	km s^{-1}	+ 3.22 ±	0.89	0.00	

Whether or not we include the terms a_R, a_λ representing the kinematics of a spiral arm system, the value of R_0 is close to the value
found in Paper II, although the standard deviations of the determinations
are now somewhat larger, suggesting that a constant velocity is not a
good representation of the stellar data. It is clear that the determination of R_0 directly from the optical data is not very dependent upon
the particular form of the velocity field assumed in defining the other
parameters.

(b) Deviation of the vertex

Inclusion of a 'deviation of the vertex' in the random velocity
components of fictitious radial velocities does not have any systematic
effect on the determination of R_0 by direct analysis.

(c) Systematic departures from circular motion in the optical data

If there are systematic departures from circular orbits among the
objects of our data set, to affect the determination of R_0 they must be
of a form different from any yet considered by us in Papers I and II,
since we have found that failure to include R_0 as a variable parameter
does not produce significant changes in the determined values of any of
the other parameters.

(d) Errors arising from the extrapolation of radio and/or optical data

A strictly linear relationship between ρ'_m and $(1-\sin l)$ would imply
that over the relevant range of R, $\tau = C + DR$, where C and D are independent of R. The linearity is consistent with, but does not imply, the
law τ independent of R. Gunn et al interpret their results in terms of a
circular velocity field with $V = V_c$, independent of R. It should be noted
that if this velocity law is only approximately true, V_0 might differ
from V_c. Under these circumstances, the mean slope of the regression line
gives V_0, not V_c, since from equation (3) we have

$$x(l) = V_0(1-\sin l) + (V_c - V_0) \tag{4}$$

Since (according to Gunn et al) the non-zero value of the intercept is
attributable to dispersion, it follows that V_c cannot be determined from
the radio observations.

Faber and Gallagher (1979) show velocity curves for other galaxies, many of which have systematic departures from the simple law V=constant, with slopes ~ 10 km s^{-1} kpc^{-1} and spatial 'wavelengths' of several kpcs. Now Figure 2, a schematic representation of our Galaxy, shows that there is a very limited overlap of location of the stellar (optical) and interstellar (radio tangent-point) objects, and that the zone of overlap covers a very limited range of R. The figure also reminds us that possible kinematic effects of spiral arms may affect the radio and optical regions differently.

A sufficient condition for finding a false value of R_0 from inter-comparison of radio and optical data is that the mean slope of the $V(R)$ curve in the neighbourhood of the Sun be different from the mean slope of the $V(R)$ curve averaged over larger ranges of R. It is quite possible for this to happen, while preserving the general overall linearity of both $x(l)$ and $x(R)$.

To test whether such an effect would be significant, using variations of $V(R)$ of the magnitude and character shown for other galaxies by Faber and Gallagher, we carried through the radio and optical comparison procedure on fictitious radial velocities based upon a variety of assumed velocity fields of the form

$$V = V_c + C{\sin\atop\cos}(2\pi R/\Lambda) \qquad\qquad C, \Lambda \text{ independent of } R$$

with known R_0. In general, large errors were found in the values of R_0 deduced from the intercomparison of $x(l)$ and $x(R)$.

For example, taking V=250+17.7cos$(2\pi R/4.44)$ and R_0=10.0 kpc yielded a 'determined' value of R_0=6.9 kpc.

Since the method described in this paper is in principle equivalent to using Oort's Constant A to represent the optical data, similar doubt is cast on any intercomparison of radio and optical data to find R_0.

/// $\equiv$ Region of stars with known r and ρ

$\lessgtr$ $\equiv$ Region of interstel-lar tangent-points

--- $\equiv$ Spiral arm

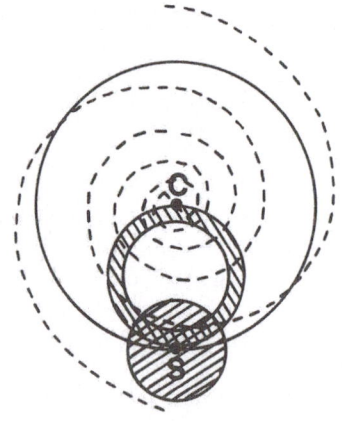

Figure 2.   Schematic diagram of Galaxy

(e) Differences between the velocity fields of the stars and the interstellar matter

The errors in R_0 under (d) arise because the radio and optical data
sample different regions of the Galaxy. A more fundamental problem would
arise if in fact the τ-velocity fields were different for the optical
and radio objects.

(i) Since the equations of Section 1 assume only that τ is normal
to the galactocentric radius vector, not necessarily that $\tau=V$, the
appropriate circular velocity, the velocity field for the (centroids of
the) stars could differ from that for interstellar matter simply as a
statistical result of a density-gradient of stars with a mean orbital
osculating eccentricity $e \neq 0$. However, this is unlikely to be a significant problem with O and B stars, where the average orbital osculating
eccentricity is small.

(ii) In another paper presented to this symposium, Ovenden, Pryce
and Shuter show that for distances restricted to a few kiloparsecs
around the Sun there are significant differences between the optical
and the radio velocity fields, and that it is the radio objects, not the
stars, which show systematic departures from circular motion. In the
light of these results, the determination of R_0 by an intercomparison of
optical and radio data seems to be unsound in principle.

8.   CONCLUSION

The determination of R_0 by intercomparison of radio and optical
radial velocities is unreliable.

The best method for determining R_0 kinematically seems still to be
the direct analysis of the radial velocities of distant stars with known
heliocentric distances, to determine statistically the convergence of
the galactic radii. The value so found

$R_0 = 10.4 \pm 0.7$ kpc

is not very sensitive to the nature of the velocity field assumed in
the analysis.

ACKNOWLEDGEMENT

This work was supported in part by a grant from NSERC.

REFERENCES

Byl, J. and Ovenden, M.W. 1978, Astrophys.J. 225, 496
Faber, S.M. and Gallagher, J.S. 1979, Ann.Rev.Astron.Astrophys. 17, 135
Gunn, J.E., Knapp, G.R. and Tremaine, S.D., 1979, Astron.J. 84, 1181

Ovenden, M.W. and Byl, J. 1976, Astrophys.J. 206, 57
Shuter, W.L.H. 1981, Mon.Not.R.Astr.Soc. 194, 851
Trumpler, R.J. and Weaver, H.F. 1953 *Statistical Astronomy* New York:
    Dover

THE VELOCITY FIELDS OF GAS AND STARS WITHIN FIVE KPC OF THE SUN

M.W. Ovenden, M.H.L. Pryce, W.L.H. Shuter
University of British Columbia

INTRODUCTION

If the velocity field around the sun is assumed to be sufficiently smooth so that it can be represented by a second order Taylor series, then the most probable value of the line of sight velocity, V_r, of an element at galactic longitude ℓ and distance projected onto the plane of the galactic disk $r = d \cdot \cos(b)$, where d is the distance to the object, can be represented as;

$$V_r = \cos(b) \cdot [c_1 \cdot \cos(\ell + c_2) + r(c_3 + c_4 \cdot \sin(2(\ell + c_5)))$$

$$+ r^2(c_6 \cdot \sin(\ell + c_7) + c_8 \cdot \sin(3(\ell + c_9)))],$$

where the coefficients are determined by a least squares fit.

The coefficients c_1, c_2 represent adjustments to the local standard of rest. If pure circular motion about the galactic centre $\ell = 0^\circ$ exists, then;

$c_4 = A$(the usual Oort constant)
$c_6 = -A/4R_\odot$ ($R_\odot$ = distance to galactic centre)
$c_8 = 3A/4R_\odot$
$c_3 = c_5 = c_7 = c_9 = 0$

and hence inspection of the coefficients gives values for A and $R_\odot$

Fits to the above formula have been done for a set of about 990 "young" stellar objects, most of which are O and B stars from a list provided by the Dominion Astrophysical Observatory - Victoria, plus a set of 112 kinematically distinct Sharpless HII regions from Blitz, Fich, and Stark, 1982, and we are at present analysing a set of about 1550 "older" A stars. Also presented here is a fit for coefficients c_3, c_4, and c_5 only for very nearby 21 cm emission from the intermediate latitude survey of Venugopal & Shuter, 1970. In all subsequent formulae, V_r is expressed in km/s and r in kpc. All observational data were

67

W. L. H. Shuter (ed.), Kinematics, Dynamics and Structure of the Milky Way, 67–72.
Copyright © 1983 by D. Reidel Publishing Company.

measured with respect to a Local Standard of Rest based on the Standard
Solar Motion with velocity components U = -10.2, V = +15.3, W = +7,9,

VELOCITY FIELD OF O AND B STARS

This sample of 988 O & B stars provided good coverage at all
galactic longitudes. The velocity field was determined to be;

$$V_r = \cos(b) \cdot [2.80 \cos(\ell+5.9^\circ) + r(-1.25 + 13.28 \sin2(\ell-3.6^\circ))$$

$$+ r^2(-.577 \sin(\ell+31.5^\circ) + .907 \sin3(\ell-2.5^\circ))].$$

This field is depicted in the plot below where the sun is at the
centre of the circle which has radius 4.8 kpc and where the galactic
centre direction is beyond the bottom of the picture. Errors have been
determined for the coefficients and are summarized in the plot where
the thickness of the bands represent a ± 2σ spread about the most
probable velocity. The velocity residuals from the fit have an rms
value of about 12 km/s.

VELOCITIES IN KM/S

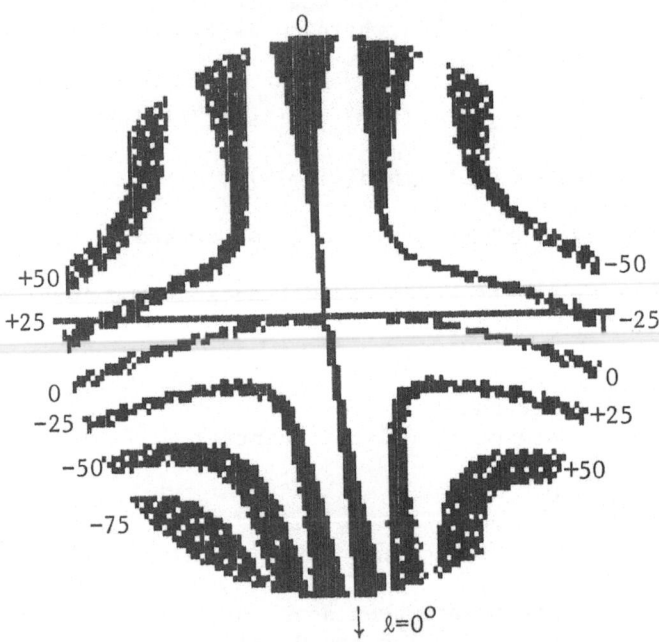

RADIUS IS 4.8 KPC

VELOCITY FIELD FOR IDEALIZED CIRCULAR MOTION

Inspection of the coefficients in the fit for the O & B stars
shows that the coefficients which describe departures from circular

motion are in most cases (with the exception of c_3) insignificant. Accordingly, an idealized velocity field which corresponds to pure circular motion has been adopted using the information obtained in the previous fit. Thus;

$$Vr = 13.5 \ r \ \sin(2\ell) + r^2(-.321 \ \sin(\ell) + .963 \ \sin(3\ell)).$$

This formula represents pure circular motion with A = 13.5 km/s/kpc, and $R_\odot$ = 10.5 kpc.

This field is plotted below on the same scale as before, with the thickness of the bands being ± 10 km/s.

VELOCITIES IN KM/S

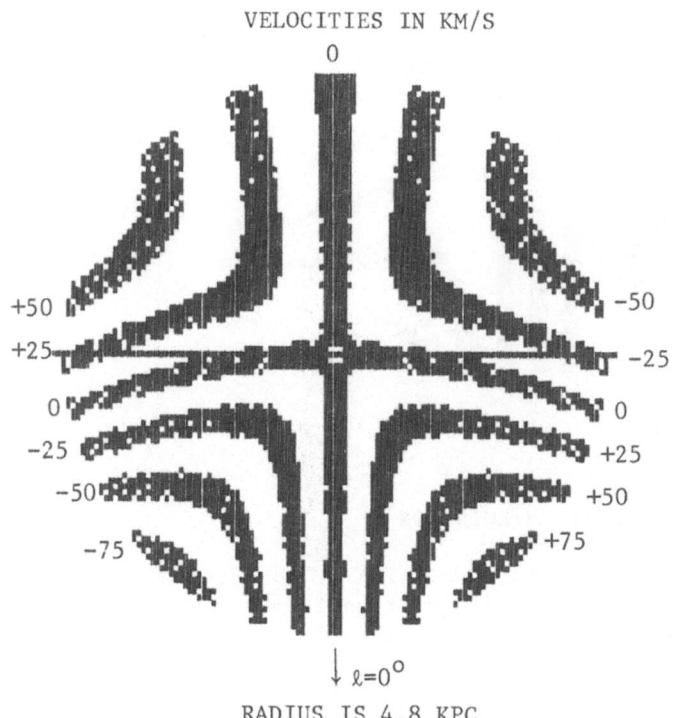

RADIUS IS 4.8 KPC

VELOCITY FIELD OF SHARPLESS HII REGIONS

112 kinematically distinct HII regions with optically determined distances were analysed and plotted in the same manner. This data represents an 'incomplete' sample in that there is no coverage of the galactic disk between longitudes ℓ = 240° - 360°. The fit gave

$$Vr = \cos(b) \cdot [2.48 \ \cos(\ell-7.1°) + r(-3.77 + 16.1 \ \sin2(\ell+7.2°))$$

$$+ \ r^2(.942 \ \sin(\ell-39.2°) + 1.111 \ \sin3(\ell+14.2°))].$$

This field is plotted below to a radius of 4.8 kpc, again with the thickness of the bands representing ± 2σ. The velocity residuals from the fit have an rms value of about 9 km/s

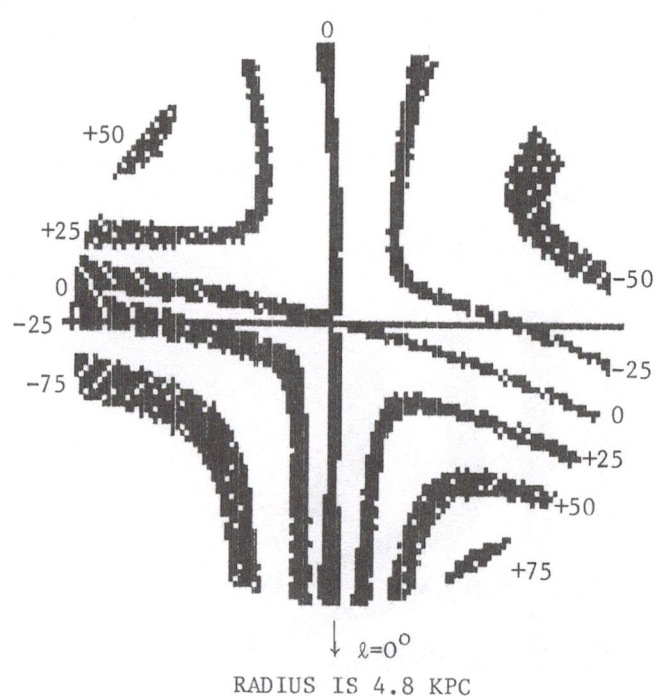

RADIUS IS 4.8 KPC

VELOCITY FIELD OF NEARBY 21 CM EMISSION

All analyses of nearby 21 cm emission or absorption that we are aware of appear to show a similar deviation from circular motion in c_5, as is apparent in the HII region fit. The fit given below was obtained by Venugopal (1969) from 21 cm emission data at intermediate latitudes with b > 0°. In this work, only the coefficients c_3, c_4, and c_5 were determined since the gas was sufficiently close that a first order Taylor expansion was adequate. He obtained

$$Vr = r(-2.4 + 16.7 \sin2(\ell+12.9°)).$$

This field is plotted below, but in this case the radius of the circle is ten times smaller than in the previous plots. The velocity code is different, and the thickness of the bands is ± 1 km/s.

VELOCITIES IN KM/S

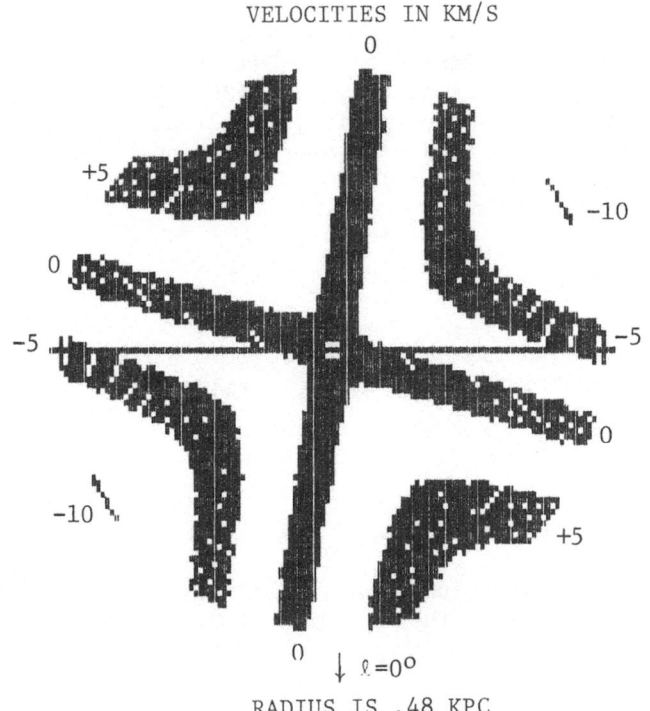

RADIUS IS .48 KPC

DISCUSSION

It is apparent that the velocity field describing the O & B stars is very close to pure circular motion, with the values of A and R_0 not too different from the standard ones. The principal departure is in the coefficient c_3 which indicates a kinematic compression which is proportional to r. Also there is a suggested adjustment of the standard LSR of about 2.5 km/s along the centre – anticentre line.

The HII data suggests a similar adjustment to the LSR and a kinematic compression term about three times larger. There is also an apparent rotation of the velocity field with respect to a circular motion field of magnitude about 15°.

The nearby 21 cm emission data again has a kinematic compression term and a field rotation, in the same sense as the above, also of about 15°.

CONCLUSIONS

It is clear from our plots here, and backed up by an extensive statistical error analysis by Pryce 1983, that the velocity fields for the nearby gas and HII regions and that of the stars are DIFFERENT.

Qualitatively, the velocity field of the O & B stars is a close approximation to that of circular motion, with A = 13.5 km/s/kpc and R⊙ = 10.5 kpc. The velocity fields of the HII regions and nearby 21 cm emission and absorption are distinctly NON-CIRCULAR, and analyses based on the assumption that the velocity fields of 21 cm data, CO data, or optical HII region data are in a good approximation to circular motion should be treated with EXTREME CAUTION!! These include determinations of R⊙, AR⊙, and rotation curves.

ACKNOWLEDGEMENTS

The authors received financial support from NSERC. We are also grateful to Edward Shuter for programming assistance.

REFERENCES

Blitz, L., Fich, M., and Stark, A.A. : 1982, Astrophys. J. suppl., in press.
Pryce, M.H.L. : 1983, This volume, pp. 73-75.
Venugopal, V.R. : 1969, Ph.D. Thesis, University of British Columbia, pp. 41-43.
Venugopal, V.R., and Shuter, W.L.H. : 1970, Memoirs R. Astr. Soc. 74, p. 1.

A COMPARISON OF THE VELOCITY FIELDS OF YOUNG STELLAR OBJECTS AND OF
SHARPLESS H II REGIONS

M.H.L. Pryce
University of British Columbia

ABSTRACT.   Assuming that the motion of Galactic objects can be
represented by a smoothly-varying velocity field on which is superposed
a random motion, it is possible to represent the radial component of the
smooth field by a formula with a small number of parameters, which can
be derived from the observations by standard statistical methods. This
has been performed for 1) a list of 989 O- and B-type stars in the solar
neighbourhood for which the distance and radial velocity are known, and
2) a list of 112 molecular clouds for which these same quantities are
known. The result is a prediction, for each class of objects, of the
most probable radial velocity at an arbitrary position in the Galaxy,
together with an estimate of its probable error. For the stellar objects,
the result deviates only marginally from what would result from circular
motion about a centre, but for the molecular clouds the result appears
to imply a distorted motion.

About two years ago, W.L.H. Shuter at UBC was analyzing the available
21-cm observations from the Galaxy, and coming to the conclusion that
they could be reconciled with the very simple model that the gas near
the plane of the Galaxy is moving in a way which deviates only slightly
from circular motion with constant speed around the Galactic centre.
As a good descriptive starting point, this model still stands, but the
value of the rotation velocity advocated by Shuter at the time appeared
to conflict with what was suggested by the kinematics of nearby stars.
Consequently, there was a good deal of animated discussion within the
astronomical fraternity here at UBC, in which I set out to act as an
arbiter. Since the observations of the hydrogen 21-cm line, and the
observations on the nearby stars sample rather different regions of
the Galaxy, apparent differences in the motions may well arise from the
effects of irregularities, possibly on a spatial scale of a few
kiloparsecs, within the Galaxy. Such irregularities, superposed on a
basically simple motion, might well lead to different estimates of the
underlying rotation velocity. I suspected that this could be a source
of the apparent discrepancy.

W. L. H. Shuter (ed.), Kinematics, Dynamics and Structure of the Milky Way, 73–75.
Copyright © 1983 by D. Reidel Publishing Company.

To test this suspicion, I borrowed a list of about a thousand
young stellar objects in the Galaxy, which Ovenden and Byl have
recently used in a study of Galactic kinematics, and I performed an
independent statistical analysis of them. They comprise O- and B-type
stars, with some Cepheids and some open clusters, close to the Galactic
plane, and mostly not much further than 3 kpc from the solar system,
for which the distances and line-of-sight velocities are reasonably well
known.

As I suspected that there might be variations between one part of
the Galaxy and another, I tried not to force too specific a model onto
the observations. I made the simplifying assumption that the stars move
according to a smoothly-varying underlying velocity field, on which is
superposed a random motion. I interpret "smoothly varying" to mean that
the Cartesian components of the velocity at any point can be expanded
as a Taylor series in terms of the Cartesian coordinates of its position.
For economy of effort, I assumed that it was sufficient to take the
Taylor series only as far as terms of the second degree. For the same
reason, I assumed that the velocity field is parallel to the plane of
the Galaxy, and depends only on the coordinates parallel to the plane.

With these simplifications, the line-of-sight component of the
velocity field is represented by a formula with nine parameters. With
nearly a thousand observations, one can use straightforward statistical
techniques to infer the most probable value of the line-of-sight
velocity field at any place in the Galaxy - and its variance.

The results of this analysis are displayed graphically in the
first diagram of Ovenden, Pryce and Shuter, 1983 (this volume), which
is a computer-generated polar diagram covering a circle of radius 4.8
kpc centred on the solar system. The filled regions correspond to
areas in the Galactic plane where the line-of-sight component of the
underlying velocity field has specified values ± two (inferential)
standard deviations.

I will refrain from making detailed comments at this point.

One would have liked to make a comparison with the 21-cm
observations, but since the latter do not contain information about the
distance to the source, this is not directly possible. Fortunately,
another quite different set of objects, for which distances and line-
of-sight velocities are available, has recently been listed by Blitz,
Fich and Stark, namely, molecular clouds associated with Sharpless H II
regions, in which carbon monoxide emission provides a measure of the
line-of-sight velocity. This list contains 112 objects. Although the
statistics are not as good as for the previous 989 stellar objects,
they are still good enough to make an interesting comparison. The
results of a similar analysis on these objects are presented in the
third diagram of Ovenden, Pryce and Shuter, 1983.

The underlying velocity fields for the two sets of objects are broadly similar, but there appear to be significant differences in detail. As an illustration of this, consider the line-of-sight component inferred for the velocity field at a point 3 kpc away from the solar system in Galactic longitude 90°. This is in a region of the Galaxy which is rather well sampled in both lists – and where the simplified formulae can be expected to yield a fairly good approximation to the "true" underlying field. For the stellar objects, the inferred line-of-sight velocity is -11.6 ± 1.6 km/sec; for the molecular clouds, it is -23.8 ± 1.9 km/sec. The difference is 12.2 ± 2.5 km/sec, which is nearly five standard deviations away from zero. Even though the basic assumptions of the analysis involve some crude approximations, there does seem to be evidence that the two classes of objects move differently. I find this surprising – even startling.

The inferred velocity field for the young stellar objects is quite close to what would be expected for circular motion about a common centre. For instance, for perfect circular motion, the circle centred on the Galactic centre and passing through the solar system is a contour of zero line-of-sight velocity. For the young stellar objects, the present analysis leads to a zero-velocity contour which, to the unaided eye, appears very circular, and with its centre definitely in a direction close to longitude zero. The corresponding contour for the molecular clouds is skewed off by about 15°.

Although I have made many simplifying assumptions in the analysis (not all of which have been mentioned in this necessarily brief account), I do believe that the data give evidence of a significant difference between the motions of the molecular clouds listed by Blitz et al, and of the young stellar objects studied by Ovenden and Byl.

A fuller account of this work, including details of the mathematical basis for the statistical analysis, is being prepared for publication elsewhere. An extension to include other types of Galactic objects, particularly A-type stars, is also being planned.

Detailed references to the papers quoted here will be found in Ovenden, Pryce and Shuter, 1983, in this volume.

DETERMINATION OF $AR_\odot$

W.L.H. Shuter
University of British Columbia

It is a well known fact, provided circular motion exists, that in the first and fourth quadrant one has:

$$\left| \partial Vm / \partial \sin(\ell) \right|_{\sin \ell \sim 1} = 2AR_\odot$$

The commonly accepted values for $2AR_\odot$ are the IAU value of 300 km/s, derived from stellar data, and the Gunn, Knapp and Tremaine (GKT) 1979 value of 220 km/s, derived using 21 cm line data. These two values are significantly different (!), and may be obtained by plotting Vm vs $\sin(\ell)$, the slope of which, at $\sin(\ell) = 1$, gives $2AR_\odot$. A recent independent analysis by the author using 21 cm data obtained by Bania and Westerhout gave a value of slightly more than 220 km/s for $2AR_\odot$ supporting GKT.

The method used in this paper in the determination involves obtaining Vm by differentiating the 2nd order Taylor Series Expression for Vr:

$$Vr = \cos(b) \cdot [c_1 \cos(\ell + c_2) + r(c_3 + c_4 \sin(2(\ell + c_5))$$
$$+ r^2(c_6 \sin(\ell + c_7) + c_8 \sin(3(\ell + c_9)))]$$

Setting $(\partial Vr / \partial r)_\ell = 0$ gives

$$r \text{ (for Vm)} = \frac{-(c_3 + c_4 \sin(2(\ell + c_5)))}{2(c_6 \sin(\ell + c_7) + c_8 \sin(3(\ell + c_9)))}$$

Substituting the expression for r into the expression for Vr, gives Vm as a function of galactic longitude. One can now plot Vm vs $\sin(\ell)$ for the range

$0.96 < \sin(\ell) < 1$, or equivalently
$76 < \ell < 90$.

The magnitude of the slope gives $2AR_\odot$. One can also get $R_\odot$, if one

77

W. L. H. Shuter (ed.), Kinematics, Dynamics and Structure of the Milky Way, 77–79.
Copyright © 1983 by D. Reidel Publishing Company.

assumes circular motion, from

$$R_\odot = r \text{ (for Vm)}/\cos(\ell).$$

RESULTS

A.)

Using O & B stars, one gets the following results:

In the Northern Hemisphere,

$$Vm = 312.7 - 312.7 \sin(\ell),$$

which gives the result that

$$2AR_\odot = 313 \text{ km/s and } R_\odot = 11.3 \text{ kpc}.$$

In the Southern Hemisphere,

$$|Vm| = 258.7 - 259.8 \sin(\ell),$$

which gives

$$2AR_\odot = 260 \text{ km/s and } R_\odot = 9.6 \text{ kpc}.$$

If one gives these results equal weight, one gets values of 286 ± 29 km/s for $2AR_\odot$, and 10.4 ± .4 kpc for $R_\odot$. A and $R_\odot$ can also be obtained from the coefficients of the Taylor expansion, assuming circular motion, by the following:

$$c_4 \simeq A, \text{ and } c_8 = 3A/4R_\odot. \text{ Thus;}$$

$$2AR_\odot = 3c_4{}^2/2c_8 = 258 \pm 26 \text{ km/s, and}$$
$$R_\odot = 3c_4/4c_8 = 9.7 \pm 1 \text{ kpc}.$$

Here one has good agreement and consistency, with values close to those recommended by the IAU.

B.)

The same type of analysis was done using the data on Sharpless HII Regions, obtained by Blitz, Fich, and Stark (1982). In the Northern Hemisphere, it was found that

$$Vm = 164.1 - 168.9 \sin(\ell), \text{ which gives}$$
$$2AR_\odot = 169 \text{ km/s},$$

while values for $R_\odot$ range from 6 kpc down to negative quantities.

On the other hand, using the Taylor expansion coefficients, one obtains

$2AR_\odot = 350$ km/s and $R_\odot = 10.9$ kpc

The results are obviously inconsistent, with the values from the coefficients being in better agreement with those derived from stellar data.

CONCLUSIONS

Determinations of $2AR_\odot$ from plots of Vm vs sin(ℓ) give incorrect results when small deviations from circular motion exist.

This is expected to apply as well as to the 21 cm determinations of GKT (1979) and Shuter (1981).

The IAU values of $2AR_\odot = 300$ km/s and $R_\odot = 10$ kpc are not too bad!

ACKNOWLEDGEMENT

This work was supported by a grant from NSERC.

REFERENCES

Blitz, L., Fich, M., and Stark, A.A. : 1982, Astrophys. J. Suppl., in press.
Gunn, J.E., Knapp, G.R., and Tremaine, S.D. : 1979, Astron. J.84, p.1181.
Shuter, W.L.H. : 1981, Mon. Not. R. Astr. Soc. 194, p.851.

VELOCITY DISTRIBUTION OF STARS AND RELAXATION IN THE GALACTIC DISK

Roland Wielen   and   Burkhard Fuchs
Institut für Astronomie und Astrophysik,
Technische Universität Berlin, Germany

ABSTRACT

The theory of the diffusion of stellar orbits predicts successfully the observed velocity distribution of nearby disk stars. We investigate the implications for the vertical scale height of galactic disks, and discuss the stability of galactic disks in which orbital diffusion operates.

1. DIFFUSION OF STELLAR ORBITS

The observed increase of the velocity dispersion $\sigma(\tau)$ of disk stars with increasing age τ indicates strongly a significant irregular gravitational field in the galactic disk (Wielen 1977). This irregular field causes a rapid diffusion of stellar orbits in velocity and positional space. The corresponding relaxation time near the sun is of the order of $2 \ 10^8$ years for young stars.

The effect of the irregular gravitational field on stellar orbits can be described quantitatively by a diffusion coefficient in velocity space. The value of this diffusion coefficient, D, has been determined empirically from the observed increase in velocity dispersion with age, $\sigma(\tau)$. The advantage of such an empirical procedure is that it avoids as far as possible uncertain assumptions on the basic physical source of the irregular part of the galactic gravitational field.

We shall always assume that the true coefficient D is isotropic in velocity space. This assumption is not only the simplest one but explains also extremely well the observed ratios between the dispersions of the velocity components perpendicular and parallel to the galactic plane.

The diffusion coefficient may depend on the velocity of a star, on time, on position etc.. We shall discuss here some typical cases, namely the constant diffusion coefficient D_0, a velocity-dependent coefficient D_1, and a velocity-time-dependent coefficient D_2. The constant coefficient D_0 represents the simplest case. Nevertheless, it fits the observa-

81

W. L. H. Shuter (ed.), Kinematics, Dynamics and Structure of the Milky Way, 81–90.

tional data already nearly perfectly. The velocity-dependent diffusion coefficient D_1 is theoretically more appropriate for gravitational encounters between stars and massive interstellar clouds (Spitzer and Schwarzschild 1953). In order to fit the observed function $\sigma(\tau)$ with such a velocity-dependent diffusion coefficient, this coefficient has to decrease with time. The time-dependence of D_2 may describe a decay with time of the irregular gravitational field.

In Fig.1, we present comparisons between the observed total velocity dispersion of nearby disk stars, $\sigma_\upsilon(\tau)$, and the fits provided by the various diffusion coefficients. For D_0 and D_2, the agreement is good. From the fits, the quantitative values of the diffusion coefficients D_0, D_1, D_2 have been determined (Wielen 1977). We list them here together with the corresponding formulae for the averaged total velocity dispersion $\sigma_\upsilon(\tau)$ of stars in a cylinder perpendicular to the galactic plane:

constant diffusion coefficient:

$$D_0 = 2.0 \ 10^{-7} \ (km/s)^2/yr \ ; \tag{1}$$

$$\sigma_\upsilon^2(\tau) = \sigma_{\upsilon,0}^2 + \alpha_\upsilon D_0 \tau \quad . \tag{2}$$

velocity-dependent diffusion coefficient:

$$D_1 = \delta_1/\upsilon \quad , \tag{3}$$

$$\text{with} \quad \delta_1 = 4.7 \ 10^{-6} \ (km/s)^3/yr \ ; \tag{4}$$

$$\sigma_\upsilon^3(\tau) = \sigma_{\upsilon,0}^3 + \frac{3}{2} \alpha_\upsilon \delta_1 \tau \quad . \tag{5}$$

velocity-time-dependent diffusion coefficient:

$$D_2 = \delta_2 \exp(-(t-t_p)/T_\delta)/\upsilon \quad , \tag{6}$$

$$\text{with} \quad \delta_2 = 3.7 \ 10^{-6} \ (km/s)^3/yr \ , \quad T_\delta = 5 \ 10^9 \ yr \ ; \tag{7}$$

$$\sigma_\upsilon^3(\tau) = \sigma_{\upsilon,0}^3 + \frac{3}{2} \alpha_\upsilon \delta_2 T_\delta (\exp(\tau/T_\delta)-1) \quad . \tag{8}$$

In these equations, υ is the total space velocity, measured with respect to the circular velocity at the position of the star ($\upsilon^2=U^2+V^2+W^2$), $\sigma_{\upsilon,0}=10$ km/s is the initial velocity dispersion, t is the time (t=0 at the formation of the disk), t_p is the present age of the disk (we adopt $t_p= 10^{10}$ yr), t_f is the time of formation of a star, and $\tau=t-t_f$ is the age of a star. The quantity α_υ is given by

$$\alpha_\upsilon = \alpha_U + \alpha_V + \alpha_W \quad , \tag{9}$$

with

$$\alpha_U = \frac{1}{2} \left(1 + \frac{4\omega^2}{\kappa^2} \right) = 1 + \frac{A}{-2B} \quad , \tag{10}$$

$$\alpha_V = \frac{1}{2} \left(1 + \frac{\kappa^2}{4\omega^2} \right) = 1 - \frac{A}{2(A-B)} \quad , \tag{11}$$

$$\alpha_W = \frac{1}{2} \qquad\qquad = \frac{1}{2} \quad . \tag{12}$$

ω is the frequency of galactic rotation, κ the epicyclic frequency, A and B are Oort's constants. The local values of α, using A=+15 (km/s)/kpc and B=-10 (km/s)/kpc, are α_U=1.75, α_V=0.70, α_W=0.50, α_υ=2.95 . The square roots of the α-values describe the axial ratios of the velocity ellipsoid, i.e. $\sigma_U : \sigma_V : \sigma_W : \sigma_\upsilon = \sqrt{\alpha_U} : \sqrt{\alpha_V} : \sqrt{\alpha_W} : \sqrt{\alpha_\upsilon}$ = 0.77 : 0.49 : 0.41 : 1 .

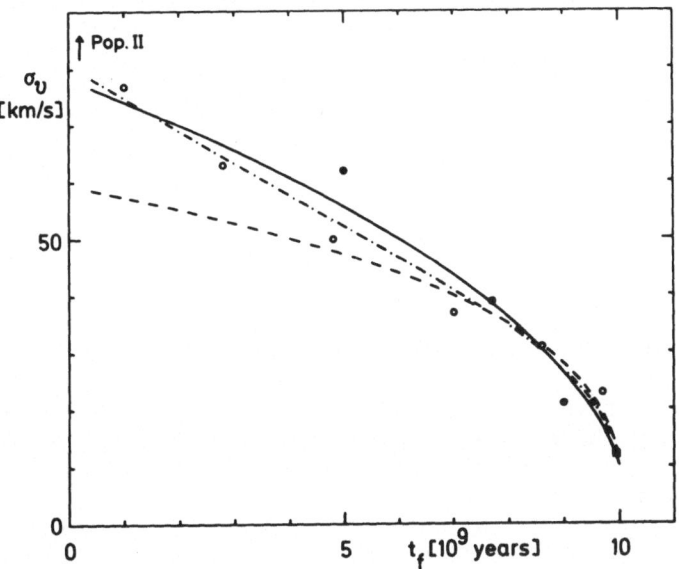

Figure 1.   Total velocity dispersion σ_υ of disk stars,
            integrated over z, as a function of the time
of formation t_f. Observed values: symbols. Theoretical
predictions for the diffusion coefficients: D_0 (full curve),
D_1 (dashed curve), D_2 (dash-dotted curve).

2. VELOCITY DISTRIBUTION OF DISK STARS

The theory of diffusion of stellar orbits does not only describe correctly the observed increase of the stellar velocity *dispersion* with age, but it gives also a rather convincing dynamical interpretation of the velocity *distribution* of nearby disk stars. In order to derive the velocity distribution, we have to solve the Liouville equation with an encounter term of the Fokker-Planck type (Spitzer and Schwarzschild 1951, Wielen 1977, Fuchs 1980). The appropriate Fokker-Planck equation is

$$\frac{\partial f}{\partial t} + W \frac{\partial f}{\partial z} - 2\omega V \frac{\partial f}{\partial U} + \frac{\kappa^2}{2\omega} U \frac{\partial f}{\partial V} - \omega_z^2 z \frac{\partial f}{\partial W} = \frac{1}{2} \left(\frac{\partial}{\partial U}(D\frac{\partial f}{\partial U}) + \frac{\partial}{\partial V}(D\frac{\partial f}{\partial V}) + \frac{\partial}{\partial W}(D\frac{\partial f}{\partial W}) \right).$$

(13)

We use here the epicyclic approximation and neglect radial and tangential gradients of the distribution function f(U,V,W,t). The velocity components U,V,W are measured with respect to the circular velocity at the actual position of the star and are directed towards the galactic center (U), in the direction of galactic rotation (V) and towards the north galactic pole (W). z is the distance from the galactic plane and ω_z the frequency of stellar motion perpendicular to the galactic plane.

For the case of the constant diffusion coefficient D_0 and for one generation of stars (born at the same time), an approximate solution of Eq. 13 is given by a Schwarzschild distribution with velocity dispersions increasing in time (Wielen 1977, Fuchs 1980):

$$f(U,V,W;\tau) = \frac{S(t_f)}{(2\pi)^2} \frac{\omega_z}{\sigma_U \sigma_V \sigma_W^2} \exp\left(-\frac{1}{2}\left(\frac{U^2}{\sigma_U^2} + \frac{V^2}{\sigma_V^2} + \frac{W^2}{\sigma_W^2} + \frac{(\omega_z z)^2}{\sigma_W^2}\right)\right) , \quad (14)$$

with

$$\sigma_U^2 = \sigma_{U,0}^2 + \alpha_U D_0 \tau , \quad \sigma_V^2 = \sigma_{V,0}^2 + \alpha_V D_0 \tau , \quad \sigma_W^2 = \sigma_{W,0}^2 + \alpha_W D_0 \tau , \quad (15)$$

and $\sigma_{U,0} : \sigma_{V,0} : \sigma_{W,0} = \sqrt{\alpha_U} : \sqrt{\alpha_V} : \sqrt{\alpha_W}$. $S(t_f)$ is the star formation rate at the time of formation t_f of this generation of stars.

In order to obtain the overall velocity distribution of nearby disk stars, F(U,V,W), we have to add up the velocity distributions $f(U,V,W;\tau)$ of all the stellar generations, i.e. to integrate f over τ. The resulting velocity distributions for each component U,V,W are shown in Figs. 2 and 3 by the smooth curves for the case of a constant star formation rate. These theoretical curves fit nicely the observed velocity distribution of the most representative sample of nearby disk stars, namely of 317 McCormick K and M dwarfs in Gliese's catalogue.

Fig. 2 gives the velocity distribution in the galactic plane (z=0). The histograms represent the directly observed statistics of the velocities of the representative nearby stars (Gliese 1969, Jahreiß 1974, Wielen 1982). In Fig. 3, we have plotted the velocity distributions for stars in a cylinder perpendicular to the galactic plane. In that case, each star observed at z=0 has to be weighted with $|W|$ in order to derive the empirical histograms for U and V (Wielen 1974). In a similar way, the W-distribution in the cylinder has been obtained from the W-distribution at z=0. Both Figs. 2 and 3 show clearly that the overall velocity distribution of all disk stars is far from being Gaussian in each component, although each generation of stars itself is described by a Schwarzschild distribution. The central peak and the outer wings of the overall velocity distribution are higher than a Gaussian with the same overall dispersion.

The theoretical velocity distributions have also been computed for time-dependent rates of star formation. As long as the star formation rate does not decrease by a factor of more than about five between the formation of the disk and now, the agreement between theory and observations is largely maintained.

From Eq.14, we have derived by integrations the z-dependence of the overall density $\rho(z)$ and of the overall velocity dispersions $\sigma_U(z)$,

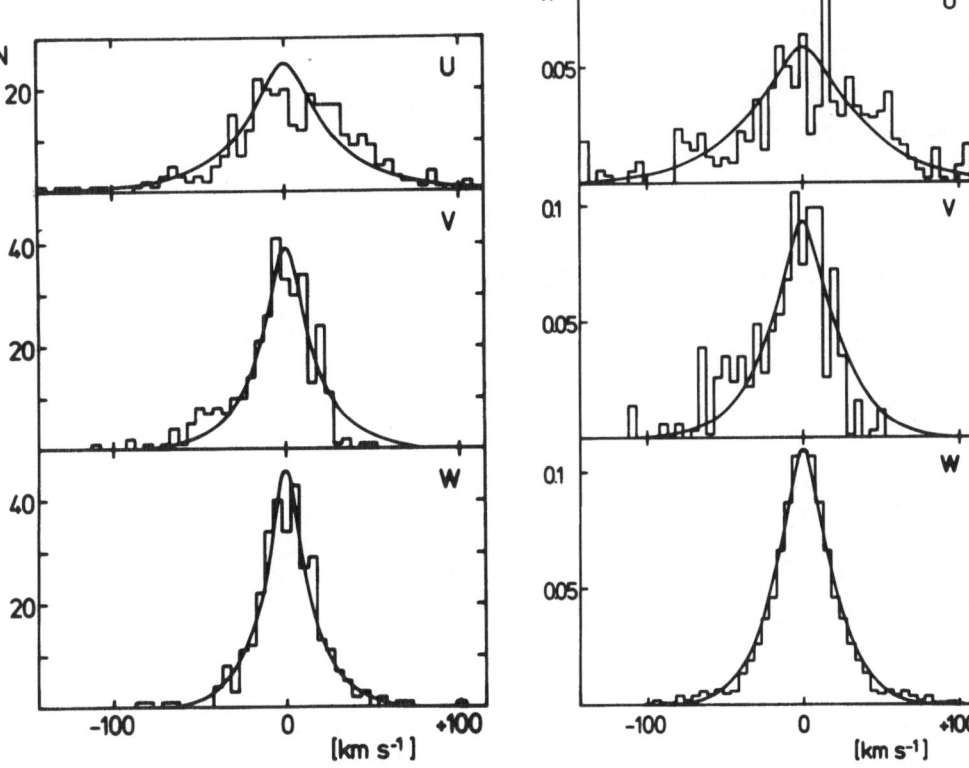

Figure 2.   Velocity dis-
            tribution of
nearby disk stars at z=0.

Figure 3.   Velocity dis-
            tribution of
disk stars integrated over z.

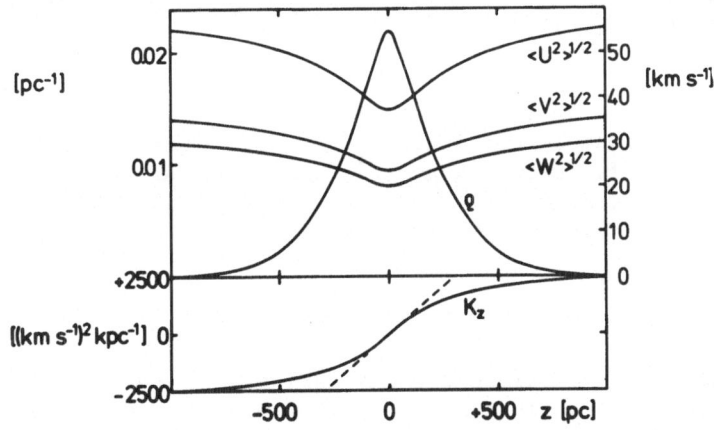

Figure 4.   Predicted z-dependence of the overall velocity
            dispersions $\sigma_U, \sigma_V, \sigma_W$, and of the overall space
density ρ, integrated over all generations of disk stars.

$\sigma_V(z)$, $\sigma_W(z)$. Unfortunately, reliable observational data for a confrontation with the theoretical results are not available at present. The theoretical predictions shown in Fig.4 are based on a constant rate of star formation and on a linear law $K_z=-\omega_z^2z$ with $\omega_z=91.4$(km/s)/kpc (Wielen 1982). A more realistic z-force and a time-dependent rate of star formation can be easily implemented.

For the cases of the velocity-dependent or velocity-time-dependent diffusion coefficients, D_1 and D_2, Eq.13 may be integrated numerically. The situation is even more complicated because impulsive encounters between stars and massive clouds produce 'diffusion coefficients' which oscillate between positive and negative values along the epicyclic orbit of a star. For motions confined to the galactic plane, the epicyclic energy E of a star is given by

$$E = U^2 + (\frac{2\omega}{\kappa})^2 V^2 \qquad . \qquad (16)$$

We consider impulsive encounters between the star and massive clouds moving in circular orbits. Each encounter conservs U^2+V^2 but changes the direction of the stellar velocity vector by an angle $\delta\theta$. Then the local expectation value $<\delta E>$ of the secular change of E within a short time interval Δt is given by

$$<\delta E> = ((\frac{2\omega}{\kappa})^2-1)(U^2-V^2)<(\delta\theta)^2> \qquad . \qquad (17)$$

Because of the assumed stochasticity of the encounters, we have used $<(\delta\theta)>=0$. The quadratic term $<(\delta\theta)^2>$ is adopted in the usual form (e.g. Spitzer and Schwarzschild 1953),

$$<(\delta\theta)^2> \propto (U^2+V^2)^{-3/2}\Delta t \qquad . \qquad (18)$$

From Eq.17, it is obvious that the impulsive encounters increase E for $|U|>|V|$, but cause a loss of epicyclic energy in those parts of the epicycle where $|U|<|V|$. If Eq.17 is averaged over an epicycle we get exactly the same result as Spitzer and Schwarzschild (1953, their Eq.18).

Vader and de Jong (1981) have essentially claimed that the velocity-dependent diffusion coefficient, D_1 or D_2, should be proportional to v^{-3} instead of v^{-1}. Their argument is incorrect because they consider erroneously the quantity U^2+V^2 as the energy instead of using the appropriate epicyclic energy given by Eq.16.

3. GLOBAL BEHAVIOUR OF THE DIFFUSION COEFFICIENT

The diffusion coefficient D may vary significantly with the distance R from the galactic center. As long as we do not know the most important source of the irregular gravitational field in the galactic disk, it is difficult to make theoretical predictions on the R-dependence of D.

Giant molecular clouds are certainly able to perturb gravitation-ally the stellar orbits. However, the abundance of such molecular clouds at the solar distance, $R_0=10$ kpc, seems to be too small for explaining the local value of D derived empirically from nearby stars. In the inner parts of the galaxy, e.g. at R=5 kpc, the frequency of giant molecular clouds is certainly much higher than at the solar distance. But even if these clouds produce at R=5 kpc a diffusion coefficient as high as ob-served at R=10 kpc (Stark and Blitz 1978, Icke 1982), we do not know (1) the source of the diffusion near the sun, and (2) the actual value of the total diffusion coefficient at R=5 kpc. Hence, it may be rather misleading to confront the diffusion coefficient determined empirically at R=10 kpc with a prediction based on the abundance of molecular clouds at R=5 kpc.

From observations it is also difficult to obtain information about the R-dependence of the diffusion coefficient D. The presently most pro-mising way of determining D(R) is the observational determination of the R-dependence of the vertical scale height H(R), either from surface pho-tometry of edge-on external galaxies or from infrared photometry of our Galaxy. H(R) is rather closely related to $\sigma_W(R)$ and hence to D(R). The available observational data seem to indicate that H(R) remains rather constant in the inner parts of a galactic disk and increases slightly with R in the outer parts.

For a thin, vertically isothermal, self-gravitating stellar disk, the density perpendicular to the galactic plane follows

$$\rho(R,z) = \rho(R,0)\text{sech}^2(z/H(R)) \quad , \tag{19}$$

with

$$H(R) = \sigma_W^2(R)/(\pi G\mu(R)) \quad , \tag{20}$$

where ρ is the space density and μ the surface density of stars, σ_W the vertical stellar velocity dispersion, and G the gravitational constant. If we assume $\mu(R)$ or $\rho(R)$ to be known, then Eq.20 provides the required relation between H(R) and $\sigma_W(R)$. Even in the presence of a halo and a co-rona, Eq.20 should be a good approximation for the disk, since the verti-cal force K_z in a thin disk is mainly determined by the disk itself.

Let us now consider various cases of an R-dependence of the diffusion coefficient, D(R), and the implied behaviour of H(R). We shall always assume that the initial velocity dispersion $\sigma_{W,0}$ and the relative distri-bution of stellar ages (especially the mean stellar age $<\tau>$) do not vary with R, and that we can use the velocity dispersion $<\sigma_W^2>$, averaged over all stellar generations, in Eq.20.

If the diffusion coefficient D is constant with respect to R, then $<\sigma_W^2>$ is also constant and H(R) increases strongly outwards because of the decreasing surface density $\mu(R)$, according to Eq.20.

A more realistic case is probably covered by the assumption that the 'constant' diffusion coefficient D_0 is proportional to the surface density μ,

$$D_0(R) = K\mu(R) \quad , \text{ with } \quad K = \text{const.} \quad . \tag{21}$$

Then, for $\sigma_{W,0}=0$, Eqs.15 and 12 imply

$$<\sigma_W^2(R)> \; = \frac{1}{2} K\mu(R)<\tau> \quad . \tag{22}$$

Inserting Eq.22 into Eq.20, we obtain

$$H(R) = K<\tau>/(2\pi G) = \text{const.} \quad . \tag{23}$$

Hence, the assumption that D_0 varies proportional to μ or ρ leads for $\sigma_{W,0}=0$ to a strictly constant thickness of the disk. For $\sigma_{W,0}>0$, the thickness $H(R)$ increases slightly with R in the outer parts of the disk according to

$$H(R) = K<\tau>/(2\pi G) + \sigma_{W,0}^2/(\pi G\mu(R)) \quad . \tag{24}$$

If we assume that the velocity-time-dependent diffusion coefficient D_2 is proportional to the surface density μ,

$$D_2(R) = L(t)\mu(R)/\upsilon \quad , \tag{25}$$

we obtain, for $\sigma_{W,0}=0$,

$$<\sigma_W^2(R)> \; = \alpha_W\alpha_\upsilon^{-1/3}<(\frac{3}{2}<L>_\tau\tau)^{2/3}>\mu^{2/3}(R) \quad . \tag{26}$$

Neglecting the slight variation of $\alpha_\upsilon^{1/3}$ with R, Eqs.26 and 20 give

$$H(R) \propto \mu^{-1/3}(R) \quad . \tag{27}$$

In this case, $H(R)$ increases always with R. The increase is more pronounced in the outer parts of the disk if $\sigma_{W,0}>0$ is taken into account, analogous to Eq.24 .

Similar derivations of $H(R)$ based on the theory of orbital diffusion have already been presented by van der Kruit and Searle (1982, following a suggestion by S.M. Fall) and by Rohlfs and Wiemer (1982). It seems to us, however, that our procedure is simpler and points out more directly the essential assumptions.

In Table 1, we compare the scale height $H(R)$, predicted from the theory of orbital diffusion, with that implied by the mass model of our Galaxy due to M. Schmidt (1965), $H(R)=\mu(R)/(2\rho(R,z=0))$. The required data of the mass model have been taken from a compilation by Wielen (1982). For the velocity dispersion at $R_0=10$ kpc, we have used the observed value derived from the McCormick K and M dwarfs in Gliese's catalogue (Wielen 1974, 1977). Table 1 shows that the velocity-time-dependent diffusion coefficient (Eq.25) represents the mass-model data quite well. On the other hand, the velocity-independent coefficient (Eq.21) seems to fit better the photometric results for $H(R)$ in edge-on external galaxies (van der Kruit and Searle 1982, Rohlfs and Wiemer 1982).

Table 1. R-dependence of H and $<\sigma_W^2>^{1/2}$ in the Galaxy

R	Mass model H	$D_0 = K\mu(R)$ H	σ_W	$D_2 = L(t)\mu(R)/\upsilon$ H	σ_W
kpc	pc	pc	km/s	pc	km/s
4	295	397	53	254	42
6	334	399	43	294	37
8	353	401	33	343	31
10	393	406	25(obs)	406	25(obs)
12	472	414	19	475	21
15	586	432	14	587	16
20	778	484	10	792	12

Table 2. R-dependence of $\sigma_{U,crit}$ and $<\sigma_U^2>^{1/2}$ in the Galaxy

R	$\sigma_{U,crit}$	$D_0 = K\mu(R)$ σ_U	$D_2 = L(t)\mu(R)/\upsilon$ σ_U
kpc	km/s	km/s	km/s
4	89	87	69
6	78	71	60
8	64	58	53
10	52	48(obs)	48(obs)
12	42	40	42
15	31	30	35
20	20	21	27

Table 3. Time-dependence of $<\sigma_U^2>^{1/2}$ at $R_0 = 10$ kpc

t	D_0 (Eq.1) σ_U	D_2 (Eq.6) σ_U
10^9 yr	km/s	km/s
0	8	8
1	15	28
2	20	34
3	24	37
4	28	39
5	31	41
6	33	41
7	36	42
8	38	42
9	40	42
10	43	42

In order to stabilize a self-gravitating disk against gravitational instabilities due to axisymmetric perturbations, the radial stellar velocity dispersion σ_U has to be larger than a critical value (Toomre 1964),

$$\sigma_{U,crit} = 3.36 \; G\mu/\kappa \quad . \tag{28}$$

The condition $\sigma_U \geq \sigma_{U,crit}$ should be fulfilled at each radius R and for all times t. In Table 2, we compare $\sigma_{U,crit}$ with σ_U predicted from the diffusion coefficients adopted in Eqs.21 and 25, using again the observational data for σ_U at $R_0=10$ kpc and μ, κ, ω from the mass model. For the velocity-independent diffusion coefficient (Eq.21), σ_U is always rather close to $\sigma_{U,crit}$. In Table 3, we indicate the variation of the mean value of σ_U for all the stars formed up to the time t as a function of t at $R_0=10$ kpc, assuming a constant rate of star formation and the diffusion coefficients D_0 and D_2 according to Eqs.1 and 6. Since D_2 leads to a value of σ_U nearly constant in time (except for the earliest stage), D_2 is more favourable than D_0 for stabilizing the disk over its whole lifetime. However, it is very likely that the disk is strongly stabilized in the radial direction by the gravitational field of the halo and the corona. This lowers the critical value for the velocity dispersion in the disk considerably at all radii and for all times. Hence the low values of σ_U predicted by the theory of orbital diffusion for the early stages, are probably not in conflict with the requirements for stabilizing galactic disks also in the more distant past.

REFERENCES

Fuchs,B. : 1980, Diss.Math.-Naturwiss.Fak.Univ.Kiel.
Gliese,W. : 1969, Veröffentl.Astron.Rechen-Inst.Heidelberg No.22.
Icke,V. : 1982, Astrophys.J. 254, p.517.
Jahreiß,H. : 1974, Diss.Naturwiss.Gesamt-Fak.Univ.Heidelberg.
Rohlfs,K., Wiemer,H.-J. : 1982, Astron.Astrophys. (in press).
Schmidt,M. : 1965, Stars and Stellar Systems 5, p.513.
Spitzer,L., Schwarzschild,M. : 1951, Astrophys.J. 114, p.385.
Spitzer,L., Schwarzschild,M. : 1953, Astrophys.J. 118, p.106.
Stark,A.A., Blitz,L. : 1978, Astrophys.J. 225, p.L15.
Toomre,A. : 1964, Astrophys.J. 139, p.1217.
Vader,J.P., de Jong,T. : 1981, Astron.Astrophys. 100, p.124.
van der Kruit,P.C., Searle,L. : 1982, Astron.Astrophys. 110, p.61.
Wielen,R. : 1974, Highlights of Astronomy 3, p.395.
Wielen,R. : 1977, Astron.Astrophys. 60, p.263.
Wielen,R. : 1982, Kinematics and Dynamics of the Galaxy. Chapter 8.4 of
     Landolt-Börnstein, Group VI, Vol.2,Subvol.2c. Eds. H.H.Voigt and
     K.Schaifers, Springer-Verlag, Berlin (in press).

THE LARGE-SCALE STRUCTURE OF ATOMIC HYDROGEN

F. J. Kerr
University of Maryland

Abstract

Recent 21cm surveys with the Parkes 18m and 65m telescopes are discussed, together with a new compilation of 21cm data right around the galactic plane. The Parkes 18m and Hat Creek 26m surveys have been used to study the distribution of HI in the outer Galaxy based on a circular rotation model. The surface density distribution is discussed, and also the warping and the thickness of the HI layer. The total mass of HI in the Galaxy is found to be $4.8 \times 10^9 M_\odot$.

This paper will not be an overall review of the large-scale structure of atomic hydrogen, because time is not available. Also, the authors of the other recent study (Kulkarni, Blitz, and Heiles) are all attending this workshop, and Kulkarni will discuss their results. This paper will therefore concentrate on the recent work of the Maryland group.

1. Galactic equator plot

We have made a new presentation of high resolution 21-cm data right around the galactic equator, in the form of a detailed array in the longitude-velocity plane. The data come from three Maryland surveys, (i) $\ell = 241°-349°$, by Kerr, Bowers and Henderson (1981) with the Parkes 64-meter telescope, (ii) $\ell = 349°-11°$, by Sinha (1979) with the Green Bank 43-meter telescope, and (iii) $\ell = 11°-231°$, by Westerhout (1973, 1982) with the Green Bank 92-meter telescope. A color version of this overall ℓ-v plot will be published shortly and the compilation is also available on magnetic tape, with profiles every 3 arcmin. In addition to the main spiral features, a very large amount of detail can be seen in this combined plot of the hydrogen distribution, including a substantial number of "holes" in the hydrogen in limited longitude and velocity ranges.

W. L. H. Shuter (ed.), Kinematics, Dynamics and Structure of the Milky Way, 91–96.

2. Outer region

The outer regions of the Galaxy have been receiving special
attention in two recent large-scale studies, because the problem of
the distance ambiguity is avoided there. One of these studies is
based on a completely sampled Southern Milky Way neutral hydrogen
survey, which was carried out at Parkes several years ago by F.J. Kerr,
P.F. Bowers, and M. Kerr, using the 18-meter telescope and covering
a range ℓ = 240° to 350° in galactic longitude, and b = -10° to +10°
in galactic latitude. The velocity resolution was 2.1 km s^{-1}, and
the spatial resolution 48 arcmin. These data have been combined with
the analogous northern survey by Weaver and Williams (1974) with the
Hat Creek 26-meter telescope in the outer region study of Henderson,
Jackson and Kerr (1982). In this analysis, a flat rotation curve was
assumed outside the solar circle, because the observational evidence
reported for a rising curve (Blitz 1979) is limited so far to a rather
small range of galactocentric angles, and other galaxies seem to show
flat curves. Also, optical depth effects were taken into account in
this study.

The related study by Kulkarni, Blitz, and Heiles (1982), which
uses a new analysis of the data of Weaver and Williams, will be
described by Kulkarni in the next paper in this workshop.

2.1 Density pattern

The overall density pattern for the HI in the outer Galaxy was
examined by integrating over the whole latitude range and then
plotting out the density projected on to the galactic plane.
Henderson et al. show that the pattern is dominated by spiral
features on both sides of the diagram, including the well-known
Perseus arm, Carina arm, and the "Outer arm". A strong feature
near R = 11 kpc and φ(galactocentric angle) = 340° extends over
a wide latitude range, and is probably a relatively local feature
with a non-circular velocity, receding from the Sun.

The bilateral symmetry which would be expected from a two-armed
"grand design" is clearly not present. In addition, the HI extends
further out in galactocentric radius in the third and fourth
quadrants of longitude (the "southern" side) than in the first and
second quadrants. This interpretation depends on the correctness
of the assumption of circular motion. Another striking feature
of the pattern is the unbalanced appearance near the omitted regions
around ℓ = 0° and 180°. As pointed out by Shuter (1981) and by
Henderson et al. (1982), there would be a more symmetrical
appearance if the local standard of rest were postulated to move
outwards by 7-10 km s^{-1} with respect to the center of mass of the
Galaxy. Alternatively the motions might be elliptical, or indeed
the structure could be quite asymmetrical.

The mean projected density of the HI as a function of R can be obtained by summing over φ for a given R. This has been done for the two intervals φ = 30° to 150°, and φ = 210° to 330°. These summations show again the large departure from bilateral symmetry, when distance estimates are based on circular motion. The total mass of HI outside R = 11 kpc is found to be 1.12 x 10^9M for 150° > φ > 30° and 1.49 x 10^9M for 330° > φ > 210°. After allowing for the mass in the inner region, and in the gaps near φ = 0° and 180°, Henderson et al. (1982) derived a figure of 4.8 x 10^9M for the total HI mass in the Galaxy, of which 81% lies outside R $\cong$ 11 kpc. These mass estimates are all very dependent on the assumptions of circular motion and a flat rotation curve.

2.2 Warp

The shape of the outer warp of the Galaxy has been more precisely delineated in the recent studies, principally because of the greater sensitivity of the newer surveys. (See Figure 1.) Again we have a striking departure from symmetry. For the north, the warping increases progressively and rapidly with R, whereas in the south the centroid of the HI distribution reaches a maximum negative value of -850 pc, near R = 17 kpc and φ = 260°, and then returns to the plane at larger R. The axis of the warp is 260° to 80° in φ, which is inclined about 10° to the Sun-center line, and points approximately in the direction of the Large Magellanic Cloud. There is an interesting "scalloping" effect in the outermost part of the warp at R > 24 kpc, which appears on both sides of the map.

2.3 Layer thickness

The recent studies refine the long-known result that the thickness of the neutral hydrogen layer increases rapidly with R in the outer regions, approaching a value of 2 kpc near the outer edge of the Galaxy. Henderson et al. point out that the apparent thickness is less near the Sun than in the far part of the Galaxy beyond the central region, but this result, which occurs on both sides of their plot, is probably an artefact of the observations. The calculations for the regions of the Galaxy at the greatest distance are affected by HI emission from nearby regions, either as "stray radiation", or through local material at noncircular velocities being attributed to greater distances.

2.4 Terminal velocities

We have also taken a new look at the terminal velocities in the line profiles referring to the outer regions on the two sides of the Galaxy. The newer, more sensitive data show that the southern terminal velocities are about 10-15 km s^{-1} higher than those for the northern side (Jackson and Kerr 1981). This result may be influenced by differences in the warp on the two sides, but a straightforward interpretation suggests a value of 250 km s^{-1} for the circular

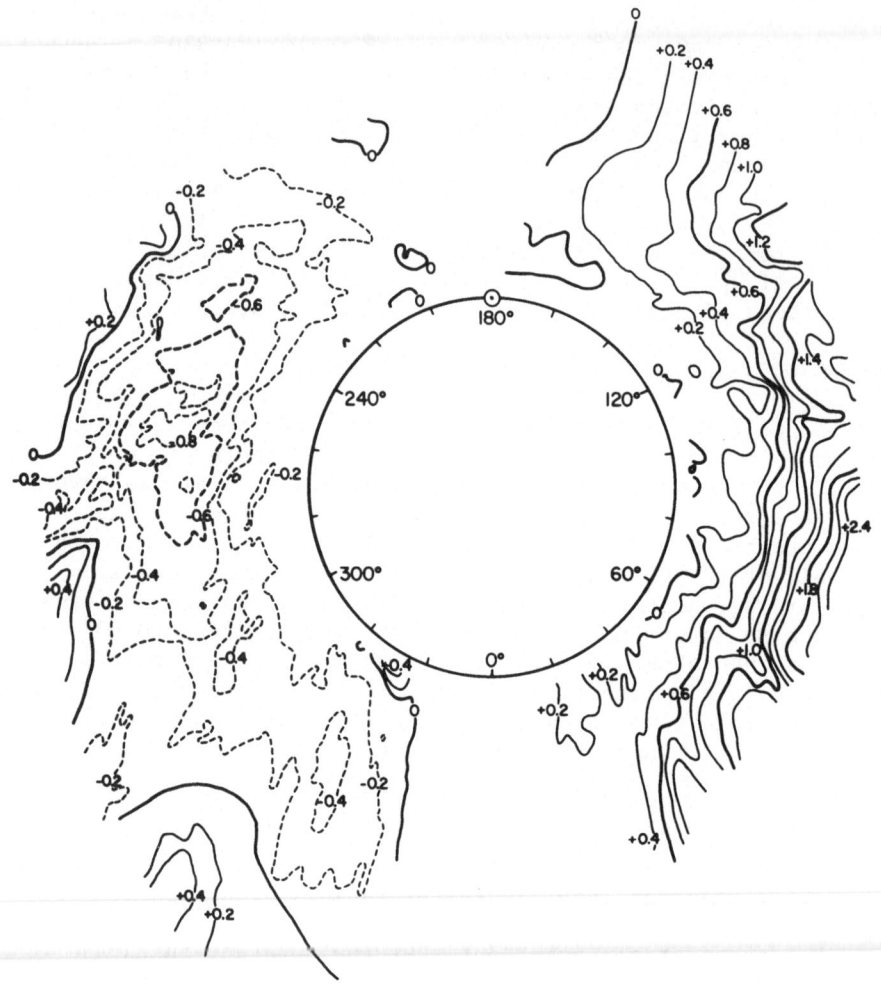

Figure 1. Contour map of the centroid of the neutral hydrogen
distribution in the outer part of the Galaxy. Contours
are labeled in kiloparsecs from the galactic plane.

velocity at the Sun, Θ_o, and also indicates that the hydrogen density
probably falls off more slowly than the exponential previously assumed.

3. Inner region

Analysis of observational data on the HI in the inner region of
the Galaxy is more difficult because of the distance ambiguity.
David L. Ball at Maryland has been looking at the data from our two
recent southern HI surveys (Kerr, Bowers and Henderson 1981; Kerr,

Bowers and Kerr 1982). He has been carrying out a modeling study,
based on a range of density-wave parameters; this study has used
both the linear and nonlinear theories, and introduces the z-dimension
so that comparisons can be made with the full latitude range of the
observational data. An important conclusion from this study is that
the Galaxy is more irregular than used to be thought. Each individual
feature must be fitted separately to an appropriate model, with
different parameters, rather than to a single grand design.

Figure 2 gives a preliminary longitude-velocity diagram of the
main spiral features in the Southern Milky Way. This type of diagram
is directly derived from the observational data, without having to
contend with velocity ambiguities. The main characteristics of the
diagram which are due to spiral features interior to the Sun are
labeled A to E. The features labeled I (intermediate) and 0 (outer)
are due to gas outside the solar circle at R > 10 kpc, and will not
be discussed further here.

Feature A, the Norma spiral arm, is a somewhat weak feature of
very narrow latitude width ($\sim1°5$), which is appropriate for a
trailing spiral arm feature at this longitude (tangent at $\ell = 328°$).
The near side, A_1, is weaker and more diffuse than the far side, A_2.

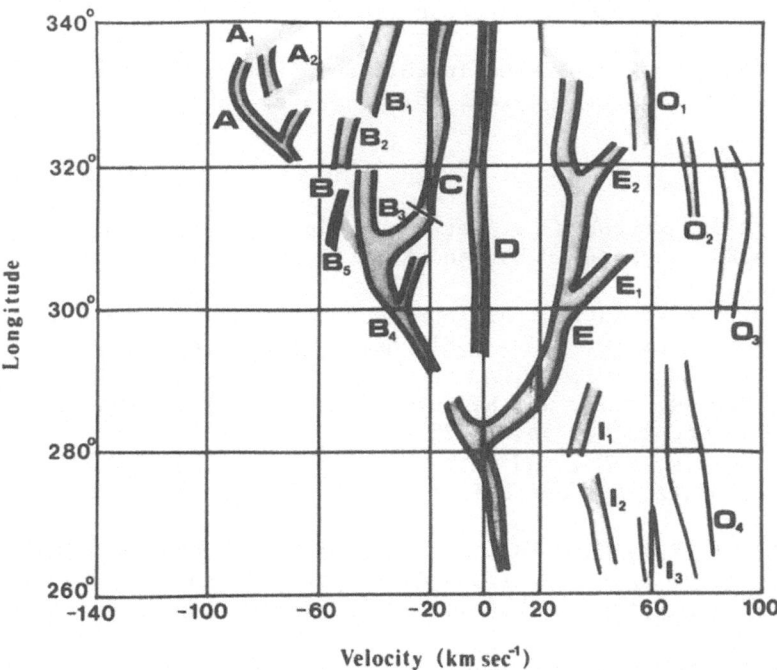

Figure 2. Longitude-velocity diagram of the main spiral features
           in the Southern Milky Way.

The Local Feature, labeled D in the diagram, is easily seen at
zero velocity and is strongly present at all latitudes $\leq 10°$. It is
also characterized by a narrow velocity width.

The most interesting and complex feature in this region is labeled
B and C; this is really a complex feature with two major parts. The
first part, B, consists of pieces of the near side of a trailing
spiral arm loop, B_1 through B_3, which is tangent at $\ell = 305°$ and has
a latitude width of $3°-4°$. Feature B_4 is due to velocity crowding
for lines of sight further from the galactic center than the main
portions of the feature B. The shoulder at $\ell \sim 312°-315°$ (feature B_5)
has been interpreted by some investigators as another spiral arm,
tangent at this longitude and with a fairly high pitch angle. In
Ball's interpretation, the best single spiral pattern which approximates
the A-B combination is a two-armed low pitch angle ($6°-8°$) pattern.

The second part of the B-C composite, feature C, is a narrow
velocity-width feature, similar to D. It is not however the far
side of the Sagittarius arm (B), as it has the ℓ-v shape and latitude
width appropriate to more nearby material. It is in fact seen best at
latitudes away from the plane, and was therefore not so apparent in
earlier studies which were more restricted in latitude.

Feature E is the Carina spiral arm, which is almost all outside
the solar circle, particularly the two highly inclined spurs, E_1 and
E_2, which were also noted by Jackson (1976).

The Maryland work described in this paper was supported by
National Science Foundation grant AST 77-26898.

References

Blitz, L.: 1979, Astrophys. J. Letters, 231: pp. L115-L119.
Henderson, A.P., Jackson, P.D., and Kerr, F.J.: 1982, Astrophys. J.,
    in press.
Jackson, P.D.: 1976, Astron. Astrophys. Suppl. 25, pp. 449-452.
Jackson, P.D., and Kerr, F.J.: 1981, Bull. Amer. Astron. Soc., 13,
    p. 538.
Kerr, F.J., Bowers, P.F., and Henderson, A.P.: 1981, Astron.
    Astrophys. Suppl. 44, pp. 63-75.
Kerr, F.J., Bowers, P.F., and Kerr, M.: 1982. In preparation.
Kulkarni, S., Blitz, L., and Heiles, C.: 1982, Astrophys. J. Letters,
    in press.
Shuter, W.L.H.: 1981, Mon. Not. Roy. Astron. Soc., 199, pp. 109-113.
Sinha, R.P.: 1979, Astron. Astrophys. Suppl., 37, pp. 403-463.
Weaver, H.F., and Williams, D.R.W.: 1974, Astron. Astrophys. Suppl., 17,
    pp. 1-249.
Westerhout, G.: 1973, Second Maryland-Green Bank 21-cm Line
    University of Maryland.
Westerhout, G., and Wendtlandt, H.: 1982, Astron. Astrophys.
    Suppl., in press.

LARGE-SCALE STRUCTURE OF HI IN THE OUTER GALAXY

Shrinivas R. Kulkarni (1)
Leo Blitz (2)
Carl Heiles (1)
(1) Department of Astronomy, University of California,
    Berkeley
(2) Astronomy Program, University of Maryland,
    College Park

Abstract — The HI in the outer Galaxy is reanalyzed using the CO
rotation curve of Blitz, Fich and Stark (1982) and the full latitude
extent of the gas from the Weaver and Williams (1974) survey. HI with
a surface density > 0.1 M$\odot$/pc^2 is found to a distance of 30 kpc from
the centre. Three distinct, well defined spiral features of roughly
constant surface density are seen, two of which extend at least 20–25
kpc along their length and can be traced to 20 kpc from the centre. If
they are logarithmic spirals, the major arms have a pitch angle of
22–25°. A nearly circular corrugation is seen as a deviation from the
large scale warping at R~11 kpc. The outermost parts of the Galaxy show
remarkable scalloping with a large azimuthal wave number (m~10). The
scale height of the gas shows an almost linear increase from the solar
vicinity to R~30 kpc. The increase implies that the large mass in the
outer Galaxy implied by the rotation curve does not reside in a thin
disk. The vertical distribution of the gas is shown not to be well
described by either a gaussian or an exponential. However the form of
the vertical distribution is independent of radius between R = 12 kpc
to 25 kpc.

The analysis of the distribution of HI within the solar circle
(R<R$_o$) is plagued by the distance ambiguity problem leading to problems
in interpretation. In contrast, for HI beyond the solar circle the
radial velocity is uniquely related to distance; this makes the
analysis of HI beyond the solar circle very attractive and fruitful.
Knowledege of the rotation curve outside the solar circle (Blitz, Fich
and Stark 1982, hereafter BFS) makes it possible to reanalyze the HI
distribution in terms of quantities intrinsic to the gas such as
density and scale height rather than observational quantities such as
antenna temperature and velocity. We have used the HI survey of Weaver
and Williams (1974) and an approximation to the BFS rotation curve to
determine the distribution of the following four quantities:
(a) σ, mass surface density. σ is the mass of HI contained in a

97

W. L. H. Shuter (ed.), Kinematics, Dynamics and Structure of the Milky Way, 97–104.
Copyright © 1983 by D. Reidel Publishing Company.

cylinder 1 pc^2 in area and perpendicular to the galactic plane.
(b) ρ_0 , the volume density in the midplane. ρ_0 is the volume density
of the HI in the mean plane defined locally i.e. $\rho_0 = \rho(<z>)$.
(c) $<z>$, the z-distance of the mean HI plane from the b=0^0 plane. In
other words this is the first moment of $\rho(z)$.
(d) B, the scale height of the HI layer. B is the rms scale height of
the HI layer i.e. B = $<(z-<z>)^2 >^{1/2}$.

The BFS rotation curve is valid upto R$\sim$19 kpc. Beyond that we have
used a flat rotation curve with a linear velocity of 285 km/sec. We
modified the formal BFS rotation curve so that it smoothly joined to
the flat rotation curve and further removed some rather steep
gradients. We have used the standard IAU values for R_0 and θ_0 of 10
kpc and 250 km/sec respectively. We have neglected the optical depth
corrections for three reasons :
(a) we expect this correction to be severe only very close to the plane
(b) in view of the wide range of temperatures of cold HI, it is not
clear to us that a blanket correction based on $T_K\sim$ 125^0 K is
meaningful.
(c) Burton (1971) has pointed that the HI profile is more sensitive to
velocity perturbations than density perturbations.

Much of the present work will soon appear in the Astrophysical
Journal (Letters). To save duplication we will discuss rather briefly
the results reported there and stress a lot on results not reported
there.

1. Surface density

Several large scale features are seen in Figure 1. The continuity
of the features across the anti-center is made on the basis of
velocity. Below is a brief description of the three prominent
features.

Feature A: $l = 15^0$, R = 10 kpc to $l = 110^0$, R = 20 kpc
           length$\sim$25 kpc; pitch angle$\sim$ 22^0 , σ = 4-8 M$\odot$/pc^2
   Feature B: $l = 60^0$, R = 10 kpc to $l = 250^0$, R = 20 kpc
           length$\sim$20 kpc; pitch angle$\sim$25^0; σ = 4-8 M$\odot$/pc ^{2}
   Feature C: $l = 90^0$, R = 10 kpc to $l = 250^0$, R = 11 kpc
           length$\sim$4 kpc; pitch angle$\sim$28^0; σ = 4-8 M$\odot$/pc ^{2}

Feature A has sometimes been called the Outer arm, but for
consistency we refer to it as the Cygnus arm since outside the solar
circle it occurs primarily in the constellation of Cygnus. Feature B
is the classical Perseus arm and C is the Orion arm. The surface
density of the Orion arm has been underestimated due to our neglect of
off-plane gas; σ, above has been corrected for that. Note that no such
correction has been applied in Fig. 1. Burton (1971) has pointed out
that perturbations in the velocity field can lead to large variations

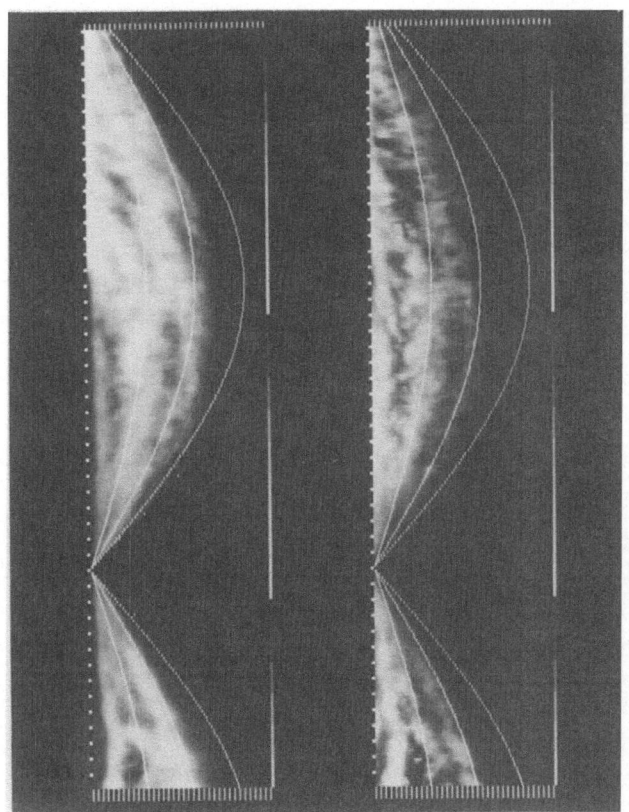

Figure 1: In Figures 1 and 2 we show grey scale plots of ρ_o, σ, $+<z>$ and $-<z>$. In each case longitude is the ordinate increasing from top to bottom; velocity is the abscissa increasing in absolute value to the right. Velocities in the first two quadrants are negative and positive in the third quadrant. In each figure, lines of constant galacto-centric radius are given at R = 11, 13, 15, 20, 25 and 30 kpc. The longitude is indicated every 5° from $l=10^o$ to $l=250^o$. The velocity scale is indicated by the tick marks on the horizontal axis; each tick mark is 5 km/sec. In all figures the strips forming the left boundary of the figures are calibration rectangles. The absolute value of the quantities always increases with the brightness of the calibration rectangles .

The figure on the right side is an l-v diagram of the surface density and that on the left, of volume density. In the left figure the faintest calibration rectangle has a value of 0.02 cm^{-3} subsequent to this the value of the calibration rectangles increases in steps of 0.04 cm^{-3} In the right figure the nine calibration rectangles have values of 0.25, 0.75, 1.5, 2.25, 3.0, 3.75, 4.5, 5.25 and 6.0 $M\odot/pc^2$ in order of increasing brightness.

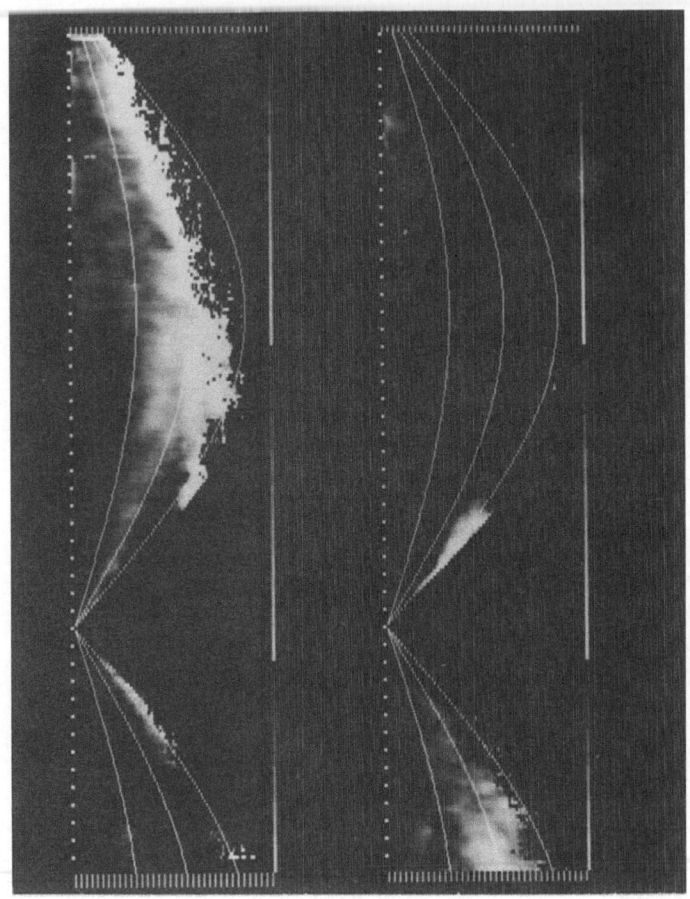

Figure 2: The positive deviation of the median plane, +<z>, of HI from
b = 0° is shown in the left hand figure and the negative deviation,
-<z> in the right hand figure. The faintest calibration rectangle has
a value of 100 pc increasing in steps of 200 pc.

in the deduced volume density. Random perturbations should occur as
relatively small scale features in Figure 1. However, the spiral
features of Figure 1 occur over large length scales and it is difficult
to understand how these could be produced without corresponding
perturbations in denstiy. Furthermore, high density molecular clouds
and their associated HII regions are also closely correlated with the
HI arms (Blitz, Fich and Kulkarni, 1982; Kutner and Mead, 1981). It
seems reasonable to conclude that the HI arms are primarily due to
density enhancements.

2. Mean plane

Figure 2 shows the well known galactic warp. The degree of
warping is fairly small to $R \lesssim 18$ kpc and becomes much more pronounced
at larger distances. The diffuse gamma ray emission in the second and
third quadrants follow the HI warp (Mayer-Hasselwander, et al. 1982);
the same effect has been noted by Phillips et. al (1981). This implies
that a significant cosmic ray flux exists at galactocentric distances
as great as 20 kpc.

In external galaxies, warps become most pronounced where the
stellar disk becomes difficult to detect (van der Kruit, 1981). In our
galaxy, Fich and Blitz (1981) do not find HII regions beyond R~20 kpc.
Thus there is strong reason for us to conclude that the onset of severe
warping coincides with the edge of the stellar disk. Independent
evidence comes from the bending wave theory of Bertin and Mark (1980).
In this theory, the amplitude, A(R) of the bending wave, is related to
the disk mass surface density as
$$A(R) \propto R^{-\frac{1}{2}} \sigma^{-1} \qquad (1)$$
A(R) is the maximum vertical displacement at a given R i.e. extremum of
<z(R)>. Since the warping becomes severe at R~ 18–20 kpc we conclude
that the stellar disk in our galaxy ends at about R ~ 20 kpc.

Two new features in the gas distribution appear in these figures.
The, first, a corrugation, is an approximately circular feature at R ~
11 kpc, which in contrast to the warp, is below the IAU plane in the
first and second quadrants and above in the third quadrant. The second
new feature is a scalloping of the outer edge of the HI gas. The outer
edge of the galaxy seems to have positive displacement with respect to
the IAU planefrom 1=10⁰ to 1=150⁰; depressed below the IAU plane, from
1=150⁰ to 1=170⁰; is elevated from 1=190⁰ to 1=220⁰ and depressed
beyond 1=220⁰. This scalloping can be traced for three more azimuthal
wavelengths in the southern data of Henderson, Jackson and Kerr (1982)
which also shows its regularity.

3. Scale height:

The run of the azimuthally averaged scale height with the

galactocentric radius R, is shown in Figure 3. Data in the longitude range 90° to 250° has been used in this figure. We find :
(a) the scale height increases linearly from R = 10 to 25 kpc.
(b) the scale height of the nearby gas is very small. This is due to our neglect of gas beyond $|b| > 10°$. Using data from off-plane surveys we have computed the approximate scale height of the nearby gas; this is shown by dashed lines in Fig. 3. The dashed line is in agreement with the determination of the local values of the gaussian scale heights found by Falgarone and Lecquex (1973).
(c) the scale height of the gas increases more rapidly in the first quadrant as compared to the gas in the second and third quadrants.

The scale height of the gas is determined by (a) surface mass density,σ_*,the scale height, H, of the stars and (b) Q^2 which is the ratio of the total gas pressure to the gas density. Locally H $\gg$ B and in this limit we have the following relation :

$$B^2 \propto HQ^2/\sigma_* \qquad\qquad (2)$$

From (2) we see that the scale height of the gas can be used to probe the mass distribution of the stars, provided we know the dependence of Q^2 on R. In the next paragraph we would like to argue that $Q^2(R)$ is known to better than a factor of 2.

The total pressure on the gas is due to (a) the turbulence in the gas which is manifested by a line of sight velocity dispersion of 5-8 km/sec; that this is roughly constant is evidenced in the recent velocity-field map of NGC 3938 (van der Kruit and Shostak, 1982); (b) the pressure due to cosmic ray particles; (c) the magnetic field pressure. In the modelling of galactic gamma-ray emission, a cosmic ray flux independent of z seems to have given excellent fit to the data (Mayer-Hasselwander et al. , 1982). From a comparision of synchrotron, gamma-ray emission and HI column density (refer to Brown's review talk, this book), one derives $<B> \propto n^{0.5}$ in which case the magnetic pressure is $\propto$ n and thus the magnetic pressure behaves like the gas pressure with dispersion independent of R. Judging from the extensive discussion following Brown's talk, this result is at best controversial. Hence there is at worst an uncertainty of a factor of two in $Q^2(R)$.

A flat or a rising rotation curve combined with an increasing scale height leads us to conclude that a large fraction of mass implied by a flat or a rising rotation curve must reside outside a thin disk. To see this we will make the simpifying assumption that the rotation curve is flat. This implies that $\sigma_* \propto R^{-1}$. In order to relate B and σ_* we will use equation (2) above. The use of this equation is justified in the next paragraph. From Fig 3 and the two relations just presented we conclude that H (R = 30 kpc) is about 30 kpc. Thus if we insist on keeping all the mass within a disk, the disk becomes very thick at the edge of the galaxy. In other words the mass implied by a flat rotation

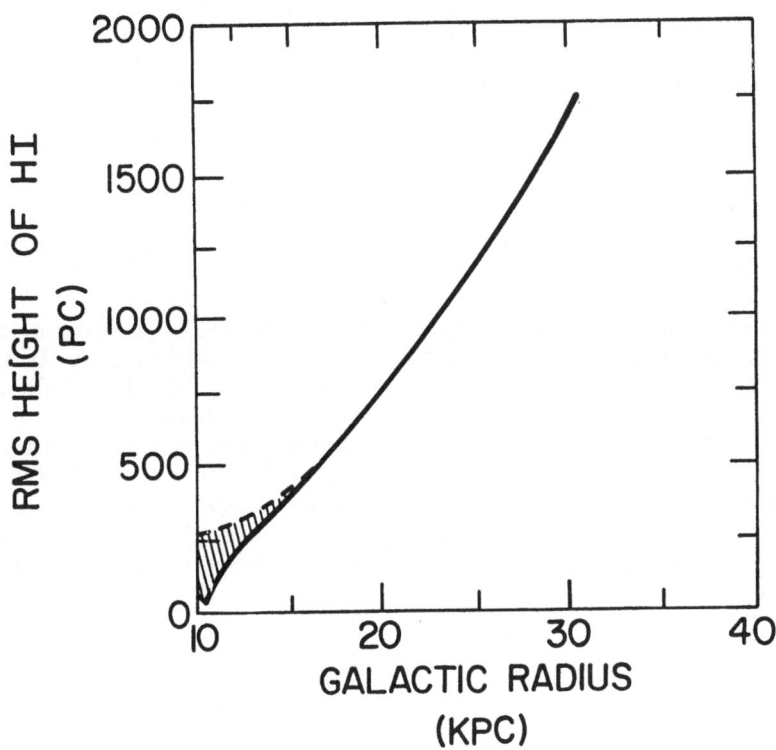

Figure 3: The scale height of the outer HI layer defined by B (see text) azimuthally averaged over $90° < 1 < 250°$. The hatched area is the region within which we estimate the errors in the scale height are greater than 10% due to our neglect of the intermediate and the high latitude gas. The dashed line gives the corrected values from the high latitude extension measured at a number of longitudes and velocities.

curve is not resident in a thin disk.

The shape of the vertical distribution of the HI layer is determined by the star layer. The vertical distribution is gaussian if H >> B and an exponential for the case H << B. Thus in principle one can use the shape of the vertical layer to determine the distribution of the stars relative to the gas layer. To examine the form of the vertical distribution of the gas, we have computed the ratio of C to B. Here C is the z-distance between the 40th centile and the 60th centile mark of the z- (z) curve; this ratio is 0.5 for a gaussian and 0.31 for an exponential. We find that this ratio is about 0.43 for R = 12 kpc to 25 kpc. The vertical gas distribution is, therefore, not well described by either a gaussian or an exponential. The important fact to note is that this ratio is essentially independent of galactocentric radius (we have not analyzed gas closer than R = 12 kpc due to the survey limitation of b > 10°). This means that the star layer is always bigger than the gas layer everywhere. So the star layer, just like the gas layer thickens.

How critical is the shape of the rotation curve in determination of the four quantities considered above? In order to answer this we have constructed l-v diagrams using (a) the Schmidt rotation curve, (b) a flat rotation curve with linear velocity of 250 km/sec and (c) a rotation whose linear velocity is between curve (b) and the rotation curve adopted in this work . We find that qualitatively all the l-v diagrams are similar; thus they differ from each other only in (roughly) a scale factor. In particular, there are great similarities between the features in l-v diagrams based on curves (b), (c), (d) and our adopted rotation curve.

REFERENCES

Blitz, L., Fich, M. and Stark, A.A. 1982, submitted to Astrophys. J. Suppl.
Blitz, L., Fich, M. and Kulkarni, S. 1982, Scinece, in preparation
Burton, W.B. 1971, Astron. Astrophys. 19, 51
Falgarone, E. and Lecquex, J. 1973, Astron. Astrophys. 25, 253
Fich, M. and Blitz, L. 1979, BAAS 11, 706
Henderson, A.P., Jackson, P.D. and Kerr, F.J. 1982, Astrophys. J. submitted
Knapp, G.R., Tremaine, S.D. and Gunn, J.E. 1978, Astron. J. 83, 1585
Kutner, M.L., and Mead, K. 1981, Astrophys. J. (Letters) 249, L15
van der Kruit, P.C. 1981, Astron. Astrophys. 99, 298
van der Kruit, P.C. and Shostak, G.S. 1982, Astron. Astrophys. 105, 351
Mayer-Hasselwander, H.A., et al. 1982, Astron. Astrophys. 105, 164
Phillips, S., Kearsey, S., Osborne, J.L., Haslam, C.G.T. and Stoffel, H. 1981 Astron. Astrophys. 103, 405
Weaver, H. and Williams, D.R.W. 1974, Astron. Astrophys. Suppl. 17, 1

MORE HI SHELLS AND SUPERSHELLS or A NEW EXPLANATION OF "NONCIRCULAR
MOTIONS" IN THE GALAXY

Carl Heiles
Astronomy Department, University of California, Berkeley

The **discovery** of HI shells and supershells in distant parts of
the galaxy (Heiles, 1979) utilized the Weaver/Williams (WW; 1973)
HI survey of the galactic plane, which is restricted to $|b| < 10°$. These
shells have diameters ranging up to 2 kpc, and some expand with
velocities up to 24 km/sec. If produced by stellar winds and
supernovae, a very large--perhaps unreasonably large--number of
massive stars in a young cluster is required.

I am currently preparing a list of newly-discovered HI shells
and supershells. This list is derived from the combination of the
WW and Heiles/Habing (1974) HI surveys, which eliminates the
artificial boundary at $|b| = 10°$. This list contains several dozen
new shells. Many are also located at large distances.

One of the most impressive shells is centered near $(1,b) = (180°, -2°)$,
with diameter about $40°$. It is easily visible over the entire
velocity range -90 to -38 km/sec, and even can be seen down to -20
km/sec. It exhibits some changes in size and appearance with velocity.
It is so large in angle, and has such a large velocity, that it has
been mistaken for a large-scale galactic feature such as an arm or a
spur. It forms the negative velocity gas near the galactic anticenter
that has stimulated some workers to invoke large noncircular motions
in the outer parts of the galaxy.

Figures 1 and 2 show grey-scale representations of the original
data at two velocities, -45 and -71 km/sec, which exhibit this shell.
Figure 3 shows a single sketch that depicts the location of the ridges
of the shell at velocities between -90 and -45 km/sec. This shell is
unusual, but not unique, because it is visible over a very large
velocity range, and because its systemic velocity is high and
apparently unrelated to galactic rotation.

W. L. H. Shuter (ed.), Kinematics, Dynamics and Structure of the Milky Way, 105–108.
Copyright © 1983 by D. Reidel Publishing Company.

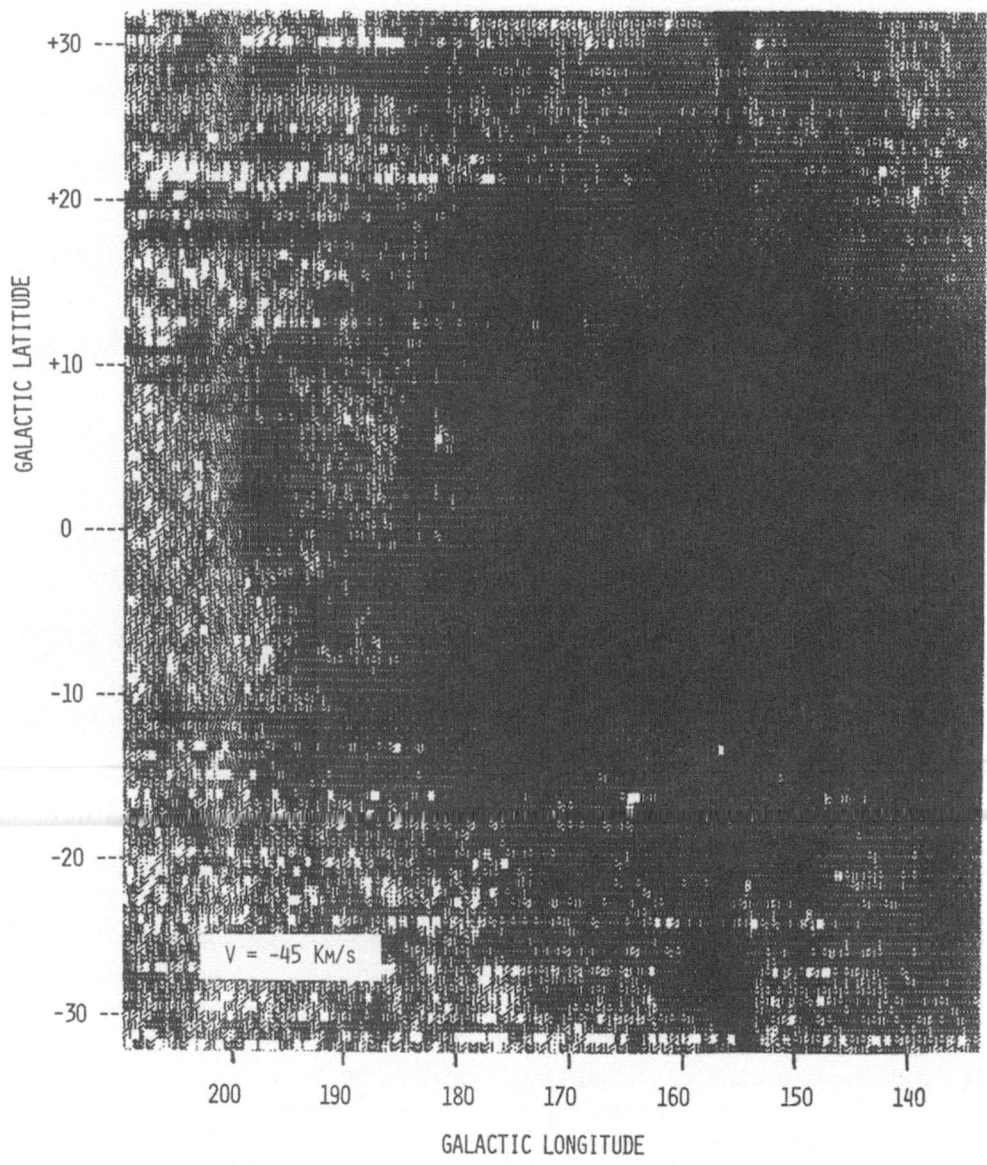

Figure 1. Grey-scale representation of shell brightness temperature at velocity –45 km/sec.

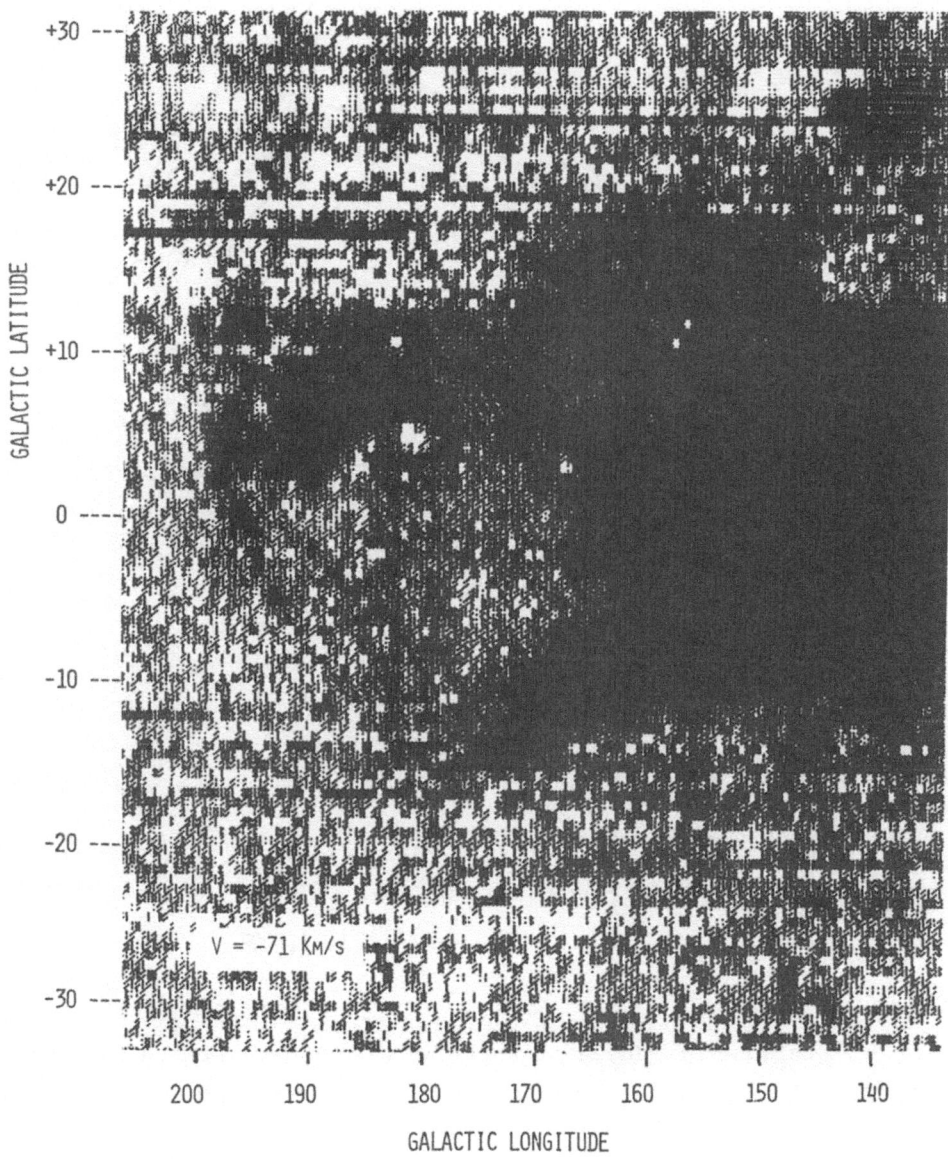

Figure 2. Grey-scale representation of shell brightness temperature
at velocity -71 km/sec.

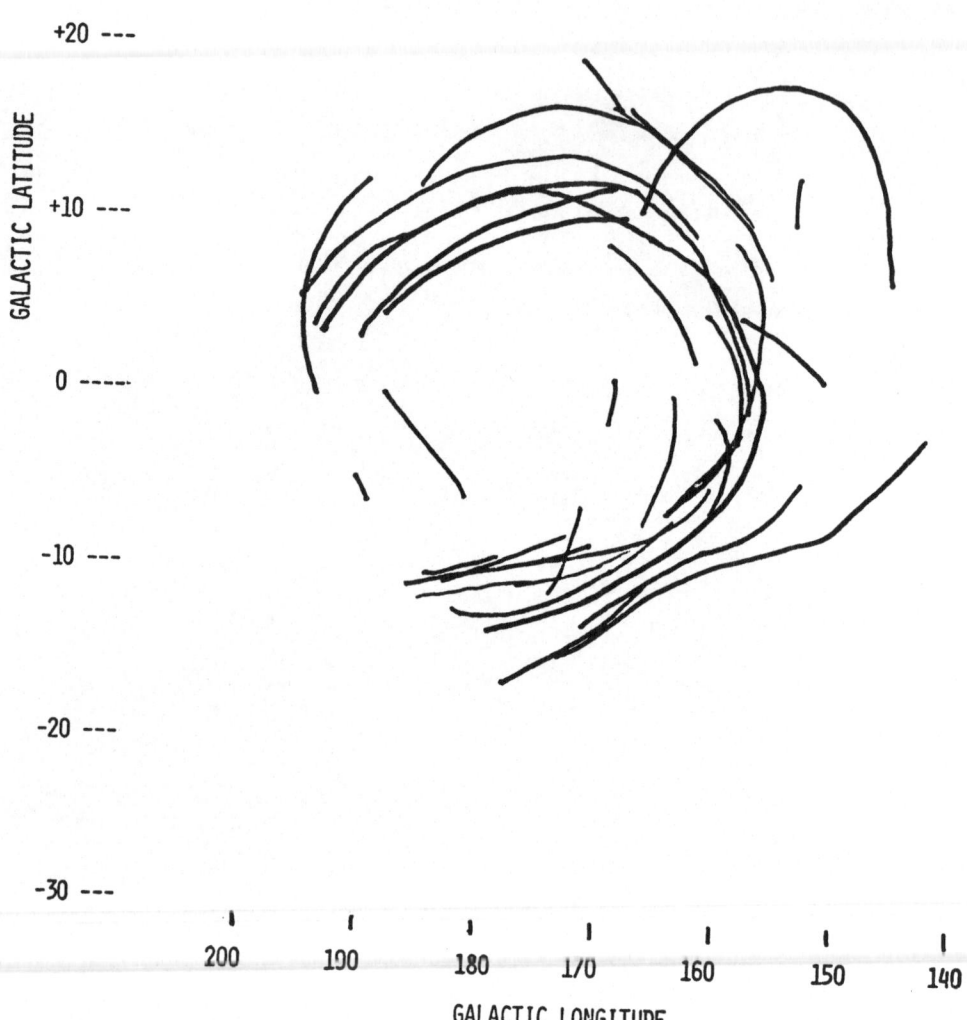

Figure 3. Locations of ridges of shell at velocities between −90 and −45 km/sec.

REFERENCES

Heiles, C. 1979, Astrophys J. 229, pp.533-544.
Heiles, C. and Habing, H.J. 1974, Astron. & Astrophys. Suppl. 14, pp.1-556.
Weaver, H.F. and Williams. D.R. 1973, Astron. & Astrophys. Suppl. 8,
     pp.1-504.

LOW LATITUDE ABSORPTION SPECTRA

J. M. Dickey[1], S. R. Kulkarni[2], J. H. van Gorkom[1],
J. M. Benson[1], and C. E. Heiles[2]

[1]National Radio Astronomy Observatory*
[2]University of California, Berkeley

Absorption spectra of the 21 cm line of neutral hydrogen at low latitudes constitute a new tracer of galactic structure, quite different from HI emission spectra because the absorption coefficient is proportional to n/T whereas the emission coefficient is proportional only to density n and independent of temperature T. So absorption surveys are sensitive to cool regions only (T $\lesssim$ 250 K) and avoid the blending of different phases of the interstellar medium which confuses emission spectra. Line widths in absorption are typically only two to three km s^{-1}, whereas emission lines may be 10 to 20 km s^{-1} wide. At low latitudes where galactic rotation provides kinematic distance estimates, the narrower line widths translate into higher spatial resolution; in a typical direction the emission linewidth corresponds to about 100 pc in z, whereas the absorption linewidth is only about 20 pc in z. In this way 21 cm absorption is like CO emission, but unlike CO it does not require high densities (n $\gtrsim$ 500 cm^{-3}) for excitation.

In addition to its value as a tracer of cool atomic gas, a 21 cm absorption survey is crucial to determine how much interstellar HI is concealed inside optically thick clouds. This is a problem with emission surveys that has long been recognized: it is impossible to correct for self-absorption to determine the true column density without both emission and absorption measurements. Rough estimates (Baker and Burton 1975, Kerr and Westerhout 1965) put this correction factor at 1.5 to 2 depending on galactic radius, i.e., one-third to one-half the HI has been missed in emission surveys. To understand the total mass density of the disk and the atomic to molecular ratio we must get more secure estimates for this factor in many directions. This talk describes new observations, some of which are still being reduced, which will increase our sample of low latitude absorption spectra.

Sensitive surveys of galactic 21 cm absorption toward extra-galactic sources (e.g., Crovisier et al. 1978, Dickey et al. 1978) have been possible only at high latitudes because the angular structure of the emission at low latitudes is too fine for even the largest single

W. L. H. Shuter (ed.), Kinematics, Dynamics and Structure of the Milky Way, 109–114.
Copyright © 1983 by D. Reidel Publishing Company.

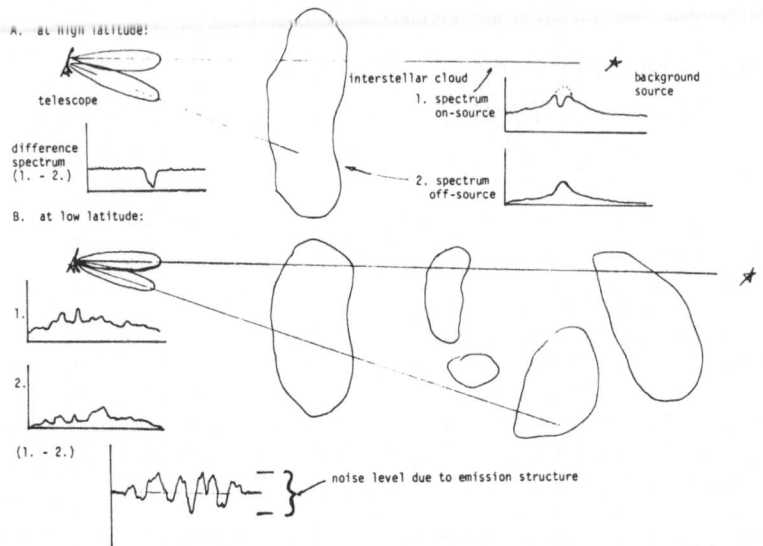

Figure 1:   An illustration of the difficulty with measuring absorption
             using a single dish telescope.   Emission structure smaller
             than the telescope beamwidth introduces spurious noise.

dish telescopes to resolve (Figure 1). Interferometer studies can
achieve higher resolution (e.g., van Gorkom et al. 1982), but high
sensitivity is difficult to achieve because the usable telescope gain
is low. We have developed two observing systems which are well suited
to measuring sensitive absorption spectra at low latitudes. The first
uses the two single dish telescopes at Green Bank, the 300-foot and the
140-foot, together as an interferometer (Dickey and Benson 1982). The
second uses the VLA, with each arm phased-up as a single dish. The
result is six interferometers, each the equivalent of a pair of Parkes
telescopes. This system has sufficient sensitivity to obtain good
absorption spectra toward background sources as weak as 200 mJy, which
are quite abundant (one per six square degrees). With this system we
have undertaken a survey of the HI absorption properties of the
galactic disk.

     Before showing new data it is worth reviewing a few points of the
single dish results which apply to the disk. These are shown in
Figure 2. Figure 2a shows the distribution of random velocities of
absorption lines detected at high latitudes (from Crovisier 1978).
There are clearly two components in this distribution: a Gaussian with
width ~6 km s^{-1}, and a less abundant population with rms velocity ~15
to 20 km s^{-1}. The presence of these two populations complicates the
interpretation of n(z) in terms of K_z, as discussed below. Figure 2b
shows the result of absorption studies toward pulsars (from Dickey
et al. 1981, cf. Weisberg 1978), which are the only sort of background
source feasible for single dish studies at low latitudes. The two lines
are not fits to the data, but the predictions of high latitude surveys

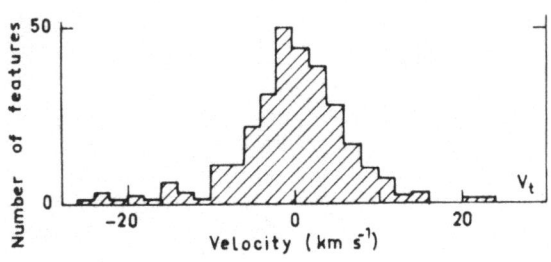

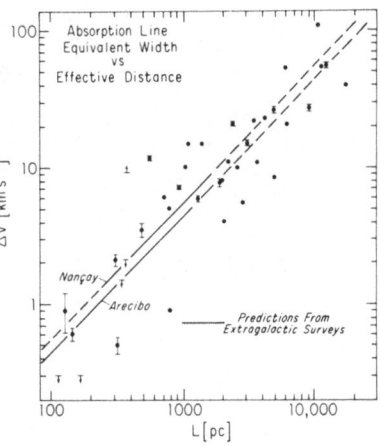

Figure 2:
Results from single dish observations.
In the left panel (a) is shown the
random velocity distribution of clouds
at high latitudes (Crovisier 1978).
In the right panel (b) is shown the results of pulsar absorption
studies (Dickey et al. 1981) which imply that the absorptive
properties of the solar neighborhood are not atypical for the disk.

which cover only the region within about 1 kpc from the sun (Crovisier
1981). It is clear that the mean opacity of the disk as sampled by the
pulsars as far as 10 kpc away is similar to that measured nearby.

What are not seen at high latitudes but are common in the plane are
molecular clouds. Figure 3 shows a spectrum (from Dickey and Benson
1982) taken at Green Bank toward a bright continuum knot in W3. The
broad, saturated absorption covering $-35 > v > -50$ km s^{-1} corresponds
to an intense CO emission line. The optical depth at these velocities
is so large ($\tau > 4$) that a tremendous amount of HI is implied

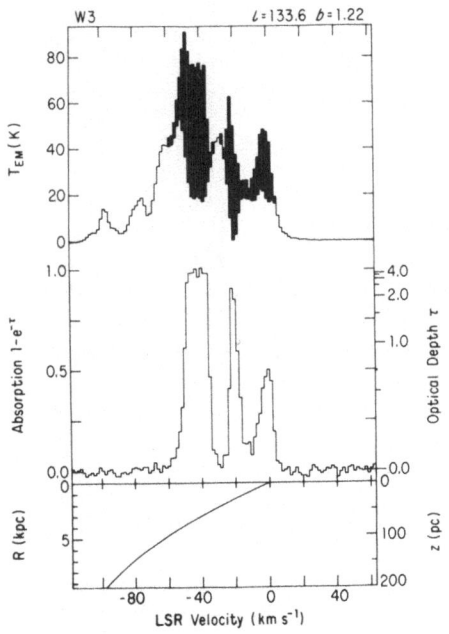

Figure 3:
Emission and absorption
spectra toward a compact
continuum knot in W3
(Dickey and Benson 1982).
The optical depth in the
molecular cloud is extremely
high. Much atomic gas may
be concealed from emission
surveys in clouds like this.

Figure 4:
Estimates for
the correction
factor to the
column density
to account for
self-absorption
in various
directions.

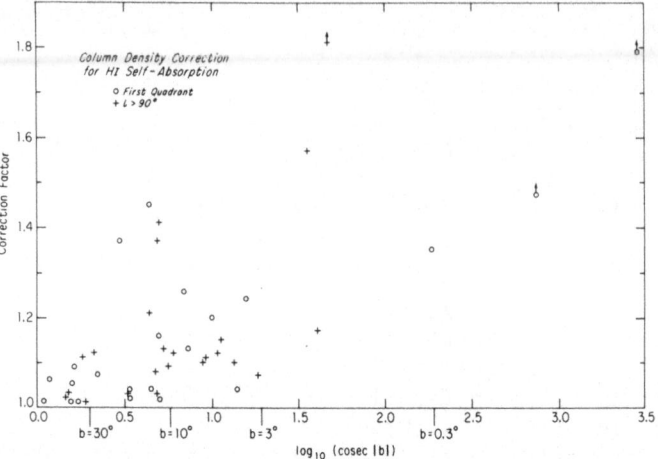

coexisting with the molecules within and around the molecular cloud.
This atomic gas concealed in molecular clouds with high 21 cm optical
depths constitutes an extra correction to the atomic surface density
presented by Burton and Gordon (1978). The factors to apply to emission
surveys (e.g., Weaver and Williams 1973) to correct for self-absorption
in the HI are presented in Figure 4 for the directions studied from
Green Bank. It is clear that at high and intermediate latitudes
($|b| > 10°$) the emission surveys miss only 5% to 25% of the gas,
roughly as expected from Gordon and Burton (1976). This is also true at
low latitudes, except in directions which pass through molecular
clouds. In those cases the correction may be 1.5 to 2, i.e., half the
total gas has been missed. To get a global estimate for the atomic gas
surface density will require the larger survey now underway.

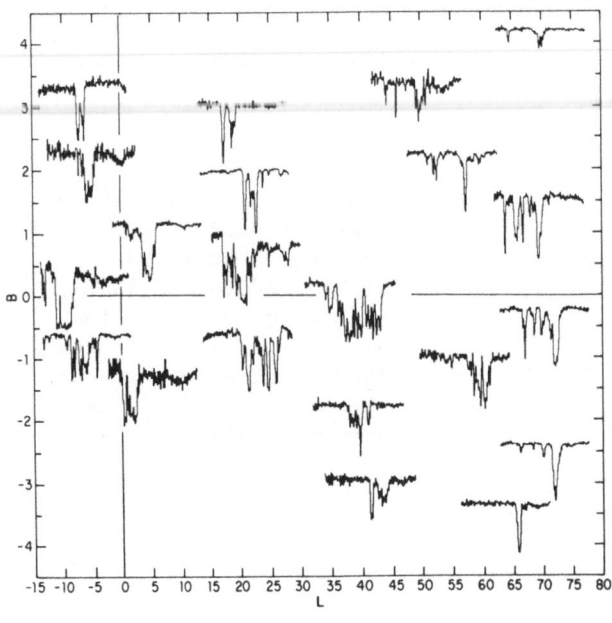

Figure 5:
Some absorption
spectra from the
VLA arranged in
galactic coordi-
nates roughly
corresponding
to the lines of
sight.

Using the VLA we have observed absorption toward 80 sources within about 12° of the galactic plane. These spectra are still being reduced, but some preliminary results are available. Figure 5 shows some spectra in the first quadrant. The richness of the profiles is clear; as many as 15 separate lines can be distinguished in some directions. Unlike emission spectra, there are gaps between the absorption lines where the optical depth returns to zero, which shows that cold gas is concentrated in distinct regions of space. There is a clear z effect when following a given range of longitudes through the plane: distant features (i.e., high velocities) drop off more quickly with increasing |b| than do nearby, low velocity structures.

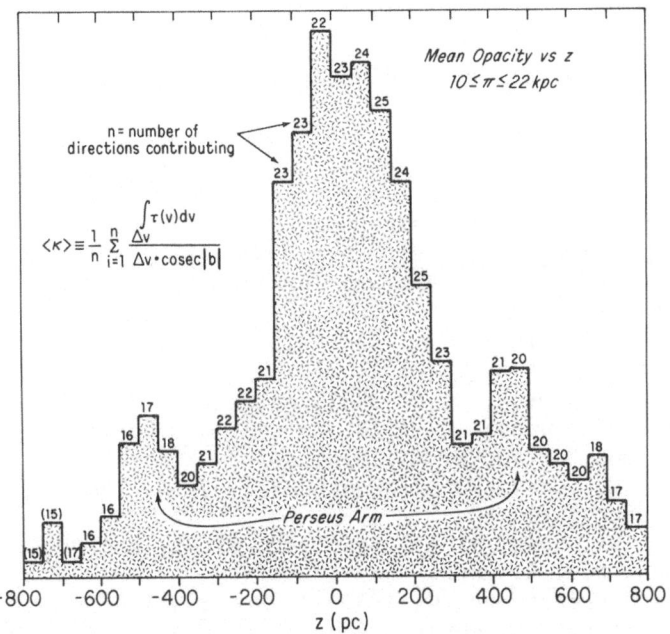

Figure 6: The distribution of absorbing gas in z computed using kinematic distances from Blitz (1979). The non-circular motions of gas in the Perseus arm (100° $\lesssim$ ℓ $\lesssim$ 140°) cause spurious distance estimates for these directions, so the excess gas at ±500 pc in fact is at lower z.

Using the rotation curve of Blitz (1979) we can look at quadrants II and III and translate velocity to distance and hence to z. Accumulating about fifty spectra gives Figure 6, which plots the average opacity as a function of z in the outer galaxy. The number of directions contributing to the estimate is shown for each 50 pc z interval. Near the plane there is clearly a Gaussian shaped z distribution with scale height about 150 pc, as expected from high latitude studies (Crovisier 1978, Falgarone and Lequeux 1973), but there seems to be an exponential tail in z extending higher than 500 pc. This may be due to the shape of K_z (e.g., Oort 1965), but its

interpretation is complicated by the shape of the random velocity
distribution (Figure 2). When the data are fully reduced we can hope to
get a better model of the z distribution of the absorbing gas.

We have not yet obtained an l-v diagram for the optically thick
gas, but that is one of the objectives of this project. In addition we
hope to study the distribution of spin temperatures of HI, and to
quantify the use of absorption spectra for distance measurements (e.g.,
van Gorkom et al. 1980). In conclusion, the HI absorption is quite
different from emission at low latitudes; it constitutes a new tracer
of galactic structure, i.e., the distribution of cold neutral gas.

REFERENCES

Baker, P. L. and Burton, W. B. 1975, Ap. J. 198, 281.
Blitz, L. 1979, Ap. J. 231, L115.
Burton, W. B. and Gordon, M. A. 1975, Astron. Astrophys. 63, 7.
Crovisier, J., Kazes, I., and Aubry, D. 1978, Astron. Astrophys. Suppl.
    32, 205.
Crovisier, J. 1978, Astron. Astrophys. 70, 43.
Crovisier, J. 1981, Astron. Astrophys. 94, 162.
Dickey, J. M., Salpeter, E. E., and Terzian, Y. 1978, Ap. J.
    Supp. 36, 77.
Dickey, J. M., Weisberg, J. M., Rankin, J. M., and Boriakoff, V.
    1981, Astron. Astrophys. 101, 332.
Dickey, J. M. and Benson, J. M. 1982, Astron. J. 87, 278.
Falgarone, E. and Lequeux, J. 1973, Astron. Astrophys. 25, 253.
Gordon, M. A. and Burton, W. B. 1976, Ap. J. 208, 346.
van Gorkom, J. H., Goss, W. M., and Shaver, P. A. 1980,
    Astron. Astrophys. 82, L1.
van Gorkom, J. H., Goss, W. M., Seaquist, E. R., and Gilmore, W. S.
    1982, Mon. Not. R. astr. Soc. 198, 757.
Kerr, F. J. and Westerhout, G. 1965, in Stars and Stellar Systems,
    Vol. V: Galactic Structure, eds. A. Blaauw and M. Schmidt
    (Chicago: University of Chicago Press).
Oort, J. 1965, Galactic Structure, eds. A. Blaauw and M. Schmidt,
    (Chicago: University of Chicago Press), p. 455.
Weaver, H. and Williams, D.R.W. 1973, Astron. Astrophys. Suppl. 8, 1.
Weisberg, J. M. 1978, Ph.D. thesis, University of Iowa.

*National Radio Astronomy Observatory is operated by Associated
 Universities, Inc., under contract with the National Science
 Foundation.

THE DISTRIBUTION OF MOLECULAR CLOUDS IN THE GALAXY

David B. Sanders[*]
Astronomy Program
State University of New York at Stony Brook
Stony Brook, NY 11794

1.   INTRODUCTION

During the last decade millimeter wave observations of the CO mole-
cule have proved to be one of the most useful probes of the large scale
structure and kinematics of the galactic disk.   The earliest CO observations indicated a large concentration of molecules in the galactic
center and at a distance of 4 to 8 kiloparsecs often refered to as the
'galactic ring'.   The shape of the CO radial distribution is similar to
nearly all other tracers of population I material (γ-rays, HII regions,
SNR ...), but quite different from the flat HI distribution.   Estimates
of H_2 mass from CO column density have led most observers to conclude
that H_2 is significantly more abundant than HI in the inner galaxy.

This paper summarizes new , more extensive CO observations which
now reveal the full extent of the molecular disk –scale height, mean
displacement, volumn density – and accurately computes the CO surface
emission at all radii, $0 < R < 16$ kpc.   Conversion of CO intensity to
H_2 density reveals that the inner disk of the Milky is dominated by
molecular hydrogen not HI.

2.   STONY BROOK – MASSACHUSETTS CO SURVEY

Two primary observing programs were begun in 1977 as part of the
Stony Brook–Massachusetts CO survey of the galaxy (1977-1981): 1) a 1000
point sample at 1 degree spacing in ℓ from –4 to 70 degrees and 12 arc
minute spacing in b from –2 to 2 degrees, designed to determine the Z
distribution of gas 2) a 1000 point strip map near the CO midplane at
2-4 arc minute spacing in ℓ from 10 to 52 degrees designed to measure
cloud properties on scales as small as 10 pc.   Two smaller observing
programs were also completed: 1) ^{12}CO and ^{13}CO observations of the
Southern Hemisphere (356 to 330 degrees longitude) at $b = 0^o$ and 2)
^{13}CO observations at $b = 0^o$, taken every degree in ℓ between 0° and 60°,
which determined that the radial distribution of the more optically thin
isotope was identical to that found for ^{12}CO.

*Present Address: Five College Radio Astronomy Observatory, Univ. of Mass.

W. L. H. Shuter (ed.), Kinematics, Dynamics and Structure of the Milky Way, 115–125.
Copyright © 1983 by D. Reidel Publishing Company.

Prior to 1979 the majority of the CO spectra were obtained with the
NRAO 36ft telescope (66" beam). Subsequent observing was done on the 14m
FCRAO antenna (45" beam). A complete accounting of the survey results
are given in Sanders (1981). Analysis of data gathered prior to 1979
can be found in Solomon, Sanders and Scoville (1979), Solomon, Scoville
and Sanders (1979) and Solomon and Sanders (1980). The following sections present the radial distribution parameters for the molecular gas –
scale height, mean Z displacement, surface density – which are then used
to compute the surface intensity I_{CO}, that an extragalactic observer
would measure for the Milky Way.

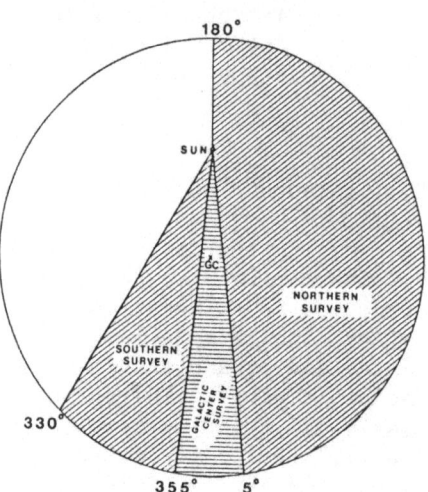

Figure 1 illustrates the survey
coverage. The bulk of the data is confined to the Northern disk where full Z
coverage allows a determination of the
vertical distribution of the molecular
gas. The Southern data is confined to b
= 0°. The galactic center (R < 2kpc),
is discussed separately primarily
because of the lack of a kinematic model
for this region of the galaxy where
large non-circular velocities suggest a
complex kinematics. Results from analysis of North, South and galactic center
regions have been combined to give a
global model for the CO distribution.
Finally we derive the face-on ^{12}CO
integrated intensity I_{CO}, to compare the
Milky Way disk with recent extragalactic
CO observations.

Figure 1. Azimuthal coverage
schematic for the Galactic
Survey

3.   THE AXISYMMETRIC DISTRIBUTION

The axisymmetric CO distribution is summarized in Figures 2 – 4.
The disk model assumes that the vertical distribution of gas is a
gaussian in Z given by (Talbot and Arnett, 1975)

$$\rho(Z) = \rho_o \; \exp\left(-\frac{(Z-Z_c)^2}{(1.2Z_{1/2})^2}\right) \qquad . \tag{1}$$

CO intensity is converted to H_2 density using the relation

$$N(H_2) = 3.64 \times 10^{20} \int T_A^*(^{12}CO) \; dv \qquad cm^{-2} \tag{2}$$

determined by relating CO integrated intensity to optical extinction
measurements in nearby dark clouds (Frerking, Langer and Wilson, 1981 –
see Sanders, 1981). Equation 2 is equivalent to the expression

$$n(H_2) = 0.12 \; J(^{12}CO) \qquad cm^{-3} \qquad . \tag{3}$$

* Operated by Associated Universities, Inc., under contract with the
  National Science Foundation

Substituting the emissivity J [K·km s^{-1}kpc^{-1}] for ρ in equation 1 allows a complete description of the axisymmetric disk in terms of the three parameters J_o, Z_c and $Z_{1/2}$.

Figure 2 shows the emissivity scale height of the molecular gas increasing with radius. A linear fit to the data from 3 to 9 kpc gives

$$Z_{1/2}(R) = 46 + 4.2(R-3) \qquad (pc) \qquad\qquad (4)$$

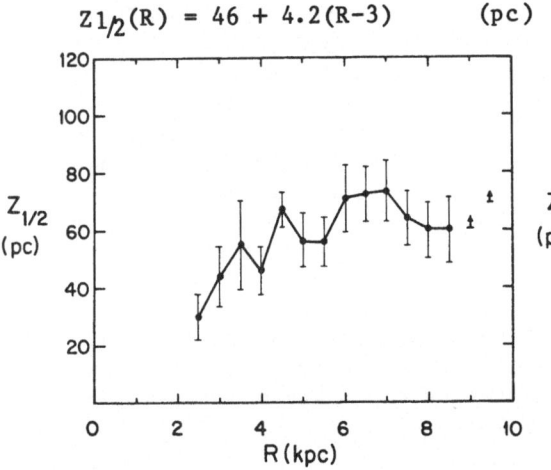

Figure 2. The scale height(HWHM)Z1/2, of ^{12}CO emission.

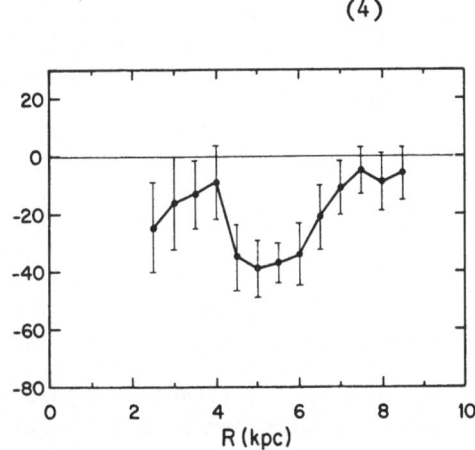

Figure 3. The mean-plane displacement Z_c, of ^{12}CO emission.

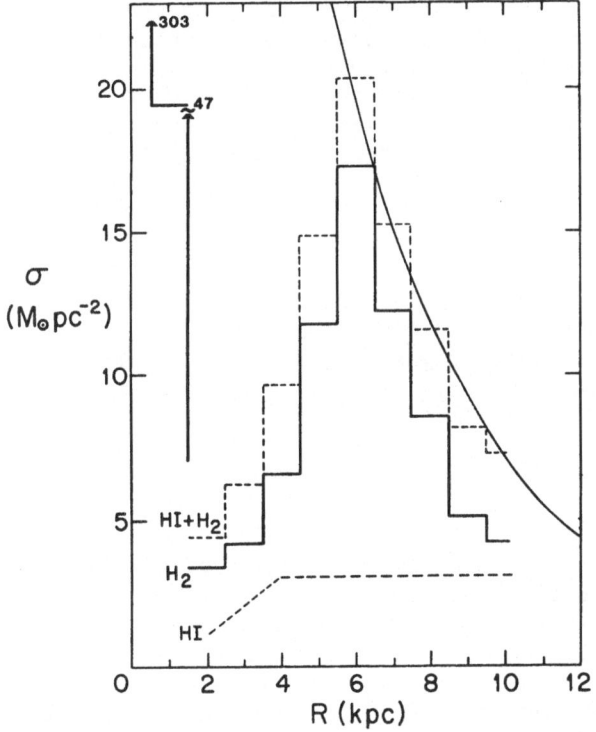

Figure 4. The mass surface density of molecular hydrogen and HI as a function of galactocentric radius.

Figure 4 shows that the mid-plane of the CO distribution is located at
-20pc for radii 3 to 9 kpc . The 'plane' of the molecular disk exhibits a corrugated structure in radius with oscillations of ±20 pc about
the mean. The origin of this warp is not clear. Other tracers of the
disk gas - HI (Quiroga, 1974), HII (Lockman, 1979) - show similar
displacement from b = 0° indicating that the warp is a fundamental property of the disk. A map in R and Θ of the vertical displacement of the
mid-plane can be found in Sanders (1981). The surface density of molecular gas shown in Figure 4 is the product of the mid-plane emissivity
J_o, and the scale height from Figure 2. For all radii R < $R_\odot$,
H_2 dominates HI. 84% of the hydrogen gas is in the form of H_2 between
4 and 8 kpc. The CO peak at 6 kpc is nearly a factor of 4 larger than
values at 4 and 8 kpc. This increase in emission is much larger than
the errors associated with the data points thus the reality of the CO
peak is of little question. There is however a significant amount of
azimuthal variation in the emissivity at a given radius eventhough the
mean is relatively well determined. At a resolution of 1 kpc for 5 < R
< 8 kpc, the azimuthal variation was typically a factor of 3. For
example, at R = 6kpc, σ (H_2) varies between values of 9 and 27. This
would correspond to the variation in CO surface emission measured by an
extragalactic millimeter telescope surveying the 6 kpc annulus face-on,
with an equivalent beam diameter of 1 kpc.

Comparing Figure 4 with previous determinations of the surface density from Northern Hemisphere data alone (Solomon, Sanders and Scoville,
1979, Gordon and Burton, 1976), one finds that incorporating the
Southern Hemisphere data has the effect of slightly decreasing the
sharpness of the peak between 4 and 8 kpc and giving a more gentle taper
to the distribution towards larger and smaller radii. The total mass of
H_2 determined from the composite distribution is identical to that found
in the earlier studies. Figure 4 also shows an exponential curve with
scale length 4 kpc determined from a least squares exponential fit to
the total gas surface density given by

$$\sigma_{HI+H_2}(R) = 19.5 \ e^{-(R-6)/4.0} \qquad M_\odot pc^{-2} \quad \text{for } R > 6kpc \qquad (7)$$

Models for the distribution of total mass by Caldwell and Ostriker
(1981) also show an exponential scale length near 4 kpc. The equality
of the two scale lengths implies a constant gas fraction (12%) at radii
beyond 6 kpc. An exponential disk also serves as a good approximation
to the H_2 distribution according to the relation

$$\sigma_{H_2}(R) = 17.3 \ e^{-(R-6)/2.7} \qquad M_\odot pc^{-2} \quad \text{at } R > 6kpc \qquad (8)$$

Our most recent survey data from longitudes 50 to 180 degrees
(Sanders and Solomon, 1982) which determined the emissivity primarily at
R >10kpc indicates that Equation 8 is a good approximation to the surface density out to at least R = 16kpc. Table 1 gives the total amount
of molecular gas inside and outside the solar circle and compares H_2
with the most recent measurements of HI. The CO emission in the outer
galaxy recently reported by Kutner and Meade (1981),if correct, is simply
the exponential tail of the CO radial distribution given by equation 8.

Table 1.

Total Disk Mass of Atomic and Molecular Hydrogen
$0 < R < 20$ kpc   ($R_\odot = 10$kpc, $V_\odot = 250$km s^{-1})

| | Galactocentric Radius, R(kpc) | | | |
	0-1.5	1.5-10	10-20	Total
m(H$_2$)	$5 \times 10^8 M_\odot$	$2.7 \times 10^9 M_\odot$	$0.8 \times 10^9 M_\odot$	$4 \times 10^9 M_\odot$
m(H$_2$)/M$_T$(H$_2$)	12%	68%	20%	100%
m(HI)	$1 \times 10^7 M_\odot$[a]	$0.9 \times 10^9 M_\odot$	$4 \times 10^9 M_\odot$[b]	$4.9 \times 10^9 M_\odot$
m(HI)/M$_T$(HI)	.2%	18%	82%	100%
m(H$_2$)/m(HI)	50	3.0	0.2	0.82

[a] Liszt and Burton (1981)
[b] Kerr (1983)

4.   THE GALACTIC CENTER

The intense CO emission from the galactic center is illustrated in Figure 5.   CO emission at $\ell < 4^\circ$ is typically a factor of 10 higher in total integrated intensity than the largest values in the ring.   The Z coverage provided in the current survey reveals an extensive nuclear structure in CO with emission extending out to 10° longitude which appears to be confined to a tilted disk of radius 1.5 kpc (see Figure 6).   Much of this emission was missed in earlier surveys which were confined to b = 0°.   Liszt and Burton (1980) have found a similar tilted arrangement for HI in the nucleus although the HI tilt of 29° is greater than the 7° tilt found for CO.   The total mass of H$_2$ at R < 1.5kpc is $5 \times 10^8 M_0$, a factor of 50 times larger than HI.

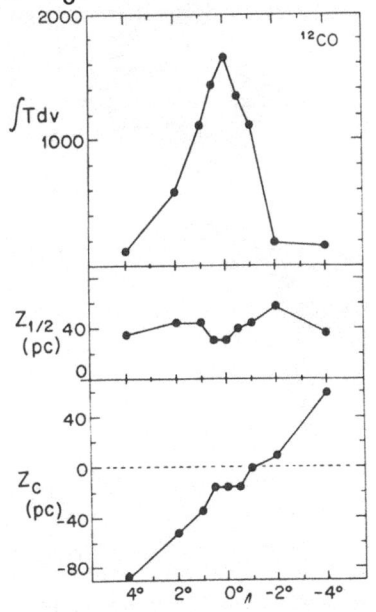

Figure 5. Integrated intensity, scale height and mean Z of ^{12}CO emission in the galactic center

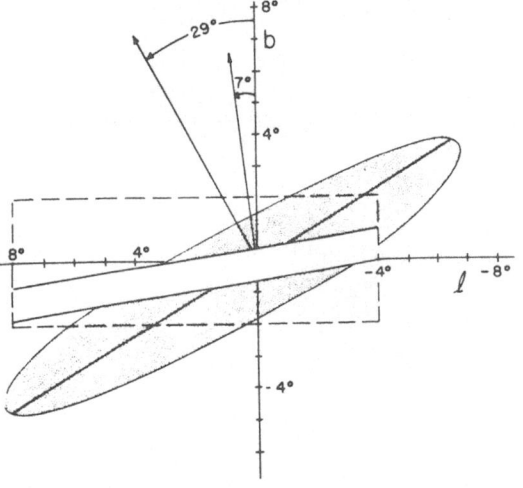

Figure 6. The two-dimensional (ℓ,b) projection of ^{12}CO emission (barlike structure) from the galactic center compared with the HI tilted disk (shaded region) of Liszt and Burton(1980)

5.   DEVIATIONS FROM AXIS SYMMETRY

While the average radial properties of the H_2 distribution are
determined by Figures 2-4, there is still a significant amount of
variation in the emissivity as a function of azimuth. We have used a
high resolution ($\Delta\ell$ = 2-4') strip survey of the galactic plane between
10 and 52 degrees longitude (Figure 7) to quantify the variation in CO
emission, to see how the variations compare with other tracers of disk
gas and to compare with putative models of spiral arms.

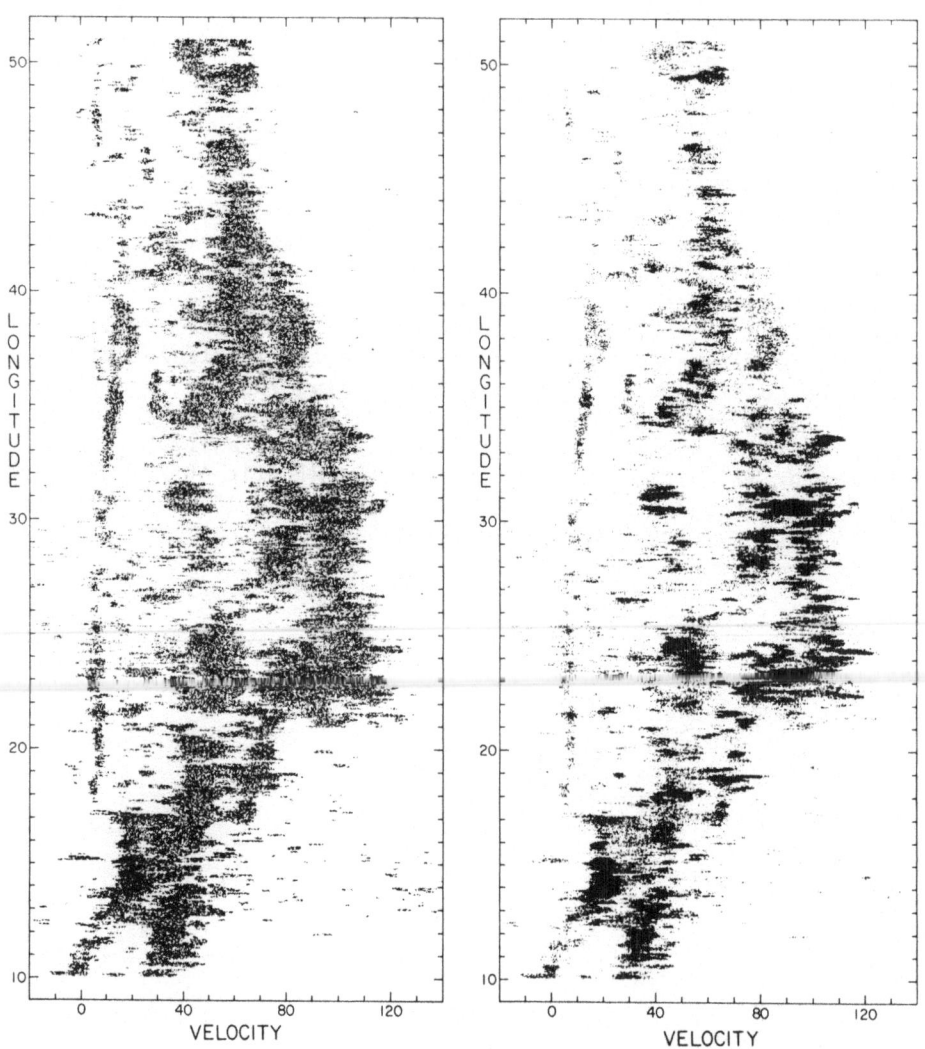

Figure 7. ^{12}CO high resolution survey ($\Delta\ell$ = 2',4') of the galactic plane
from ℓ = 10 to 52 degrees. The dot density in the left and right panels
is proportional to the square root and to the square of the antenna
temperature respectively.

Because of the increasing distance ambiguity as the velocity difference from the terminal point increases the CO contrast is determined
using only the upper 10km/s of the emission profiles. The corresponding
region of the disk is shown in Figure 8. The shaded area represents the
region of the galactic plane over which the line of sight velocity due
to differential rotation is within 10 km/s of the terminal velocity.
Figure 9 illustrates the variations in CO emissivity observed along the
tangent point strip. There is a general decrease in emission as one
moves to larger radius but the emissivity profile is far from smooth.
The smooth solid curve represents an axisymmetric distribution of clouds
with an exponential radial distribution normalized to give equal
integrated emissivity compared to the data over the longitude interval
20 to 50 degrees. The dashed curve represents a similar model normalized to best fit the data from trough regions. Five peaks are identified in the figure based on both 20' and 1° binning of the data in
longitude. Adjacent emission peaks and troughs show ratios of approximately 3:1 at 1° resolution increasing to 5:1 at 20' resolution. Much
higher contrast and a maximum of 2 to 3 peaks are expected from prior
studies of HI and HII recombination lines (e.g. Lockman, 1979). Their
counterparts are seen in CO (at $\ell \simeq 30^{\circ}$ and $\ell \simeq 50^{\circ}$), but the most prominent HI peak at $\ell \approx 50^{\circ}$ (Sagittarius arm) is the weakest CO peak. An
extragalactic telescope surveying the Milky Way along the tangent point
with a resolution of 1 kpc^{2} (typical of current extragalactic data),
would see a somewhat smoothed version of Figure 8. Regions prominent in
HII emission such as the W51 complex near $\ell \simeq 50^{\circ}$ would appear relatively weak in CO emission when compared to the $\ell \simeq 40^{\circ}$ region which contains no giant HII regions. However there would also be correlation of
CO emission with giant HII regions such as the peak at $\ell \simeq 30^{\circ}$.

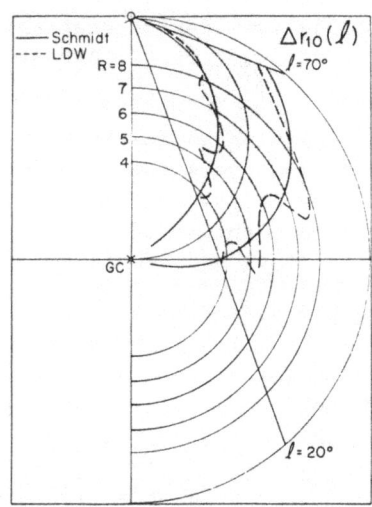

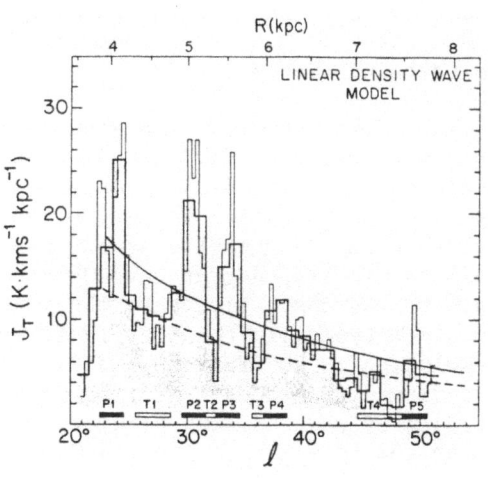

Figure 8. Schematic of sample area
in the galactic plane near the loci
of tangent points

Figure 9. The ^{12}CO emissivity
along the loci of tangent points
using a linear density wave rotation curve, Burton (1970)

6.   COMPARISON OF THE MILKY WAY WITH EXTRAGALACTIC CO OBSERVATIONS

The CO integrated intensity corresponding to a particular line of sight through the disk is simply the integral of the emissivity along the beam path. For extragalactic observations perpendicular to the Milky Way disk

$$I_{CO}(R) = \int_z J(R,z) \, dz \qquad (K \cdot km \, s^{-1}) \tag{9}$$

For a gaussian gas distribution in Z, Equation 9 simplifies to

$$I_{CO}(R) = 2.13 \, J_0(R) \, \Delta Z_{1/2} \tag{10}$$

The relationship between I_{CO} and H_2 column dinsity and surface density follows directly from equation 1:

$$N(H_2) = 3.64 \times 10^{20} \, I_{CO} \qquad (cm^{-2}) \tag{11}$$

$$\sigma(H_2) = 6.25 \times 10^{-19} \, N(H_2) = 5.82 \, I_{CO} \qquad (M_\odot pc^{-2}) \tag{12}$$

Figure 10 compares the observed quantity I_{CO} for the Milky Way with five other galaxies which are the first to be mapped in the ^{12}CO line. All show molecular gas distributions in marked contrast to the relatively flat HI. The central region of each of these galaxies (R < 10kpc) is primarily molecular. The most important interpretation of the CO measurements is that the distribution of molecules follows the distribution of the disk light (Young and Scoville, 1981; Solomon, Sanders, Barrett and deZafra, 1982). This result gives support to the finding that the scale lengths of total disk mass and the gas (HI+H_2) are approximately equal in the Milky Way and, to the implication that the star formation rate is proportional to the first power of the average gas density (Sanders, 1981).

All five extragalactic disks show a continuous rise into the nucleus, only the Milky Way disk shows a hole in CO emission. In the nucleus and in the disk outside 6kpc, the distributions for all galaxies are similar in form. While the mean values show a relatively smooth decrease in R suggestive of exponential disks (Young and Scoville, 1981; Solomon, Sanders, Barrett and deZafra, 1982), there is real scatter in the data points contributing to the mean. Azimuthal variations of factors of 3 are typically found at radii inside 10 kpc. Eventhough the 1-2kpc resolution is sufficient in M101, IC342 and NGC6946 to resolve arm-interarm differences in spiral patterns deduced from optical and radio studies of these disks, results on CO spiral arms are sofar, mixed. Young and Scoville (1981) – NGC6946, IC342; Rickard and Palmer (1981) –IC342, NGC6946, M51; and Solomon, Sanders, Barrett and deZafra (1982) – M101, report finding for each galaxy studied, some instances of relatively strong CO peaks at minima in the optical patterns and vice versa. Similar results are also found for the Milky Way from the study of CO emission along the loci of tangents.

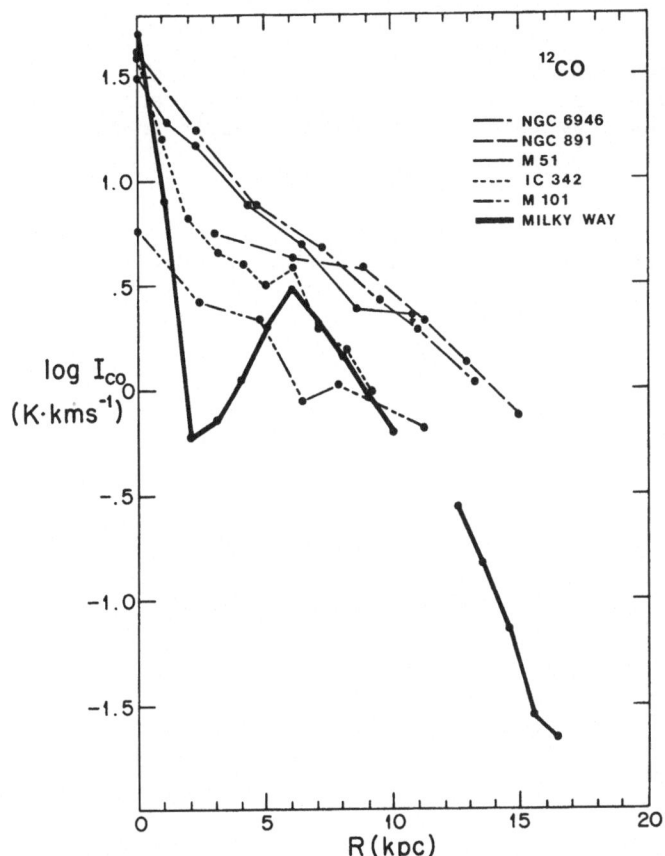

Figure 10. Radial distribution of ^{12}CO integrated intensity I_{CO}, in the Milky Way compared with five galaxies for which radial distribution data is currently available: M51, NGC6946, IC342 – Young and Scoville (1982); M101, NGC891 – Solomon, Sanders, Barrett and deZafra (1982).

The origin of the depletion in total gas inside 5 kpc for our galaxy seems to be tied to the shape of the rotation curve. To date most extragalactic CO observations have focused on a narrow range of Hubble type, in particular to the more easily observable luminous Sc's. All four face-on Sc galaxies in Figure 10 show smoothly rising rotation rotation curves in their inner parts. By contrast the relatively large bulge component of the Galaxy yields a steeply rising rotation curve in the inner kiloparsec with a keplerian falloff from 1 to 3 kpc before flattening out as the differentially rotating disk begins to dominate. The gas distribution in the inner disk must somehow reflect the influence of the bulge. Further extragalactic CO observations including galaxies of different Hubble type will surely yield gas distributions similar to the Milky Way.

8.   SUMMARY

The CO distribution in the Milky Way is summarized in Table 2 which gives the observed surface integrated intensity I_{CO}, and the derived column density of H_2 from 0 to 16 kpc galactocentric radius.

Table 2.

^{12}CO Integrated Intensity and H_2 Column Densities
Perpendicular to the Galactic Plane

R (kpc)	$I(^{12}CO)$ (K·kms^{-1})	$2N(H_2)$ (10^{20}cm^{-2})	$\sigma(H_2)$ ($M_\odot$ pc^{-2})	$\Delta M(H_2)$ (10^8 $M_\odot$)
0 −.5	52.0	374.0	303.0	2.38
1[a]	8.0	57.6	46.6	2.93
2	0.58	4.2	3.4	0.42
3	0.73	5.3	4.2	0.79
4	1.13	8.2	6.6	1.67
5	2.03	14.8	11.8	3.72
6	2.97	21.6	17.3	6.51
7	2.10	15.3	12.2	5.37
8	1.46	10.6	8.5	4.23
9	0.88	6.4	5.1	2.89
10[a]	0.62	4.5	3.6	2.26
11[a]	0.30	2.2	1.8	1.24
12	0.33	2.4	1.9	1.46
13	0.21	1.5	1.2	1.0
14	0.10	0.72	0.58	.51
15	0.06	0.43	0.35	.33
16	0.03	0.22	0.18	.18

[a] annulus emissivity poorly determined

The most important results from the current CO survey are:

1) H_2 is the dominant form of the interstellar medium at $R < R_\odot$. The total mass of H_2 and HI are approximately equal at $R < 18$ kpc but, 80% of HI is found outside the solar circle while only 20% of H_2 is found there.

2) The distribution of total disk gas (HI+H_2) and the distribution of total disk mass at $R > 6$ kpc are similar in form. Both can be approximated by decreasing exponentials with scale length near 4 kpc. This suggests that at a given R, the rate at which gas is converted into stars must proceed as approximately the first power of the average gas density.

3) The Milky Way is not alone in exhibiting a predominantly molecular inner disk. Extragalactic CO observations of 4 nearby Sc galaxies show CO distributions which follow the exponentially decreasing distribution of optical disk luminosity.

REFERENCES

Blitz, L. and Shu, F.H. 1980, Ap.J., 238, 148.
Burton, W.B. 1970, Astr. Ap., 10, 76.
Burton, W.B.,Gordon, M.A.,Bania, T.M. and Lockman, F.J. 1975, Ap.J.,
      202, 30.
Burton, W.B., Gordon, M.A. 1978, Astr. Ap., 63, 7.
Caldwell, J.A.R. and Ostriker, J.P. 1982, Ap.J., 251, 61.
Dickey, J.M. 1981, NRAO Workshop on Extragalactic Molecules, Green Bank.
Frerking, M.A., Langer, W. and Wilson, R.W. 1981 preprint.
Georgelin, Y.M. and Georgelin, Y.P. 1976, Astr. Ap., 49, 57.
Innanen, K.A. 1973, Ap. and Space Sci., 22, 393.
Kerr, F.J. 1983, this volume, pp. 91-96.
Kutner, M. and Meade, K. 1981, Ap.J.(Letters), 249, L15.
Liszt, H.S. and Burton, W.B. 1981, Ap.J., 243, 778.
Lockman, F.J. 1979, Ap.J., 232, 761.
Quiroga, R.J. 1974, Ap. and Space Sci., 27, 323.
Rickard, L.J. and Palmer P. 1981, preprint.
Sanders, D.B. 1981, Ph.D. Thesis, SUNY at Stony Brook.
Sanders, D.B., Solomon, P.M. 1982, in preparation.
Scoville, N.Z. and Solomon P.M. 1975, Ap.J.(Letters), 199, L105.
Solomon, P.M., Sanders, D.B. and Scoville, N.Z. 1979, "Large Scale
      Characteristics of the Galaxy",ed. W.B. Burton, p.35, (Dordrecht:D.
      Reidel).
Solomon P.M., Scoville, N.Z. and Sanders, D.B. 1979, Ap.J.(Letters),
      232, L89.
Solomon, P.M. and Sanders, D.B. 1980, "Giant Molecular Clouds in the
      Galaxy", eds. P.M. Solomon and M.G. Edmunds, p.41,(Pergamon Press).
Solomon P.M., Sanders, D.B., Barrett, J. and deZafra, R. 1982, preprint.
Spitzer, L. 1968, "Diffuse Matter in Space",(Wiley-Interscience, NewYork).
Talbot, R.J. and Arnett, W.D. 1975, Ap.J., 197, 551.
Young, J. and Scoville, N.Z. 1982, Ap.J., 258, in press.

KINEMATICS OF MOLECULAR CLOUDS: EVIDENCE FOR AGGLOMERATION IN SPIRAL ARMS

Antony A. Stark
Bell Laboratories
Holmdel, NJ 07733
and
Princeton University Observatory, Princeton, NJ 08540

ABSTRACT

A new survey of CO in the first Galactic quadrant has been analyzed to yield a catalog of 320 molecular clouds near the tangent velocity. These clouds have known distances, so that cloud sizes and heights above the Galactic plane can be determined. The largest clouds ($M_c > 10^{5.5}\ M_\odot$) have a reduced scaleheight relative to smaller clouds by an amount which is consistent with equipartition of energy. This can be interpreted as evidence for small clouds combining to form giant clouds in spiral arms.

INTRODUCTION

It has now been adequately shown that giant molecular clouds in at least some galaxies are found only in spiral arm segments. In M31, Boulanger et al. (1981) and Stark, Linke and Frerking (Linke 1981) have shown that the CO emission coincides with the optical spiral arm segments, and that there are no giant clouds in the interarm regions. This seems to be true also of M33. In the outer Milky Way Cohen et al. (1980) find no giant clouds in a gap 10 km/s wide between the Perseus arm and local arm; their map of the first Galactic quadrant also shows large holes in ℓ, b, v space which contain much less emission than surrounding regions. Stark (1979) has shown that these holes contain many times fewer giant clouds than would be expected in a uniform distribution, although there are small clouds present throughout. There seem to be regions in the inner Galaxy of a size 1 Kpc in galactocentric radius by $\gtrsim 60°$ galactocentric azimuth which have 10 to 20 times more giant molecular clouds than do adjacent regions of similar size. No other galaxies have yet been mapped in CO with sufficient resolution to discriminate between arm and interarm regions. None of the observed galaxies are grand design spirals, but they do have spiral patterns consisting of broken spiral arm segments. The giant molecular clouds are confined to these spiral arm segments.

W. L. H. Shuter (ed.), Kinematics, Dynamics and Structure of the Milky Way, 127–133.

In a certain sense, the giant molecular clouds are the spiral arms: in other galaxies the spiral structure is traced out by dust, IIII regions and young stars; in the solar neighborhood, these objects are invariably associated with giant molecular clouds.

If the giant clouds are associated with spiral arms, then the familiar problems relating to formation and persistence of spiral structure are transferred to the clouds. The giant clouds cannot survive as material objects for many galactic rotations and still trace the spiral arms. (The giant clouds might, of course, persist as wave phenomena rotating around the galaxy at a slightly different speed than the matter from which they are constituted.) No matter what the detailed mechanism, giant molecular clouds must be renewed or recreated by the addition of material. Small molecular clouds are the most likely reservoir of new material.

Kwan (1979) and Cowie (1980, 1981) have considered the agglomerative formation of molecular clouds. One prediction of such models is that the cloud-to-cloud velocity dispersions of massive clouds should be significantly smaller than that of smaller clouds: the clouds mutually interact, so the ensemble tends toward a uniform energy per cloud ("temperature"). For clouds smaller than $10^{5.5} M_\odot$, this is not borne out by observations in that the random velocities of $10^5 M_\odot$ clouds are about the same as those of $10 M_\odot$ clouds (Stark 1979). Either these small to moderate size clouds do not form by agglomeration, or there is some process which injects kinetic energy into the larger clouds. But, for clouds larger than $10^{5.5} M_\odot$, there is a reduction in the random velocities. This may be an indication that the very largest clouds, those which lie in the spiral arms, are built up by combining smaller clouds.

THE SURVEY

The present work is the first reported result of an extensive CO survey still in progress, a continuation of the first three-dimensional (ℓ, b, v) CO survey of Stark (1979). Survey parameters have been chosen as follows:

1) Both ^{12}CO and ^{13}CO are observed.

2) Sensitivity - 0.1 K rms in 0.68 km/s channels for ^{13}CO
                  0.3 K rms in 0.65 km/s channels for ^{12}CO.

3) Sampled on a square grid in ℓ and b, 3' between grid points.

4) Calibrated within 5% of absolute temperature scale.

5) Accurate beam pattern - the Gaussian central lobe contains $\gtrsim 87\%$ of antenna response; less than 5% of the antenna response is from Milky Way emission not in the central lobe.

6) Coverage - at least ±1° in b. So far, 16 square degrees have been
    observed in ^{13}CO: 32.5° < ℓ < 37.5° and 50.5° < ℓ < 53.5°. More
    than 6400 ^{13}CO and 1600 ^{12}CO spectra have been obtained so far.

Presently, the receiver on the 7m antenna at Crawford Hill has a DSB
receiver temperature of ~ 70 K (Linke et al. 1982); on an average
winter day the survey progresses at 2 square degrees per day in ^{13}CO or
4 square degrees per day in ^{12}CO.

        Figures 1 and 2 are illustrative samples of the survey data.
Figure 1 shows thin velocity slice containing two giant clouds and
several small clouds, as well as some emission belonging to giant clouds
which peak at other velocities. The emission from the giant clouds is
strongly concentrated toward the mid-plane. Figure 2 shows a similar
volume of ℓ, b, ν space which does not contain giant clouds; the small
clouds fill the field with fair uniformity. Note that these giant
clouds in Figure 1 cannot be illusions created by chance superposition
of clouds like those in Figure 2 - each giant cloud has at least 30
times the emissivity of the brightest cloud in Figure 2.

DATA ANALYSIS

        A portion of the survey has been analyzed to yield a catalog
of about 300 molecular clouds that are within 10 km/s of the tangent
velocity. Naturally, there are some subjective decisions to be made
in such a cataloging procedure - whether an extended object with several
intensity maxima is one large cloud or a chance superposition of smaller
clouds. The ^{13}CO data set as fewer such ambiguities than the ^{12}CO data.
A subset of both the ^{12}CO and ^{13}CO survey data were cataloged
independently by three different people using different algorithms
(Stark, Penzias and Beckman 1982). The various ^{13}CO catalogs disagreed
for only about 10% of the objects, whereas the ^{12}CO catalogs disagreed
for about 25% of the objects. The results presented here are not only
affected by these disagreements; all three ^{13}CO catalogs show a reduced
scale height for largest clouds. The algorithm used on the more
complete data set cataloged all statistically significant local maxima
of ^{13}CO luminosity in ℓ, b, ν space as separate clouds. This procedure
probably errs on the side of splitting up giant clouds, but this is
conservative because it would tend to reduce the presently hypothesized
effect. It also minimizes systematic problems caused by velocity crowding near the tangent velocity.

        The position assigned to a cataloged cloud is the position of
the spectrum with greatest peak temperature for that cloud. The size
assigned to a cloud is $2(A/\pi)^{1/2}$ where A is the area of the 0.4 K km/s
contour. Since all the clouds are near the tangent velocity, their
distance is known to be approximately that of the tangent point; this
distance was used to convert angles to physical lengths. The clouds
were divided into bins in ℓ, each one degree wide, and three size bins:
5-20 pc, 20-40 pc, and bigger than 40 pc. The root-mean-square b values
were determined for each of the 24 bins; these were further averaged over
the eight ℓ bins to give:

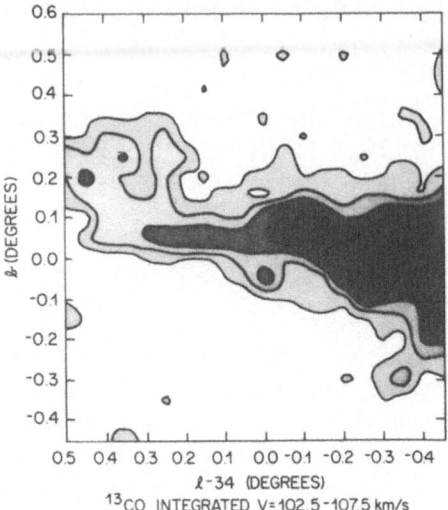

Figure 1 - ^{13}CO in a one square degree field around $\ell=34°$, $b=0°$. Contours are emmisivity integrated over the velocity range
102.5 to 107.5 km/s. Contours are at 0.4, 1, 2, 4, 10 K km/s.
The rms noise level is 0.2 K km/s.

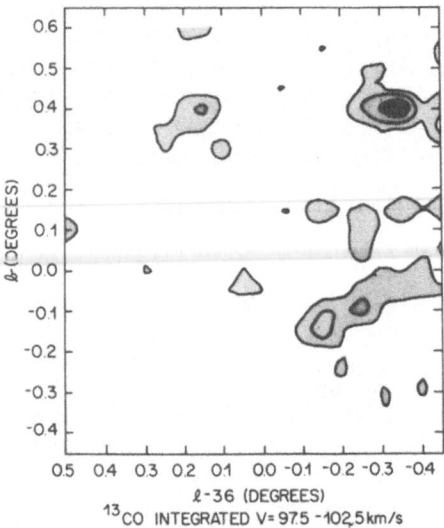

Figure 2 - ^{13}CO in a one square degree field around $\ell=36°$, $b=0°$. Contours are emissivity integrated over 97.5 to 102.5 km/s.
Contours are as in Fig. 1.

$$\overline{<Z^2>^{1/2}} \gtrsim 62 \text{ pc for 5-20 pc clouds } (M_C \lesssim 10^4 \ M_\odot)$$

$$\overline{<Z^2>^{1/2}} \gtrsim 48 \text{ pc for 20-40 pc clouds } (10^4 \ M_\odot \lesssim M_C \lesssim 10^{5.5} \ M_\odot$$

and

$$\overline{<Z^2>^{1/2}} \approx 21 \text{ pc for 40 pc and larger clouds } (10^{5.5} \ M_\odot \lesssim M_C).$$

The first two values are upper limits because of the limited coverage in b. The third value is approximate because it is only three times the spatial resolution of the survey, and because there are only 24 clouds in the sample greater than 40 pc size.

All mass estimates discussed here are made by assuming that cloud mass is a strict function of the cloud size parameter. This function is an empirical fit to the masses and sizes of a selected sample of clouds. These were in turn estimated by assuming that cloud mass is proportional to ^{13}CO surface brightness integrated over area:

$$M_C = 15 \ M_\odot (\text{K km s}^{-1} \text{ pc}^2)^{-1} \iiint T_A^*(^{13}\text{CO}) \ dv d\ell db \quad .$$

This further assumes that the metallicities, excitation temperatures, and $N(^{13}\text{CO})/N(\text{total})$ values are like those of local clouds.

DISCUSSION

In the absence of non-gravitational forces, the cloud scale-heights are a measure of velocity dispersion. (Provided cloud life-times are longer than the oscillation period for motions perpendicular to the galactic plane. This condition may only marginally be met.) The velocity dispersion, v_{rms}, of a population of clouds is linearly related to the scaleheight $<Z^2>^{1/2}$, through the central density, ρ, of the galactic disk:

$$v_{rms} = (4\pi G \rho)^{1/2} <Z^2>^{1/2} \quad ,$$

if $<Z^2>^{1/2}$ is sufficiently small (Oort 1965). Since $\rho \approx 4 \times 10^{-23}$ g cm^{-3} in the molecular ring (Caldwell and Ostriker 1981), $v_{rms} = (0.18$ km s^{-1} pc^{-1}) $<Z^2>^{1/2}$; i.e. $<Z^2>^{1/2} = 50$ corresponds to $v_{rms} = 9$ km s^{-1}.
The scaleheight vs. size relation therefore suggests a mass vs. dispersion relation like Figure 3.

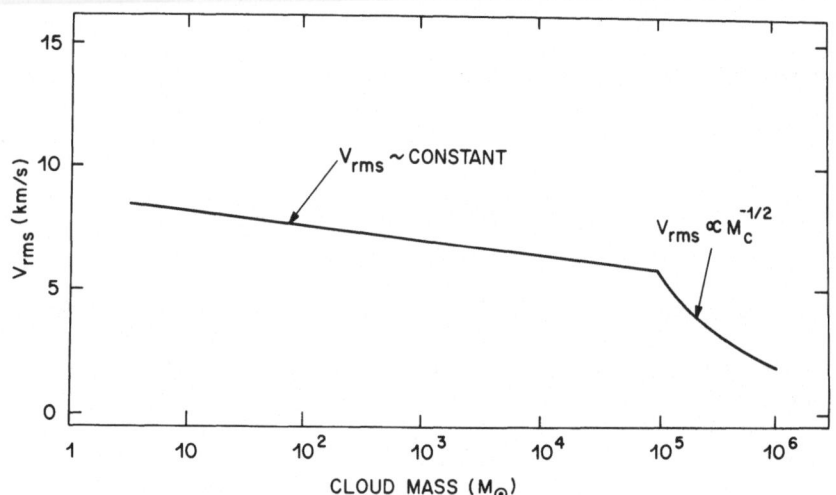

Figure 3 - Cloud-to-cloud velocity dispersion as a function of cloud
           mass (schematic).

The velocity dispersion is roughly independent of mass for clouds less
than 10^5 M_0, but is consistent with equipartition of kinetic energy for
giant molecular clouds. Equipartition results from interaction between
clouds, for example any of the mechanisms resulting in agglomerative
build-up of the largest clouds. It may, therefore, be possible to
explain the confinement of giant clouds to spiral arm segments by a
process of continual renewal, in which small clouds entering a spiral
arm collide and combine to make the giant molecular clouds.

SUMMARY

        Over 300 molecular clouds found between 5 and 8 kpc galactocen-
tric radius have been observed and cataloged; positions and sizes have
been analyzed to find a relation between scaleheight and mass. The very
largest clouds, those which are confined to the spiral arms, have a much
reduced scaleheight in comparison to smaller clouds. This may be a
manifestation of the formation of giant clouds by agglomeration.

REFERENCES

Boulanger, F., Stark, A. A. and Combes, F. 1981, Astr. Ap. 93:L1.
Caldwell, J. A. R. and Ostriker, J. P. 1981, Ap. J. 251:61.
Cohen, R. S., Cong, H., Dame, T. M., and Thaddeus, P. 1980, Ap. J. (Letters)
    239:L53.
Cowie, L. L. 1980, Ap. J. 236:862.
Cowie, L. L. 1981, Ap. J. 245:66.
Kwan, J. 1979, Ap. J. 229:567.

Linke, R. A. 1981, in "Extragalactic Molecules" ed. L. Blitz and
    M. Kutner (Green Bank, W. Va.: NRAO), p. 87.
Linke, R. A., Stark, A. A., Bally, J. and Miller, R. 1982, (in
    preparation).
Oort, J. H. 1965, in Galactic Structure, ed. A. Blaauw and M. Schmidt
    (Chicago: University of Chicago Press), p. 455.
Stark, A. A. 1979, Ph.D. thesis Princeton University.
Stark, A. A., Penzias, A. A. and Beckman, B. 1982, (in preparation).

^{13}CO IN THE GALACTIC PLANE: THE CLOUD-TO-CLOUD VELOCITY DISPERSION IN
THE INNER GALAXY

H.S. Liszt
National Radio Astronomy Observatory

W.B. Burton
Sterrewacht Leiden

We have observed galactic ^{13}CO over the region b=0°, ℓ=20° to 40°
at 3' spacings of the 36 foot telescope beam. Because of its moderate
opacity, ^{13}CO is a better probe of molecular column density than ^{12}CO.
In addition, because the ^{13}CO features are narrower and more transparent, the perceived nature of the cloud ensemble suffers less from
blending. This simplifies study of the parameters of the ensemble.
Here we confine our remarks to derivation of the parameter σ_{cc} speci-
fying the velocity dispersion of the ensemble clouds with respect to
each other.

The ^{13}CO spectra were observed in sessions dating from 1977 to 1981
at the NRAO 36 foot telescope on Kitt Peak. At the rest frequency (110.2
GHz) of the J=1 $\to$ 0 transition the HPBW of the telescope is 66". Use of
the 500 kHz filterbank gave a velocity resolution of 1.3 km s^{-1}. Integrations of 10 minute duration gave a typical 3σ noise level of about
0.2 K. Intensities are expressed in units of T_A^*; velocities are expressed relative to the local standard of rest.

Figure 1 shows a velocity-longitude map of the survey. The survey
comprises spectra at 391 positions at b=0°.0 between ℓ=20°.5 and 40°.0.
The sampling interval of 3' of longitude sets the effective angular
resolution of the survey. This relatively small interval is important
because of the use of the data for a statistical investigation of cloud
parameters. The 3' interval subtends about 7 pc at a representative
subcentral point in the observed sector; individual clouds typically
have diameters about three times this length, but of course rarely
present an entire diameter to an undersampling survey.

Earlier, Liszt, Xiang, and Burton (1981) used a subset of this
survey to derive information on the mass and size distribution characterizing the galactic molecular cloud ensemble. One of the most important
characteristics is the cloud-to-cloud velocity dispersion of the members of this ensemble. Accurate determination of this parameter is important, for example, to discussions of possible cloud-cloud accretion,
the role of clouds as sites of star formation, the statistical accele-

135

W. L. H. Shuter (ed.), Kinematics, Dynamics and Structure of the Milky Way, 135–142.
Copyright © 1983 by D. Reidel Publishing Company.

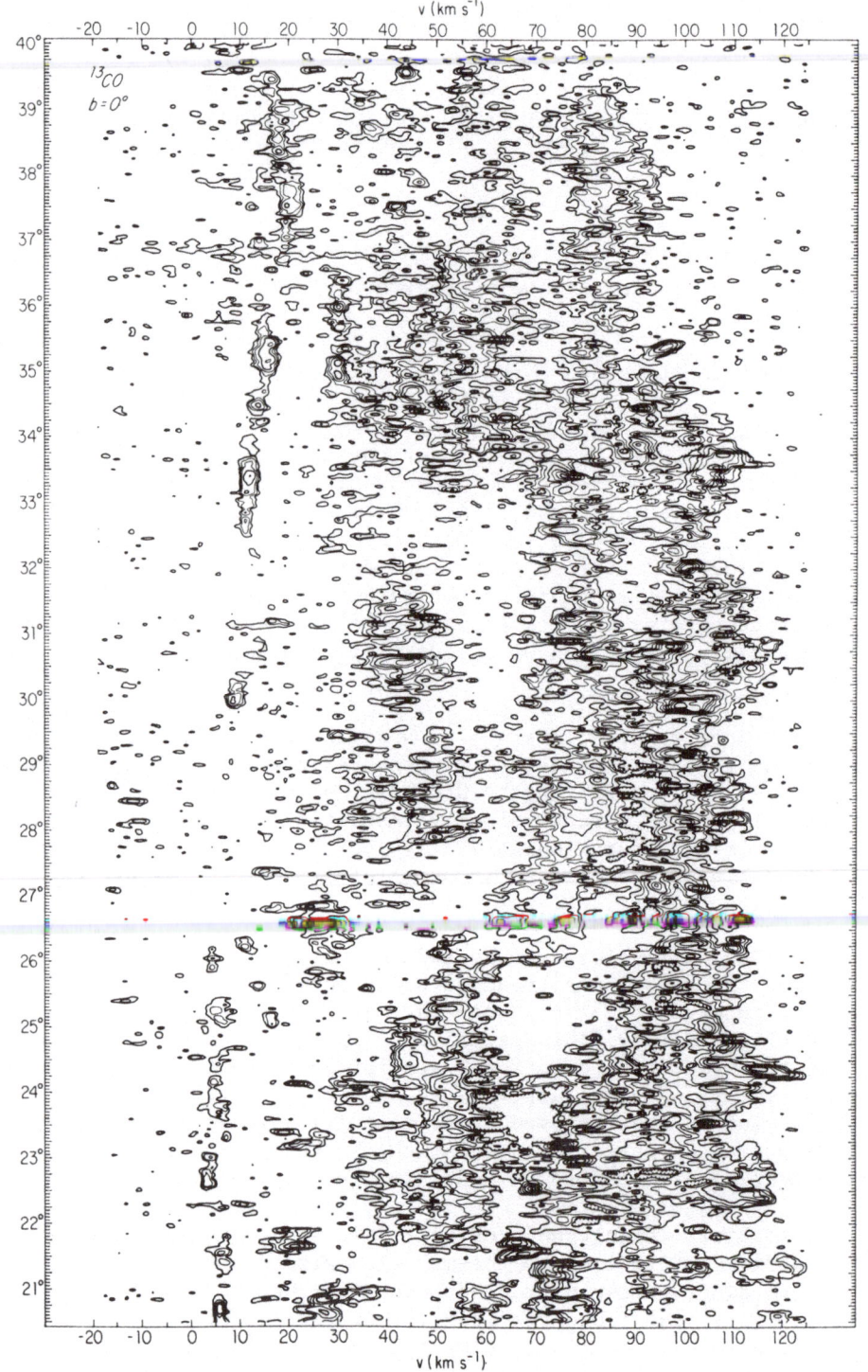

Figure 1. ℓ-v arrangement of emission from ^{13}CO in the galactic equator.
The contours give T_A^* at 0.19, 0.4, 0.7, 1.0, 1.4, 2.0, 2.5, 3.0, 4.0,...K.

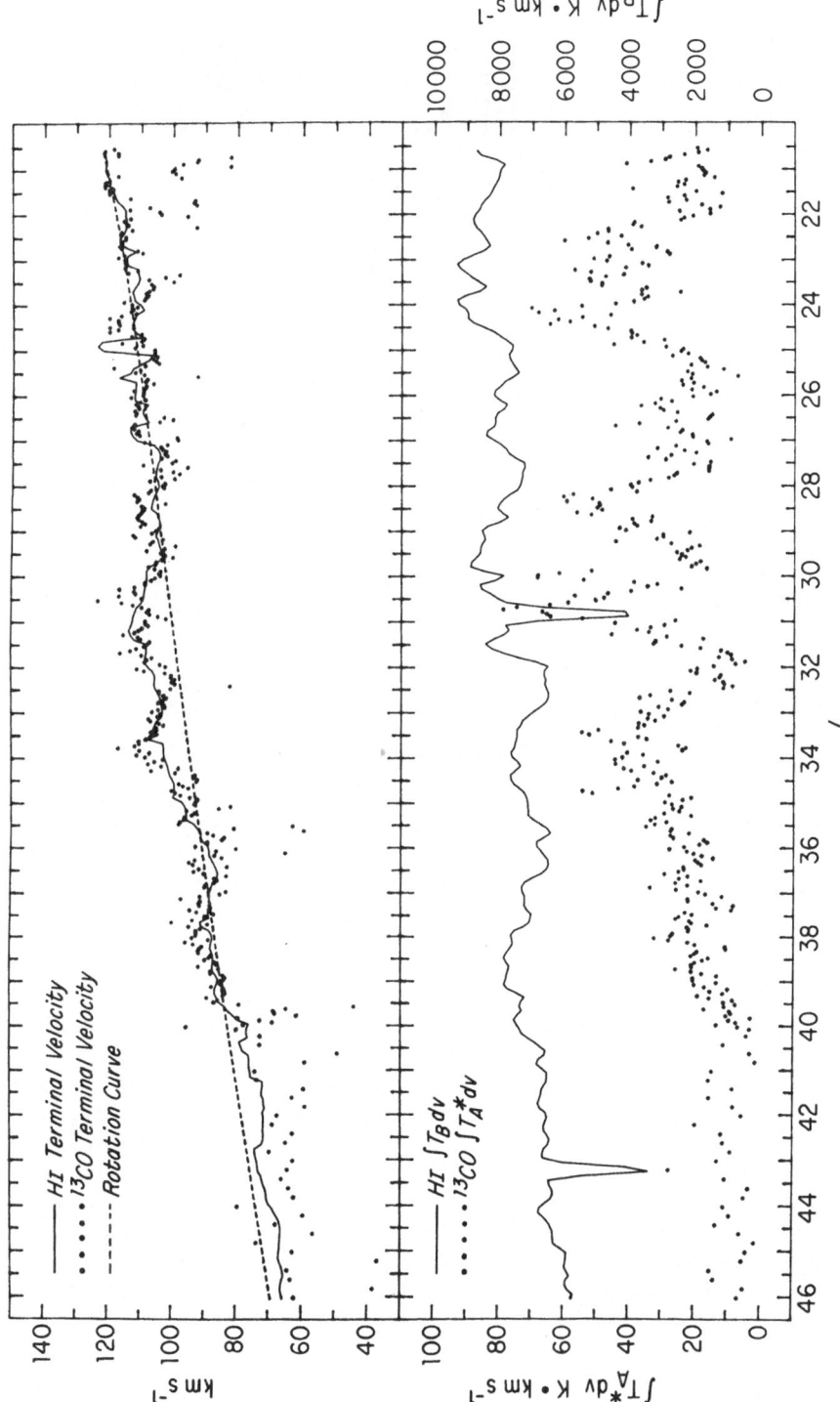

Figure 2. (Top): Variation with ℓ of the terminal velocities measured in our ¹³CO data (dots) and the HI data of Westerhout 1976 (solid line). The dashed line shows the velocity contributed from the locus $R_0 \sin \ell$ according to the perturbation-free rotation curve of Burton and Gordon (1978). (Bottom): Variation with ℓ of the integrated intensities at $v \geq 0$ km s⁻¹ for ¹³CO (dots) and HI (line).

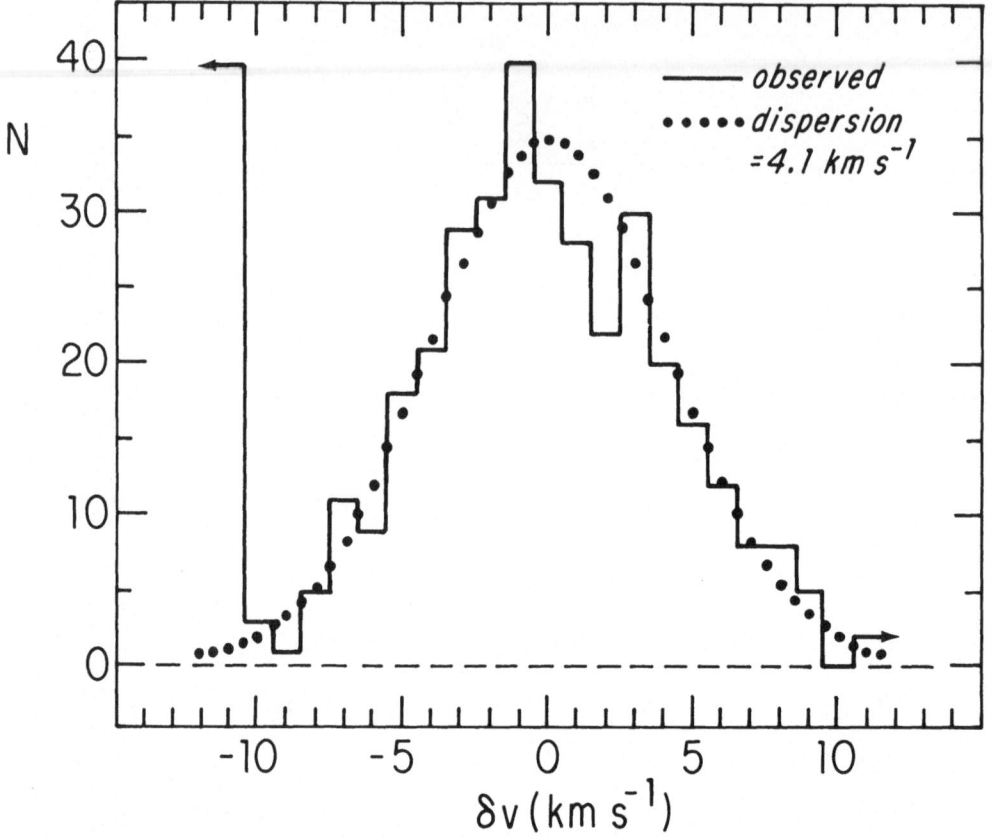

Figure 3. Histogram of the observed differences $v_t(^{13}CO) - v_t(HI)$ between the terminal velocities of the clumped molecular component and the ubiquitous neutral gas. This measure reflects directly the cloud-to-cloud velocity dispersion. The dots show a gaussian distribution of dispersion 4.1 km s^{-1}.

ration of various galactic components through encounters with clouds, the ages of clouds, and the degree of their confinement to large-scale structures.

In several earlier papers (e.g. Liszt and Burton, 1980; Liszt, Xiang, and Burton, 1981) we approached the general problems of interpreting CO survey data by simulating spectra from the galactic ensemble of clouds. The controlled conditions inherent in such modeling allow identification of the manner in which the various parameters characterizing the cloud ensemble influence spectra of the sort observed. The dominant parameters are the cloud size spectrum, the mean free path, λ, and the whole-cloud velocity dispersion, σ_{cc}.

An easily observable measure which is quite sensitive to both λ and σ_{cc} is provided by the degree of scatter shown by the terminal-velocity ridge of emission. Gas located near the subcentral-point locus, at a distance $R_0 \sin \ell$ from the galactic center, contributes emission at the positive-velocity (in the first quadrant) cut-off. If the gas is ubiquitous at this locus (as it demonstrably is for the case of HI), deviations from a smoothly-varying distribution of terminal velocities reflect kinematic irregularities only. Because the molecular cloud ensemble is composed of discrete objects, empty regions along the subcentral point locus also contribute to the scatter of the CO terminal velocities. By measuring the scatter in the molecular spectra, we can derive in a statistical manner information on both λ and σ_{cc}.

Figure 2 shows the longitude dependence of the terminal velocities measured (in the manner discussed by Burton and Gordon 1978 and Liszt, Xiang, and Burton 1981) from the present ^{13}CO data and from the HI data of Westerhout (1976). If the CO v_t were smoothed over lengths corresponding to about 1°, the CO and HI v_t would coincide, indicating that the molecular ensemble as a whole shows the same kinematics as the diffuse gas. We see no evidence in this measure, for example, for the gross migration of the clouds with respect to the HI, suggested in a series of papers by Bash (e.g. 1979) and collaborators. The substantial point-to-point jitter in the CO v_t reflects the parameters λ and σ_{cc}. Observational noise is negligible. Excursions of the individual v_t to velocities higher than the local mean reflect directly the line of sight component of σ_{cc}. Excursions to velocities lower than the local mean reflect the combined influence of σ_{cc} and possibly empty regions near a subcentral point locus and thus the mean free path parameter λ. In our stochastic modeling, we have a variety of constraints on λ. These constraints are based chiefly on the integrated properties of the profiles and the results of cloud counting. For a mean free path of 1.7 kpc we find that a value of $\sigma_{cc} = 4.2$ km s^{-1} yields agreement with the observed situation of the v_t jitter calculated from simulated profiles.

Figure 3 shows a histogram of the differences $v_t(^{13}CO) - v_t(HI)$. The distribution is skew at differences < -10 km s^{-1}, as would be expected for occasional empty subcentral regions; all such differences are lumped together in the histogram. (That there are no such excursions to differences $> +10$ km s^{-1} shows immediately that $2 \times \sigma_{cc} \lesssim 10$ km s$^-$.) The dots in the figure show a gaussian distribution of dispersion 4.1 km s^{-1}.

The sensitivity of the v_t jitter method to the parameters σ_{cc} and λ is shown in Figures 4 and 5, respectively. In these figures the dots represent the histogram of observed differences as also plotted in Figure 3. The line histograms in Figures 4 and 5 show the differences $v_t(^{13}CO) - v_{max}$ found on simulated profiles representing ^{13}CO emission from a stochastic ensemble of clouds. Here v_{max} is the maximum velocity predicted by the smoothly-varying rotation curve used in the modeling procedure; v_{max} is plotted as the dashed line in Figure 2. The error bars in the figures were estimated by some 1000 successive realizations of the random distributions used in the modeling process.

Figure 4 shows that a value of $\sigma_{cc} = 3$ km s^{-1} results in a cloud v_t scatter which is somewhat more sharply peaked than that observed;

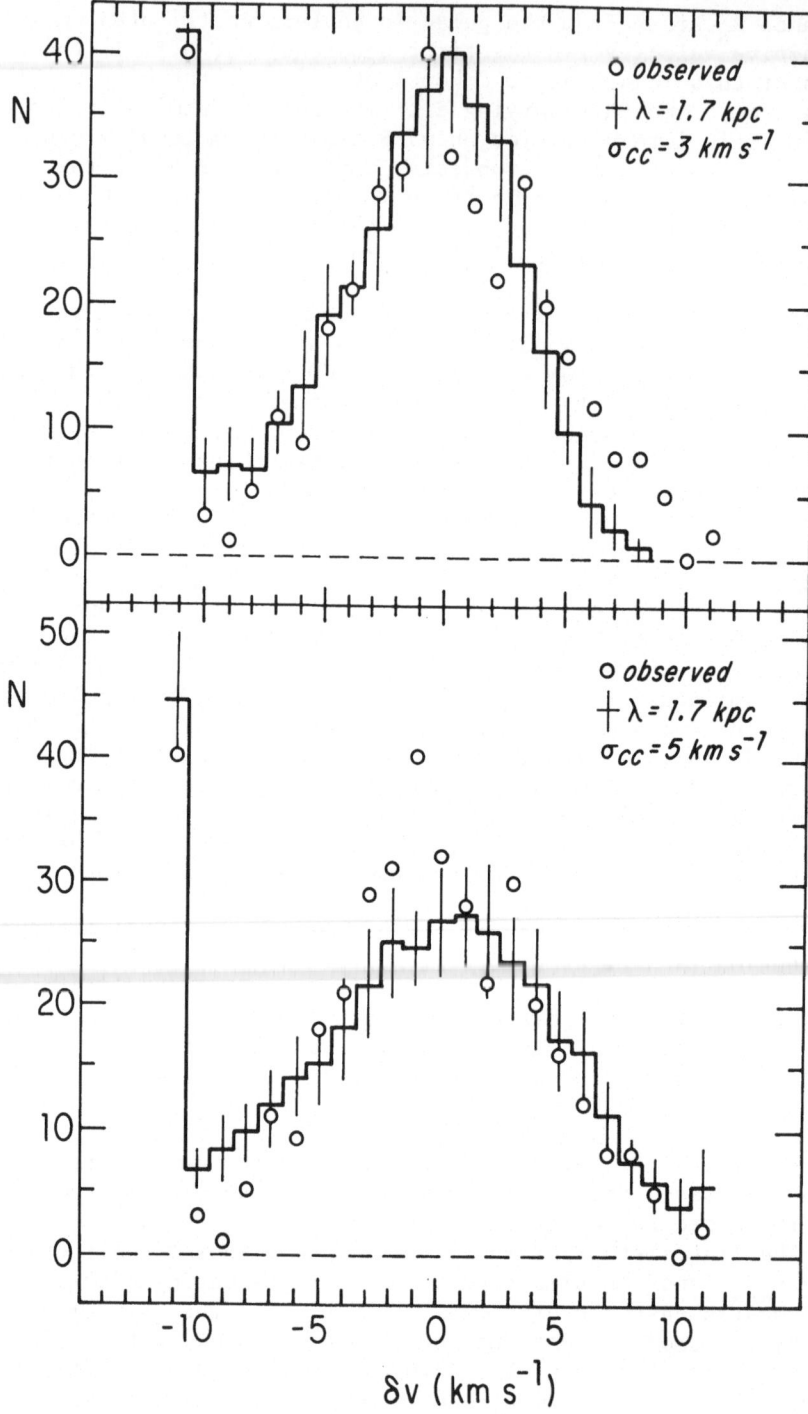

Figure 4. Comparison of the observed (dots) and synthetic-profile (line) terminal-velocity differences for two values of the velocity dispersion parameter, σ_{cc}, in the model ensemble of clouds.

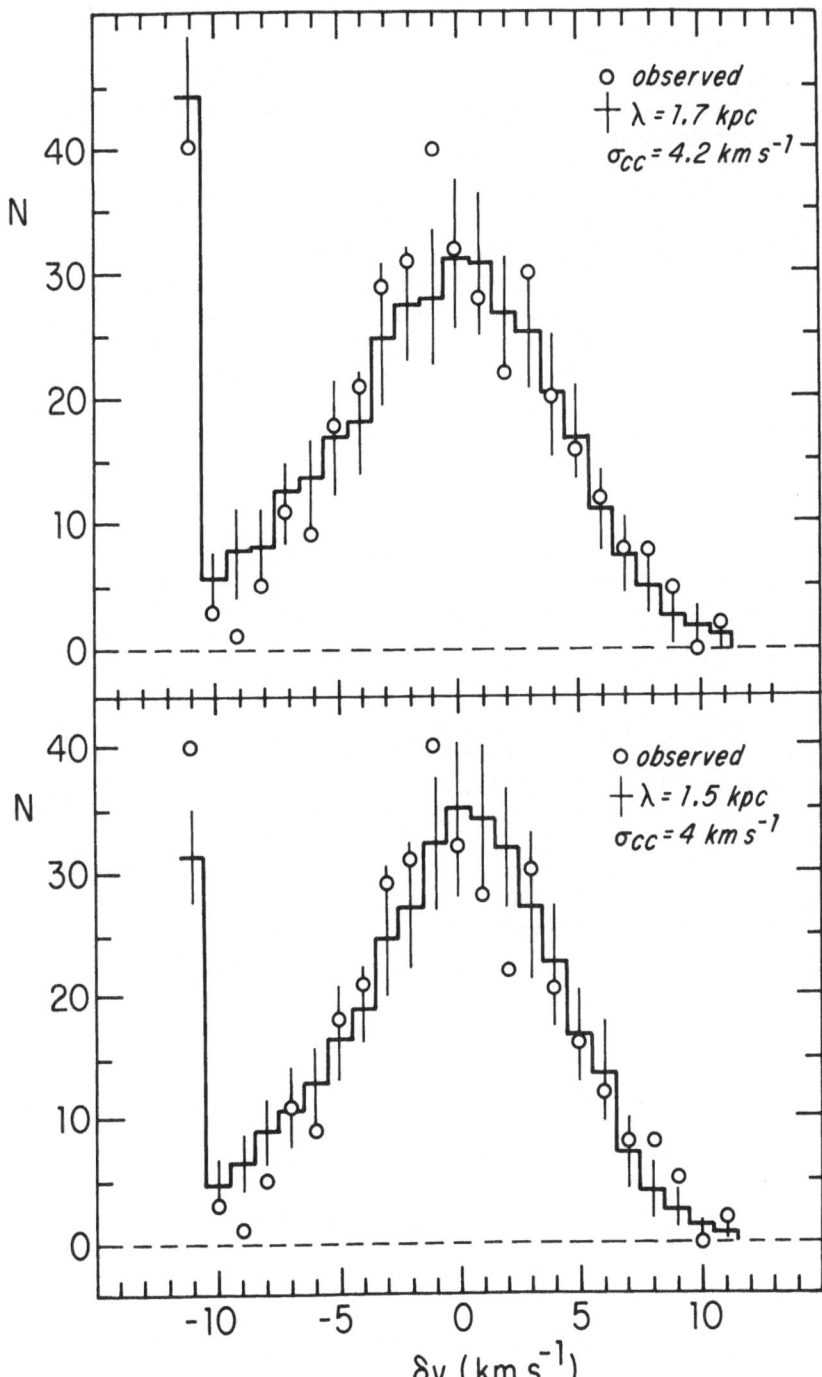

Figure 5. Comparison of the observed (dots) and synthetic-profiles (lines)
terminal velocity differences for two values of the cloud separation
parameter, λ, in the model ensemble of clouds.

σ_{cc} = 5 km s^{-1} results in a somewhat too broad distribution. Figure 5 shows that for a reasonable value of $\sigma_{cc} \sim 4$ km s^{-1}, a value of λ as small as 1.5 kpc damages the agreement with the observed situation by resulting in too few extreme negative-velocity excursions, and a too sharply peak distribution.

We conclude that the velocity dispersion of molecular clouds in the inner parts of the galaxy is 4.2+0.5 km s^{-1}. The error is an estimate based on numerical experiments to gauge the sensitivity of the method. The dispersion refers to the one dimensional motions measured in the galactic equator. Clouds in a rather large range of the Galaxy, 3.5 $\lesssim$ R $\lesssim$ 6.5 kpc contribute to the measure. Other estimates of σ_{cc}, all based either on local data or on clouds in the outer galaxy, give values approximately twice that derived here (e.g. Stark and Blitz, 1978). Further work is necessary to evaluate the nature of the different results.

Acknowledgement: We gratefully acknowledge support from the North Atlantic Treaty Organization through Research Grant No. 008.82.

REFERENCES

Bash, F.N.: 1979, Astrophys. J. 233, 524.
Burton, W.B., and Gordon, M.A.: 1978, Astr. and Astrophys. 63, 7.
Liszt, H.S., and Burton, W.B.: 1980, Astrophys. J. 236, 779.
Liszt, H.S., Xiang, Delin, and Burton, W.B.: 1981, Astrophys. J. 249, 532.
Stark, A.A., and Blitz, L.: 1978, Astrophys. J. Lett. 225, L15.
Westerhout, G.: 1976, Maryland-Bonn Galactic 21-cm Line Survey (College Park, Md: University of Maryland).

HOW CONFIDENTLY DO WE KNOW THE CO ROTATION CURVE OF THE OUTER GALAXY?

Leo Blitz
Astronomy Program, University of Maryland
Michel Fich
Astronomy Department, University of California, Berkeley

ABSTRACT

We examine the sources of error in the determination of the CO rotation curve beyond the solar circle. The largest uncertainties are due to uncertainties in R_0 and θ_0. Systematic errors in stellar distance determinations, non-circular motions and incomplete galactic coverage could be important in principle, but the evidence suggests that they do not affect the shape of the rotation curve much in practice. We conclude that unless the rotation constants have values outside the range accepted by most observers (i.e. $\omega_0 < 20$ km s^{-1} kpc^{-1} the rotation curve beyond the sun rises to 18 kpc from the galactic center. The implications for the Oort A constant are also discussed.

The work of at least three different groups has shown that the rotation curve of the outer Milky Way is approximately flat to large R (Knapp, Tremaine and Gunn 1978; Jackson, Fitzgerald and Moffatt 1980, Blitz, Fich and Stark 1980). Evidence from the latter two groups has indicated that the rotation curve may be rising by as much as 50 km s^{-1} beyond the solar circle, but the conclusions have been tentative because of the preliminary nature of the data or the analysis. Attempts to obtain information on the rise of the rotation curve from observations of other galaxies give apparently contradictory results: the Hα rotation curves continue to rise to the last measured point (Rubin, Ford and Thonnard 1980), but the HI rotation curves are generally rather flat (Bosma 1978). The detailed shape of the rotation curve not only has intrinsic interest, but provides the only available observational evidence on the form of the density distribution of a massive halo. In this paper, we examine to what degree errors and uncertainties affect the determination of the CO rotation curve of the outer Milky Way.

The basic observational data are from the catalogue of Blitz, Fich and Stark (1982) which provides stellar distances and CO velocities for a large number of optical HII regions. The data

W. L. H. Shuter (ed.), Kinematics, Dynamics and Structure of the Milky Way, 143–150.

reduction and the observational uncertainties are as given by Blitz
(1979) with the following exceptions: i) the rotation curve is
determined by fitting $(\omega - \omega_o)$ as a function of $(R-R_o)$ and then
converting to R, θ coordinates. Linear regressions, however, are fit
directly in R, θ space. ii) The small apparent motion of the molecular
cloud sample relative to the LSR has been removed in the fitting. We
have also produced curves by fitting R as a function of ω; the results
do not differ appreciably from the plots of ω as a function of R. A
fuller description of the analysis and the results will appear in a
forthcoming publication.

The best fit to the data under the assumption of circular
rotation, R_o = 10 kpc, θ_o = 250 km s^{-1} is shown in Fig. 1, which shows
the rise beginning at ~ 12 kpc. The sources of error can be grouped
into the following three categories: measurement errors, assumptions
and modeling uncertainties, and rotation constants. We discuss each
of these below.

I. MEASUREMENT ERRORS

A) <u>CO Velocities</u> The CO velocities are the most accurately
determined values in the data set, with typical uncertainties per
observation of ~ 0.5 km s^{-1}. Even though each HII region has a number
of measurements, the 1σ uncertainty in the center-of-mass velocity of
the entire complex is worse than this, about 1 km s^{-1} producing
typical uncertainties in θ of ~ 2 km s^{-1} (see Blitz 1979 for a
discussion of this point). Because other sources of error are larger
than this, improvements in the velocity measurements of the HII
region/molecular cloud complexes will not significantly affect the
rotation curve. A comparison of the CO velocities versus the Hα and
H109α velocities indicates that no significant bias is introduced with
the use of the CO velocities (Fich, Treffers and Blitz 1982).

B) <u>Stellar Distances</u>

1) <u>Random Errors</u> – The dispersion in θ from the mean curve
shown in Fig. 1 is 12 km s^{-1}. Most of this error is due to
uncertainties in the distances to the stars exciting the HII
regions. Our analysis of the CO velocities toward the galactic center
and anticenter indicates that the cloud – to – cloud velocity
dispersion is 7 ± 1.5 km s^{-1}. Thus, even if the uncertainties in the
distances were lowered by a large factor, the scatter in the θ
coordinate would be reduced only by a factor of two. Confidence in
the rotation curve at large R could be more profitably improved by
obtaining stellar distances to additional HII regions rather than by
improving the distances already at hand. In any event, the magnitude
of the rise in the rotation curve in Fig. 1 is greater than the
dispersion of the points. Improvements in the random uncertainties
therefore will not significantly lower the rotation curve.

2) <u>Systematic Errors</u> – The rotation curve could be flatter
than is shown in Fig. 1 if the distances to the outlying HII regions
were systematically overestimated. Imagine that the stars which are
used to determine the distance moduli of the more distant HII regions

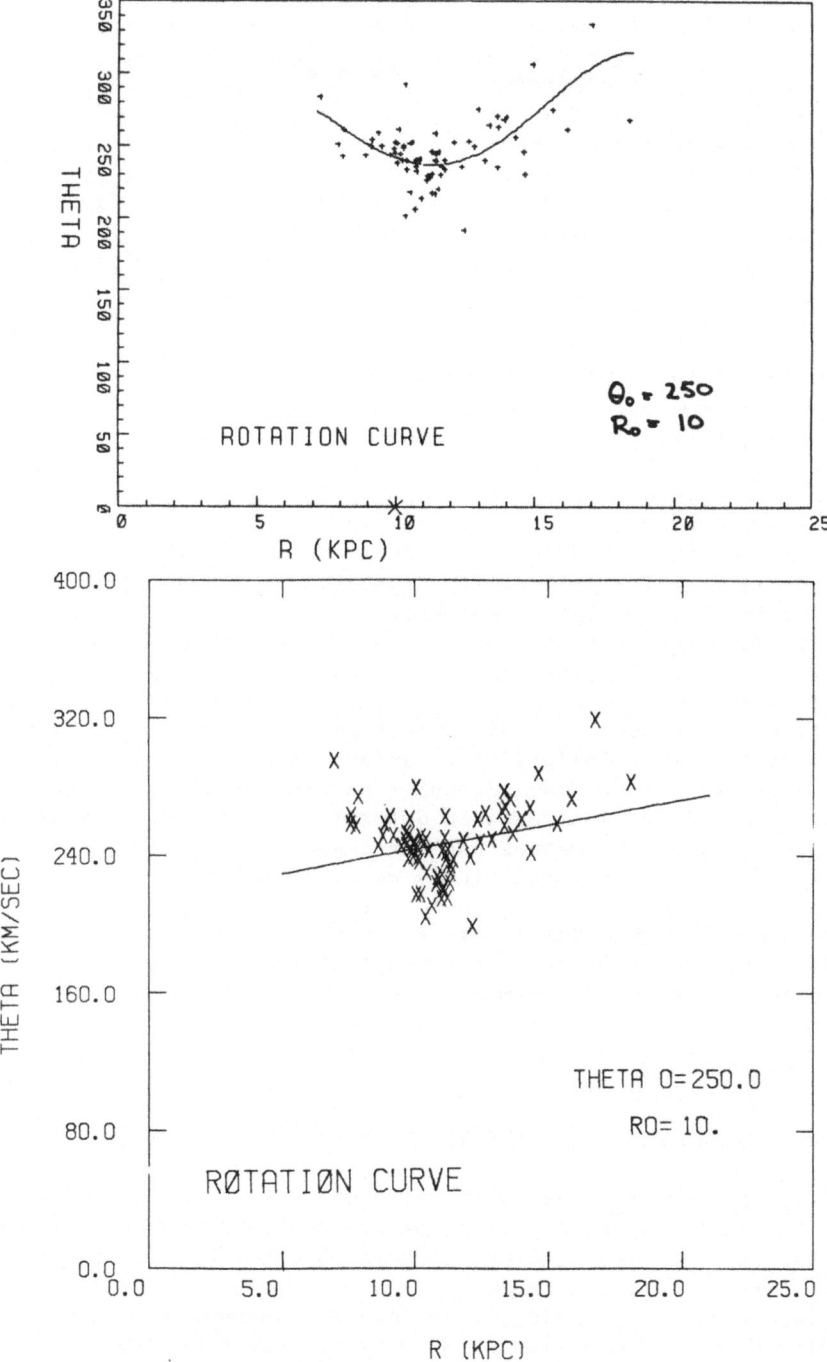

Fig. 1   a) The CO rotation curve with R_o = 10 kpc θ_o = 250 km s^{-1}.
The line is a fourth order fit to the data.   b) Same as a) except the
fit is linear.

were fainter than stars of the same spectral type used to calibrate
the upper end of the main sequence locally. The observed distance
moduli would then require a downward revision which would produce a
flattening of the rotation curve. It is not clear what could produce
such a change in the stellar properties; a metallicity decrease for
example would be expected to produce brighter stars. In any event,
the effect would have to be large. For example, to obtain a circular
velocity of 250 km s^{-1} near 18 kpc, the outermost points in Fig. 1
would require a systematic error in distance modulus of ~ 35%, or a
change in the intrinsic brightness of a given spectral type of almost
a factor of 2. Such a large effect would probably have been noticed
in extragalactic HII regions and OB stars but cannot be ruled out for
the Milky Way. It therefore seems unlikely that the rise in the
rotation curve can be affected by systematic errors in the distances,
although such errors could produce some flattening.

   C) <u>Center of Mass Position</u> In general, the molecular clouds are
not mapped, and thus their center of mass positions are often not well
determined. The mean diameters of 12 well mapped local giant
molecular cloud complexes is 45 pc (Blitz 1978). Thus, the angular
uncertainty is less than 0.5° at 5 kpc from the sun. This error
therefore contributes insignificantly (typically < 1 km s^{-1}) to
uncertainties in the circular velocities.

   D) <u>Misidentifications</u> Because of the large number of points
used to determine the rotation curve, and the certainty with which
molecular clouds can be identified with HII regions in the second and
third quadrants, we expect that misidentifications of CO clouds with
HII regions will have an insignificant affect on the rotation curve.
Misidentifications, if any, should number no more than two or three.
For the outermost HII regions used to determine the rotation curve,
the CO complexes have been mapped and the association of a complex
with a given HII region is particularly well determined.

   We conclude that measurement errors do not contribute
significantly to uncertainties in the shape of the outer galaxy
rotation curve. Improvements are best achieved by obtaining distances
to the outlying HII regions for which they have not yet been
determined.

II.   ASSUMPTIONS AND MODELING UNCERTAINTIES

   A) <u>Error in the LSR</u> The molecular cloud sample has a component
of motion of about − 4.5 km s^{-1} relative to the local standard of rest
in the radial (Π) direction (Blitz, Fich and Stark 1980). This may be
due to a variety of affects, among which are a non-circularity of the
overall galactic velocity field, or an inward component of motion
which results from the formation of molecular cloud complexes in the
spiral arms. HI absorption measurements toward the galactic center
(e.g. Radhakrishnan and Sarma 1980) suggest that any radial component
in the motion of the LSR with respect to the galactic center is less
than 1 part in 1000.

The effect of the radial motion of the molecular clouds with respect to the LSR has been removed in Fig. 1 and the effect on the rotation curve is less than 10 km s^{-1} at all points on the mean curve. As long as the other (θ,Z) components of motion of the molecular cloud sample with respect to the LSR are of the same order as the Π component, the overall shape of the rotation curve should be insignificantly affected.

B) <u>Non-Circular Motions</u> Non-circular motions such as those which have long been known in the Perseus arm (e.g. Roberts 1972) could affect the shape of the rotation curve. Such non-circular motions have been thought to be responsible for the perturbations of the HI rotation curve measured inside the solar circle (e.g. Burton 1974). Furthermore, the inward streaming of some of the HII toward the galactic anticenter is quite large ($\sim$ 30-50 km s^{-1}), and if the HII region/molecular cloud complexes shared this motion, the shape of the rotation curve could be significantly affected.

The effect of the radial streaming motions is that inward streaming causes an underestimate in the derived circular velocities. Thus, the observed non-circular motions would have the effect of requiring an even steeper rise in the CO rotation curve than is shown in Fig. 1. As mentioned in the previous paragraph, the small radial motion of the molecular cloud sample with respect to the LSR has been removed from the point plotted in Fig. 1. The value of the correction, -4.5 km s^{-1} toward ℓ = 180°, shows that either the molecular clouds do not share the large inward streaming of the HI gas, or the negative velocity gas is located at distances $>$ 18 kpc, the edge of the molecular cloud sample. Another possibility which has been raised recently by Heiles (1982) is that the HI streaming is due to a large shell of expanding atomic gas. Such a shell could not affect the overall kinematics of the ensemble of molecular clouds in the outer galaxy.

C) <u>Incomplete Galactic Coverage</u> The data used to construct the outer galaxy rotation curve covers only about 60° in galactocentric azimuth. Velocity field asymmetries on this size scale and larger have been observed in other galaxies (e.g. M101-Bosma, Goss and Allen 1981). The Milky Way shows asymmetries in the distribution of HI which could be partly due to large scale velocity field asymmetries (Henderson, Jackson, Kerr 1982). Such asymmetries could affect the determination of the overall rotation curve of the Milky Way.

By obtaining more CO and optical data, it is possible to obtain a rotation curve which samples as much as 120° of galactocentric azimuth, but not significantly more. The limitation is due to the requirement for obtaining optical distances. Beyond this 120° slice, velocity field anomalies can only be determined indirectly. However, if the velocity field is well behaved in the region where the rotation curve is measured (as appears to be the case), and no gross asymmetries appear in the HI distribution in the first and fourth quadrants, the measured rotation curve should be representative of the Galaxy as a whole.

III.   ROTATION CONSTANTS

Changing the rotation constants R_o, θ_o by as little as 20% can cause significant changes in the rotation curve. Decreasing R_o and leaving θ_o constant steepens the rotation curve (see Fig. 2), while lowering θ_o tends to flatten it. Fig. 3 shows a combination of rotation constants such that a linear fit to the data actually exhibits a decline. Note, however, the trend in the data beyond 13 kpc even for R_o = 10 kpc and θ_o = 200 km s^{-1} is up.

Although values of R_o between 7 kpc and 10 kpc and θ_o between 180 km s^{-1} and 250 km s^{-1} have been advocated in the recent literature, there are observational constraints on the <u>pairs</u> of permitted values. There has been considerable support recently for R_o = 8.5 kpc and θ_o = 220 km s^{-1}. These values leave the shape of the rotation curve nearly unchanged from that shown in Fig. 1. The reason for this is that for objects in circular rotation, $\omega-\omega_o$ = R_o^{-1} $v_r/\sin\ell$ is the observed radial velocity. For a fixed ω_o, $\omega-\tilde{\omega}_o$, scales with R_o only. The most commonly discussed rotation constants all have $\omega_o \simeq$ 25 km s^{-1} spc^{-1}, in which case the rotation curve will rise for $R-\bar{R}_o$ > 2 kpc by a common scale factor.

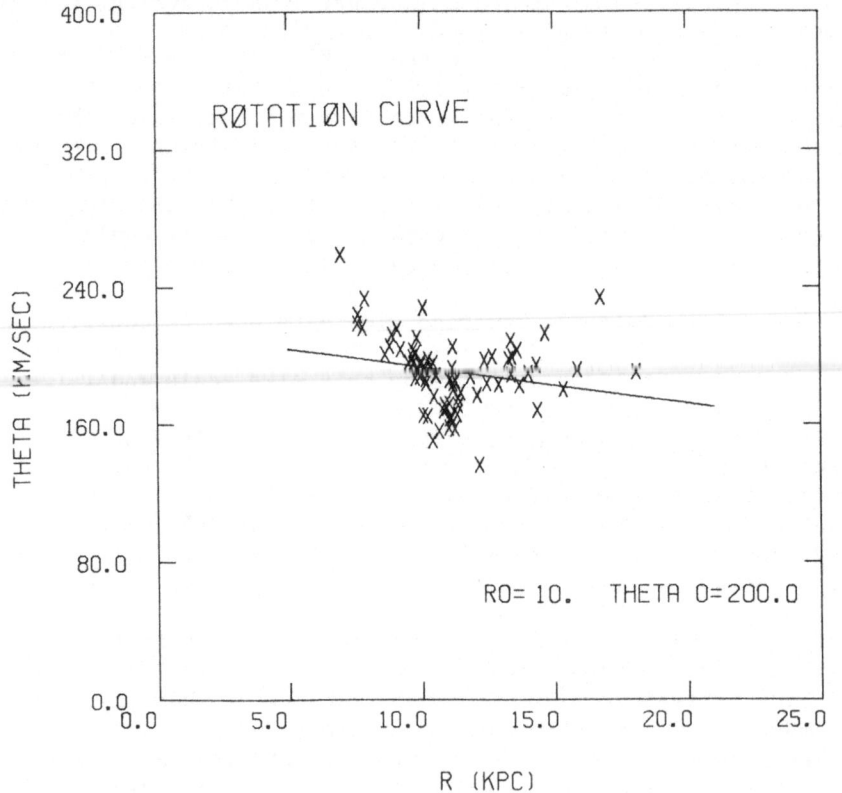

Fig. 2   Least squares fit to the data used in Fig. 1 with R_o = 10 kpc and θ_o = 200 km s^{-1}. Note that the points at R > 13 kpc continue to show an increase in θ with increasing R.

We conclude that although it is possible to find pairs of values for R_o and θ_o that will cause the rotation curve to be flat or to decline, these values are outside the range accepted by most observers. For the most commonly accepted values, the rotation curve rises by 30-50 km s^{-1} at the last measured point.

IV.   THE VALUE OF A

Stellar dynamics has reasonably well established that the local value of the A constant is 15 km s^{-1} kpc^{-1}. However, a flat rotation curve requires (for ω_o = 25 km s^{-1} kpc^{-1}) that A = 12.5 km s^{-1} kpc^{-1} and a rising rotation curve requires that A < 12.5 km s^{-1} kpc^{-1}. The apparent discrepancy between the value from stellar dynamics and a flat rotation curve can be understood if there is a local velocity perturbation.

The rotation curve data near R_o are heavily weighted by objects near the sun which have a small range of galactocentric azimuth. In fact, when the value of $-1/2 R_o \, d\omega/dR|R_o$ is evaluated from our data set, we also find that A = 15 km s^{-1} kpc^{-1}, a value which is due to

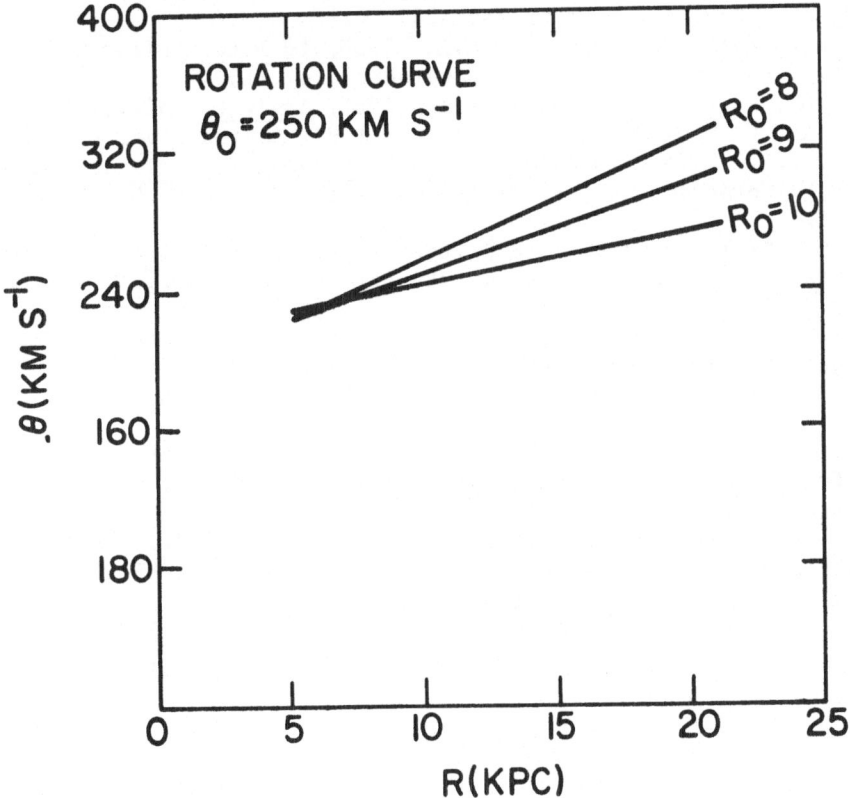

Fig. 3   Least squares fits to the data used in Fig. 1 with θ_o = 250 km s^{-1}, R_o = 8, 9 and 10 kpc.

the dip in the curve beyond R_o. If there were data near R_o around the entire solar circle, the local velocity fluctuations would presumably average out and a mean value of A somewhat less than 12.5 km s^{-1} kpc^{-1} would be obtained.

V. SUMMARY

We conclude that the largest uncertainties in the rotation curve stem from uncertainties in the rotation constants, but unless ω_o is less than about 20 km s^{-1} kpc^{-1}, the rotation curve rises beyond the solar circle at a distance of 12 kpc from the center. Other significant uncertainties could come from systematic errors in the stellar distances and incomplete galactic coverage, but there is no evidence at this time that the rotation curve is significantly different from that shown in Fig. 1.

REFERENCES

Blitz, L. 1978, Ph.D. Dissertation, Columbia University.
Blitz, L. 1979, Ap.J. (Letters), 231, L115.
Blitz, L., Fich, M. and Stark, A.A. 1980, in Interstellar Molecules, B.H. Andrew, ed. (Reidel: Dordrecht), p. 213.
Blitz, L., Fich, M. and Stark, A.A. 1982, Ap.J. (Suppl.), 49, 183.
Bosma, A., Goss, W.M. and Allen, R.J. 1981, Astron. Ap., 93, 106.
Burton, W.B. 1974,, in Galactic and Extragalactic Radio Astronomy, G.L. Verschuur, and K. Kellerman, eds. (Springer: New York), p. 82.
Fich, M., Treffers, R.R., and Blitz, L. 1982, in Regions of Recent Star Formation, R. Roger and P. Dewdney eds. (Reidel: Dordrecht), p. 201.
Heiles, C.E. 1982, paper presented at Leiden meeting on Southern galactic surveys.
Henderson, A.P., Jackson, P.D. and Kerr, K.J. 1982, Ap.J., 263, in press.
Jackson, P.D., FitzGerald, M.P. and Moffatt, A.F.J. in The Large Scale Characteristics of the Galaxy, W.B. Burton, ed. Reidel, Dordrecht, p. 221.
Knapp, G.R., Tremaine, S.D. and Gunn, J.E. 1978, A.J., 83, 1585.
Radhakinshran, V. and Sarma, N.V.G. 1980, Astron. Ap, 85, 249.
Roberts, W.W. 1972, Ap.J., 173, 259.
Rubin, V.C, Ford, W.K. and Thonnard, N. 1980, Ap.J., 238, 471.

THE DISTRIBUTION OF HII REGIONS IN THE OUTER GALAXY

Michel Fich
Astronomy Department, University of California, Berkeley
Leo Blitz
Astronomy Program, University of Maryland

ABSTRACT

We have determined the positions of a large sample of HII regions in the outer Galaxy. Assuming R_0 = 10 kpc and Θ_0 = 250 km s^{-1} we find that essentially all HII regions are within R < ~20 kpc. The dispersion in their distances from the galactic plane increases from 90 pc near the Sun to greater than 300 pc at large R. There is also a strong suggestion that the HII regions follow the warp in the plane that is seen in HI. There are some indications that the mean diameter of HII regions decreases with increasing R.

INTRODUCTION

We have used all of the known optical HII regions in the Galaxy which are visible from the northern hemisphere to determine the distribution of HII regions in the outer Galaxy. One third of these have positions which are known from optical (spectrophotometric) distances determined to their exciting stars. The remaining two thirds have kinematic distances determined using the radial velocity of CO measured in molecular clouds associated with the HII regions.

It is our intention here to show some results of this investigation without discussing details of the analysis or giving in-depth interpretation of the results. These details will appear in a future paper.

RESULTS

The HII regions used here are taken from the catalogue of Blitz, Fich, and Stark (1982). Only 242 of the 377 objects listed in that catalogue are distinct HII. regions with measured CO velocities. The remaining 135 objects include 1) those in which CO was not detected,

W. L. H. Shuter (ed.), Kinematics, Dynamics and Structure of the Milky Way, 151–158.
Copyright © 1983 by D. Reidel Publishing Company.

2) those in which it was not possible to make a definite association of
a particular CO line with the HII region, 3) HII regions which, though
catalogued seperately, are in fact part of a larger complex of HII
regions and molecular clouds and thus are not 'kinematically distinct
objects', and 4) a few supernova remnants, planetary nebulae, and a
couple of extragalactic objects.

Of the 242 'useful' objects in this catalogue 94 have distances
determined directly by spectrophotometric techniques and thus have
well-determined positions in the Galaxy. Kinematic distances were
calculated for the remaining 148 objects. The rotation curve of Blitz,
Fich and Stark (1980) has been used and is shown in Figure 1. This
curve has been extrapolated for R > 18 kpc assuming a flat rotation
curve at Θ = 300 km s^{-1} (solid line).

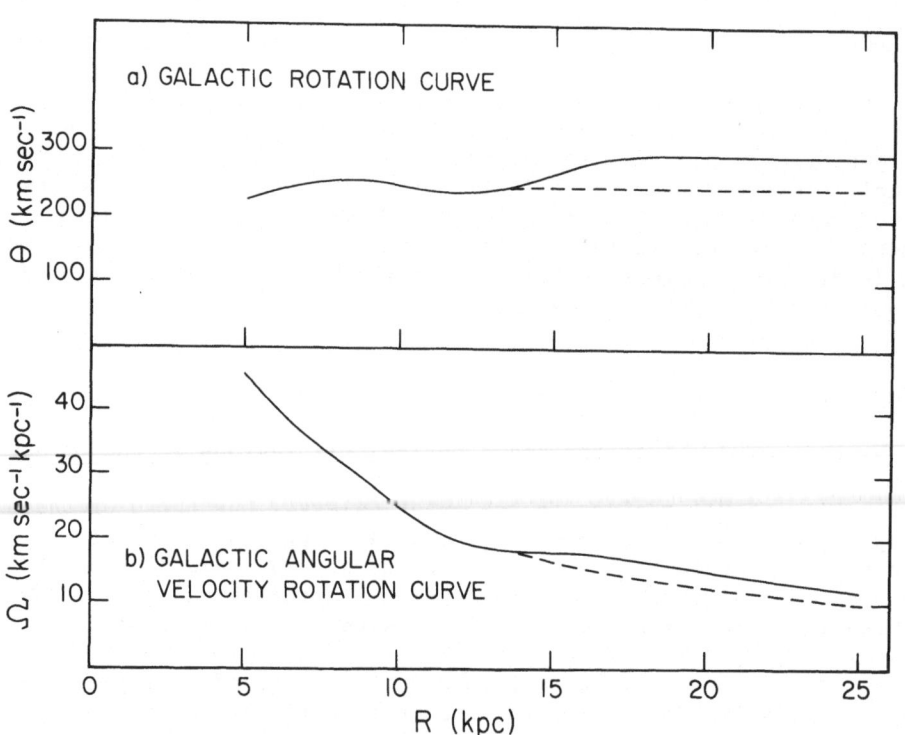

Figure 1.   Rotation curves used to calculate kinematic distances.

For comparison we repeated our calculations for a rotation curve
which rises to a maximum of Θ = 250 km s^{-1} at R=15 kpc and which remains
flat beyond that (dashed line). In both cases we have used R_0=10 kpc
and Θ_0=250 km s^{-1}. The Ω (angular velocity) vs R version of the
rotation curve which is actually used for the distance calculation is
also shown in Figure 1.

The positions of HII regions (projected onto the galactic plane) are shown in Figures 2 and 3 (for the 300 km s^{-1} and the 250 km s^{-1} curves respectively). Objects whose positions were determined kinematically are shown as filled circles while open circles denote those where optical distances were used. The distribution of HII regions locally is not shown in either figure but will be discussed in a future paper.

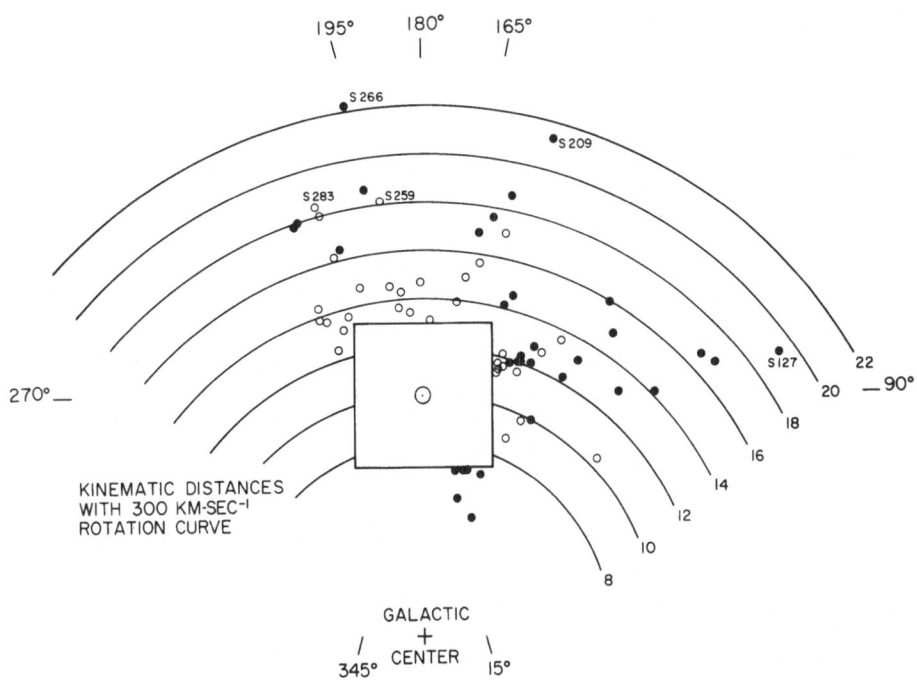

Figure 2. The positions of HII regions projected onto galactic plane.

The most obvious feature of the distributions shown in Figures 2 and 3 are the 'fingers' of HII regions pointing towards the Sun (at $\ell \cong 95^\circ$, 155°, and 210°) showing some of the local extinction pattern. In each of these fingers the most distant HII region is at R ≅ 20 kpc. These most distant HII regions are all bright and several arcminutes in angular diameter. Since extinction is low in the outer Galaxy we would expect to see other similar HII regions at up to 20 kpc from us (R ≅ 30 kpc) if there were any. That we do not see any in three different directions is therefore suggestive that we are seeing the edge of the HII region disk. This is an important quantity since it can be easily compared to other galaxies.

The primary difference between Figures 2 and 3 is that the lower rotation curve (used in Figure 3) brings the most distant objects in from R ≅ 22 kpc (Figure 2) to R ≅ 18 kpc (Figure 3). However if R_o is

smaller than assumed here all distances are less, scaling approximetely
linearly with R_0 (ie. if R_0 = 8 kpc then the most distant HII regions
could be at R = 14 kpc for the rotation curve flat at 220 km s^{-1}).

No spiral structure is visible in either figure but this is not
surprising for two reasons. First, the uncertainties in the distances
are typically larger than the typical distances between spiral arms thus
smearing out the picture. And second, the HII regions used here
represent many different types of HII regions. It might be expected
that only the youngest or the giant HII regions will trace out spiral
features.

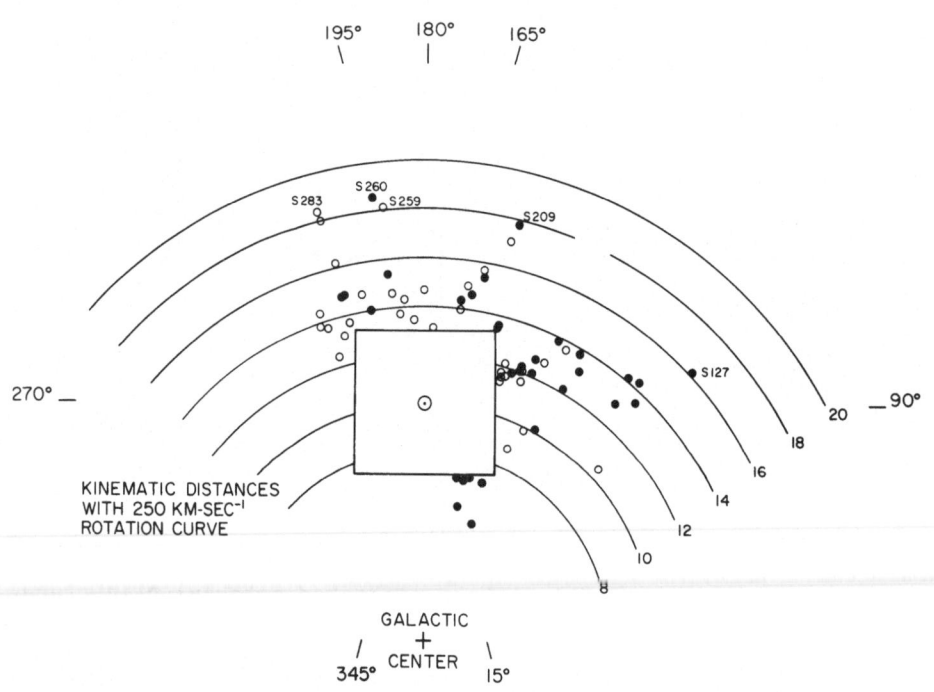

Figure 3. The positions of HII regions projected onto galactic plane.

The surface number density of HII regions is shown in Figure 4.
The cutoff at R < 6 kpc is due to the high extinction in the inner
Galaxy which prevents us from seeing any HII regions optically in the
deep interior of the Galaxy. The peak near R = 10 kpc is almost
certainly a selection effect but the long tail for R > 14 kpc is perhaps
real and of some interest. Figure 5 shows a similar result in the HII
regions of M31 (from Arp 1964) and many other galaxies also show similar
distributions in their outer edges (see Hodge 1969).

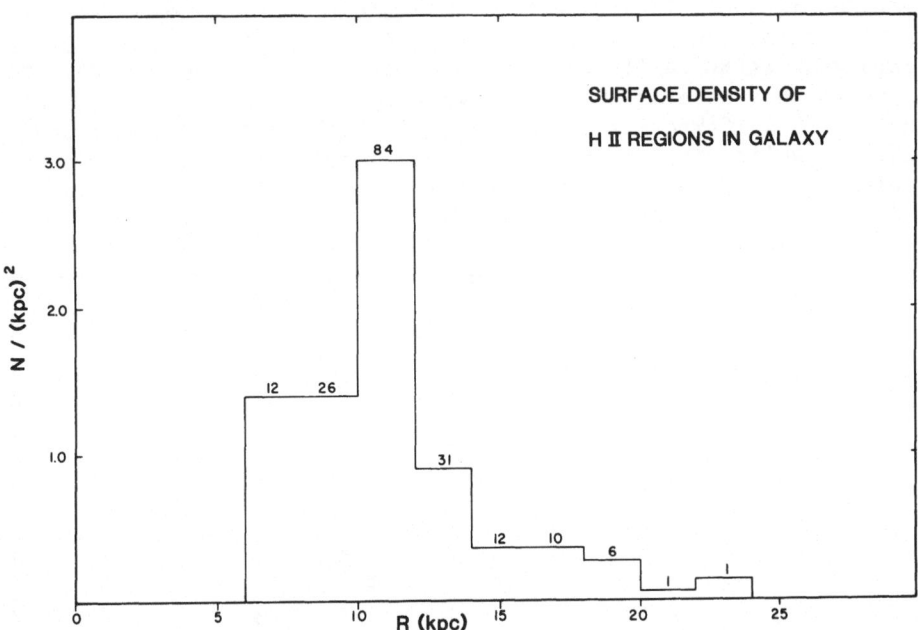

Figure 4. Surface number density of optical HII regions in outer Galaxy

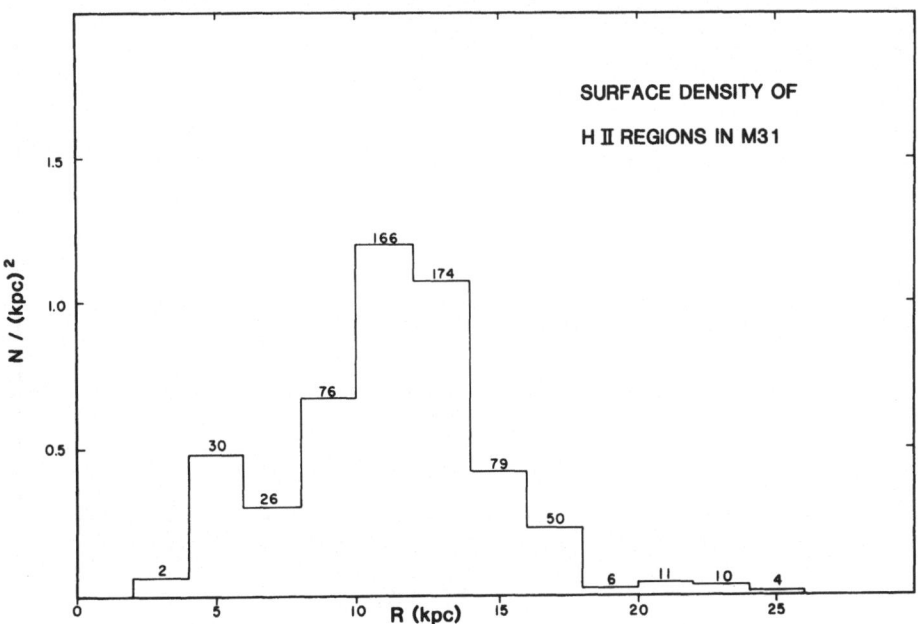

Figure 5. Surface number density of HII regions in M31

The distribution of HII regions perpendicular to the plane (the z direction) is shown in Figure 6. It is clear that the dispersion in the z direction increases with increases in R. The filled circles are those objects with ℓ<180° while those with open circles are at ℓ>180°. There appears to be a seperation in z between these two groups for those objects at R > 15 kpc. The 'northern' (ℓ<180°) objects are primarily at positive z while the 'southern' (ℓ>180°) objects are mainly found at negative z. The HI in the outer Galaxy shows this same behaviour.

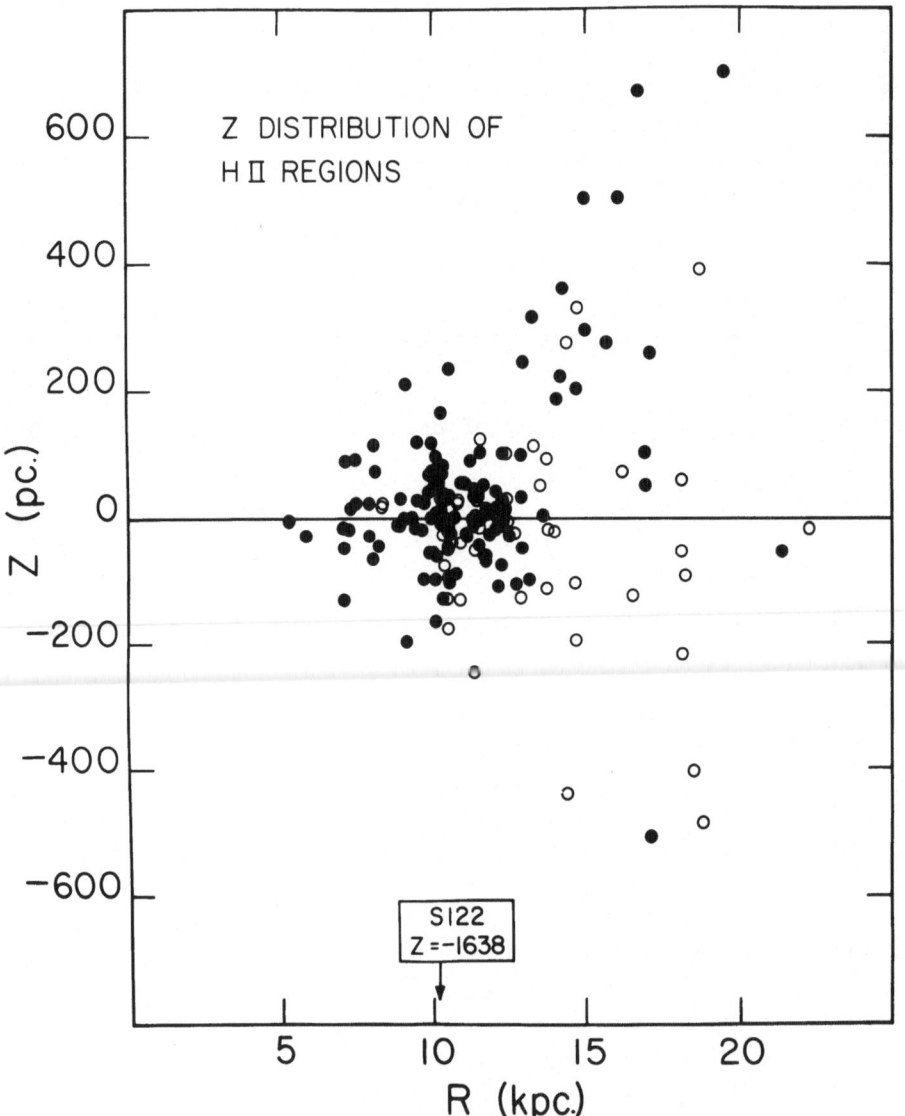

Figure 7. The z distribution of HII regions in the outer Galaxy

Table 1 shows the mean z and the z dispersion of the HII regions. One way to test the correlation between the z height of the HII regions and that of the HI is to calculate z using the HI to define the mean plane near the HII region (we use the HI maps of Kulkarni, Blitz, and Heiles 1982) instead of using the position of the Sun. The last line of Table 1 shows $\langle z \rangle$ and σ_z for the most distant HII regions when z is defined this way. Both statistics are significantly reduced when computed using the HI plane indicating that there is a correlation between the HI plane and the positions of the HII regions. The dispersion in the z positions of the optical HII regions clearly increases by more than a factor of three between R = 10 kpc and 15 kpc. Kulkarni, Blitz, and Heiles have shown that the HI increases in scale height by almost a factor of four in the same range in R.

TABLE 1

The z Distribution of HII Regions

R(kpc)	n	$\langle z \rangle$(pc)		σ_z(pc)
7.5 - 8.5	14	9.79 ±	16.0	60.0
8.5 - 9.5	10	-23.5	25.3	80.0
9.5 - 10.5	37	-0.2	14.9	90.7
10.5 - 11.5	29	-9.6	12.9	69.2
11.5 - 12.5	33	6.9	8.7	49.8
12.5 - 13.5	12	68.3	37.1	128.7
13.5 - 14.5	11	53.5	74.9	248.4
14.5 - 15.5	6	67.8	127.2	311.7
15.5 - 21.5	17	117.8	100.6	414.9
	Relative to Mean HI Plane			
15.5 - 21.5	14	30.2	84.7	316.8

Figure 7 shows the average diameter for those HII regions in this sample which have optical distances determined to them. Both the mean and median diameters for each 1 kpc wide (in R) bin are shown. There is a very strong suggestion that the diameters decrease with increasing R but because of selection effects involved in choosing this sample (these are only those HII regions seen optically) this can only be taken as a hint.

CONCLUSIONS

We have shown that HII regions only extend out to R ≅ 20 kpc in our Galaxy, following closely the distribution of the HI. Since HII regions are the signpost of massive star formation we can further say that massive star formation, and perhaps all recent star formation, occurs within the same limit.

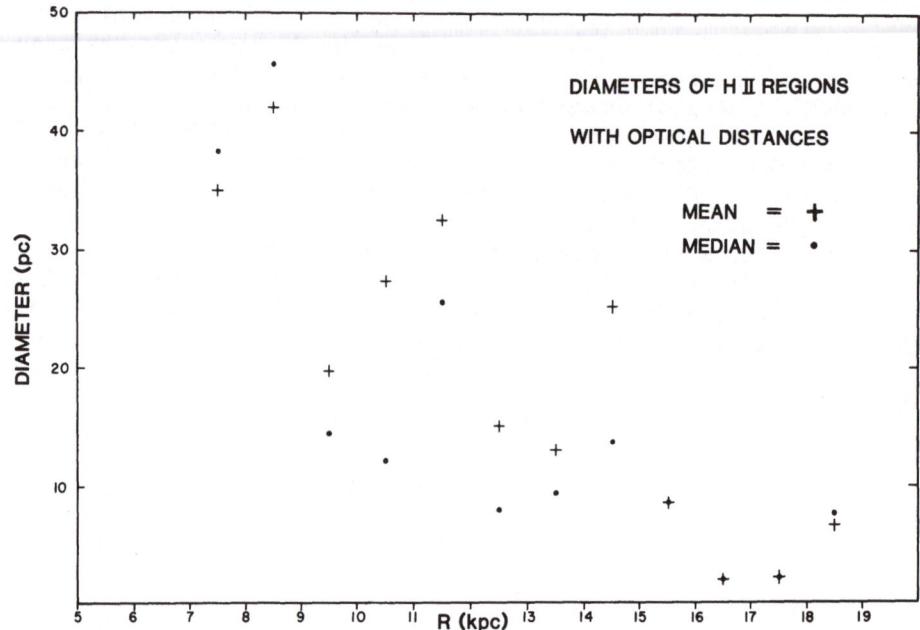

Figure 7. Average diameter of HII regions with optical distances.

REFERENCES

Arp H. 1964, Ap.J. 139, 1045
Blitz L., Fich M., and Stark A.A., 1980, IAU Symposium #87,
    "Interstellar Molecules", ed. B.H.Andrew (Reidel,Dordrecht), p213
Blitz L., Fich M., and Stark A.A., 1982, Ap.J. Suppl., 49, 183
Hodge P.W., 1969, Ap.J., 155, 417
Kulkarni S., Blitz L., and Heiles C., 1982, Ap.J.(Letters), in press

CO(J=2-1) OBSERVATIONS OF GALACTIC HII-REGIONS (x)

Jan Brand, Marco D.P. van der Bij, Harm J. Habing,
Cor P. de Vries
    Sterrewacht, Leiden (NL)
Thijs de Graauw, Frank P. Israel
    Astronomy Division, European Space Agency,Noordwijk (NL)
Herman van de Stadt
    Astronomical Institute, Utrecht (NL)

We have used three different telescopes and the ESTEC/Utrecht (sub-) mm heterodyne receiver and backend to detect CO(J=2-1) emission from molecular clouds associated with/close to galactic HII-regions. Table 1 gives specifics pertaining to the observations. System calibration was done by looking alternately at a hot and ambient temperature load. Atmospheric opacity was determined from skydips,which also gave us an efficiency factor which includes spillover and an "extended source beam efficiency". Our receiver uses room temperature Schottky barrier diode mixers and a backward wave oscillator as local oscillator. The backend consists of two 256-channel filterbanks with a resolution of respectively

TABLE 1

telescope	2.5 m du Pont Las Campanas	3.6 m ESO La Silla	1.4 m CAT,ESO La Silla
Altitude (m)	2400	2450	2450
Efficiency (%)	45	55	35
HPBW (arcmin)	2.3	1.9	5.5
Atmospheric opacity at zenith	0.29-0.71 (av: 0.51)	0.14-0.43 (av: 0.24)	0.01-0.78 (av: 0.18)
Double sideband system temperature (K)	2200	2200	1750
Observing date	April,May 1980	May 1980	June-October 1981

(x) Partly based on observations made at the European Southern Observatory,La Silla (Chile)

159

W. L. H. Shuter (ed.), Kinematics, Dynamics and Structure of the Milky Way, 159–163.

1 MHz and 250 kHz, corresponding to a velocity resolution of 1.3 kms^{-1} and 0.33 kms^{-1} respectively, at the observing frequency of 230 GHz.

With the ESO 1.4 m Coudé Auxilliary Telescope (CAT) at La Silla (Chile) we observed 16 points in molecular clouds associated with galactic HII-regions. The results are listed in table 2, together with some comparison data. Three of the sources are particularly interesting because they show signs of self absorption (RCW 36, 38 and 74C). The most likely candidate in this matter is RCW 38 where the velocities of the higher excitation molecules cluster around the velocity of the dip. In the case of RCW 36, these molecules have velocities close to that of the higher-velocity peak indicating that we may be seeing two separate clouds. RCW 74C only shows a very shallow dip.

At Las Campanas (Chile) we have mapped a 14x14 arcmin area around the radio continuum source G327.3-0.5. This radio source is very close to the optical HII-region RCW 97. With a spacing between the grid points of 2 arcmin, the map has been slightly undersampled. The spectra obtained with the higher resolution filterbank are shown in figure 1. Some interesting features are found from these observations:
(i) in many of the spectra from the lower resolution filterbank (not shown in the figure) a second, low-intensity, broad feature is visible, with a range in velocity from -70 kms^{-1} to -30 kms^{-1}. It is not yet clear what the nature of this feature is;
(ii) there is a shift in the velocity of the peak of the line of 4 kms,$^{-1}$

TABLE 2

| source designation | | observed position (1950.0) | | CO(J=2-1) data | | | comparison data | | | Notes |
G	Other	α h m s	δ ° ' "	T_A^*(K)	V_{LSR}(kms^{-1})	ΔV(kms^{-1})	(b) $V_{H109\alpha}$(kms^{-1})	(c) V_{CS}(kms^{-1})	ΔV_{CS}(kms^{-1})	
208.9-19.4	Orion-A	05:32:47.0	-05:24:21	35.6	7.4	4.9	-2.8	9.7	2.3	
265.1+1.5	RCW 36	08:57:38.0	-43:33:24	5.5,9.4	4.1,6.6	-	2.8	(6.9)	(3.6)	d
267.9-1.1	RCW 38	08:57:25.0	-47:19:18	9.1,5.2	-2.2,3.9	-	1.8	-1.3	10.1	e
301.3 0.7	RCW 57/NGC3603	11:09:47.0	-61:02:36	9.9	-22.4	10.1	-23.4	-22.2	9.2	
305.4+0.2	RCW 74C	13:09:21.0	-62:18:54	7.6,6.8	-41.3,-36.8	-	-39.1	-37.7	4.7	f
306.3+0.2	RCW 75	13:16:30.0	-62:15:00	6.4	-29.7	4.0	-	-	-	
322.2+0.6	RCW 92	15:14:50.0	-56:28:00	7.7	-56.9	5.8	-51.8	-55.1	3.8	a
327.3-0.5		15:49:13.0	-54:26:30	12.2	-48.4	8.4	-48.8	-46.8	5.7	g
332.8-0.6	RCW 106	16:16:25.0	-50:47:30	5.2	-57.6	8.2	-57.2	-57.2	2.9	
333.0-0.4		16:16:52.0	-50:33:00	9.5	-55.3	10.3	-53.8	(-55.8)	(14.7)	a
336.5-1.5	RCW 108	16:36:20.0	-48:45:36	12.7	-24.7	4.9	-24.9	-	-	
348.7-1.0	RCW 122	17:16:40.0	-38:54:06	16.6	-13.6	6.9	-12.8	-12.6	4.6	a
351.4+0.7	RCW 127E/NGC6334	17:17:35.0	-35:43:42	20.8	-7.2	11.0	-3.2	-5.0	5.0	
351.6-1.3		17:25:56.0	-36:37:54	13.7	-11.0	7.7	-12.2	-11.7	6.9	a
353.1+0.6	W 22/NGC6357	17:22:18.0	-34:19:54	7.8	-5.8	10.2	-4.1	-	-	
15.0-0.7	M 17(SW)	18:17:26.5	-16:14:54	15.8	17.4	7.8	18.4	20.0	4.5	

Notes to the table:
a: baselines have been removed by using preferentially only the outer parts of the filterbank. Due to the presence of
standing waves and/or other irregularities in the spectra, this may imply a T_A^*-value that is too low or too high (apart
from the general 15% to 20% uncertainty in the calibration).

b: recombination line velocity from Wilson et al. (1970)   (beamwidth 4.0)

c: CS data from Gardner and Whiteoak (1978)   (beamwidth 1.5)

d: double peaked; V_{LSR}(dip)=5.0 kms^{-1}; T_A^*(dip)=2.7 K

e: double peaked; V_{LSR}(dip)=1.8 kms^{-1}; T_A^*(dip)=0.6 K

f: double peaked?; V_{LSR}(dip)=-38.4 kms^{-1}; T_A^*(dip)=5.1 K

g: also mapped (14'x14' grid) with 2.5 m du Pont

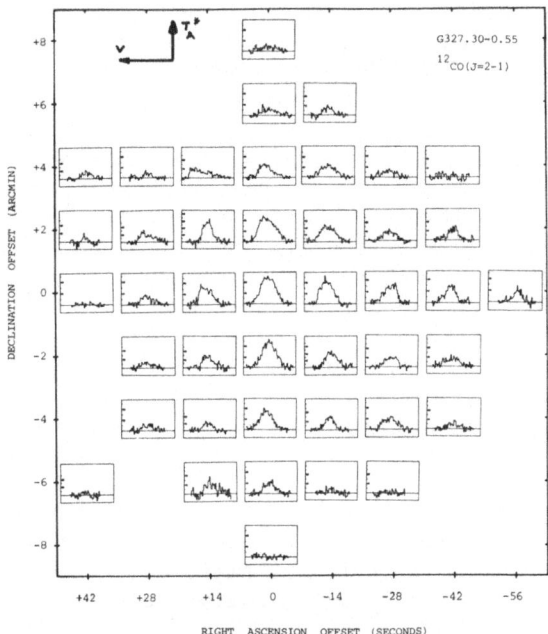

FIG.1 SPECTRA OBSERVED IN G327.30-0.55.
(THE SCALES OF THE SPECTRA ARE NOT IDENTICAL)
OFFSETS ARE WITH RESPECT TO RA=15^{h}49^{m}16^s, DEC=-54°28'24" (1950.0)

from SE to NW along a line which is at an angle of ~60° with respect to the galactic plane. However, since ^{12}CO is not a good density indicator it is not certain whether this can be interpreted as rotation of the bulk of the gas, until a similar gradient has been found in e.g. ^{13}CO or NH$_3$ lines;

(iii) in CO, the corrected antenna temperature T_A^* peaks over an elongated area. This area coincides with a secondary peak found at 1415 MHz continuum (Retallack, 1980);

(iv) the linewidths are largest (~9 kms^{-1}) at the position of the main radio peak indicating it is a region of relative kinematic turbulence; at this position the visual extinction is large, $A_V \gtrsim 20^m$ (Sherwood et al., 1980; Frogel and Persson, 1974; Persson et al., 1976) and signposts of star formation are present, such as OH-masers (Caswell and Robinson, 1974) and 10μm continuum emission (Frogel and Persson, 1974);

(v) at least six or seven of the profiles in figure 1 show a clear asymmetry. At five positions the profile has an enhanced blue side and at one or two positions it has an enhancement on the red side. On basis of the presently available material the dividing line between these two types cannot be drawn uniquely (due to the small number of asymmetric profiles).

The explanation for these features may be similar to that of Lichten (1982) for NGC 2071: a rotating cloud (disk?) tilted at a projected angle of ~60° with respect to the galactic plane, with gas outflow along the rotation axis. The axis runs from NE to SW and is inclined in such a way that the NE end is directed towards the observer (thus giving rise to a larger number of enhanced blue wings than red ones). A rotation period of 5×10^6 years is derived for a uniformly rotating disk, the rotation axis of which is tilted by $\sim 75°$ with respect to the plane of the sky.

With the ESO 3.6 m at La Silla (Chile) we have made a strip scan in R.A. of W 48. This region encompasses three radio sources; in one of these denser knots can be distinguished. The sources are situated some 100 pc below the galactic plane at a distance of 3 kpc and have no associated optical HII-region. Combining other observations (Wilson et al., 1974; Goss et al.,1971),W 48 appears to be situated,together with the SNR W 44,in one molecular cloud of dimensions 120x60 pc at a velocity of $\sim +43$ kms^{-1}. In figure 2 we show a R.A. - velocity plot. The narrow feature at +14.5 kms^{-1} can be identified with what is sometimes called "Lindblad's local expanding ring" (Cohen et al.,1980). It is not clear why this feature splits at $\Delta RA \approx -48^S$. The broad feature between +30 kms^{-1} and +55 kms^{-1} consists of very complex line profiles and appears to be a blend of two or more emission lines. Part of the structure (the depressions at +40.7 kms^{-1} and +46.0 kms^{-1}) seems to be due to absorption,possibly by colder foreground material. The emission structure extending between $\Delta RA \approx -54^S$ and $\Delta RA \approx 0^S$ could be a separate cloud.

In this paper,preliminary results have been presented. A more detailed discussion of these observations is in preparation by the same authors.

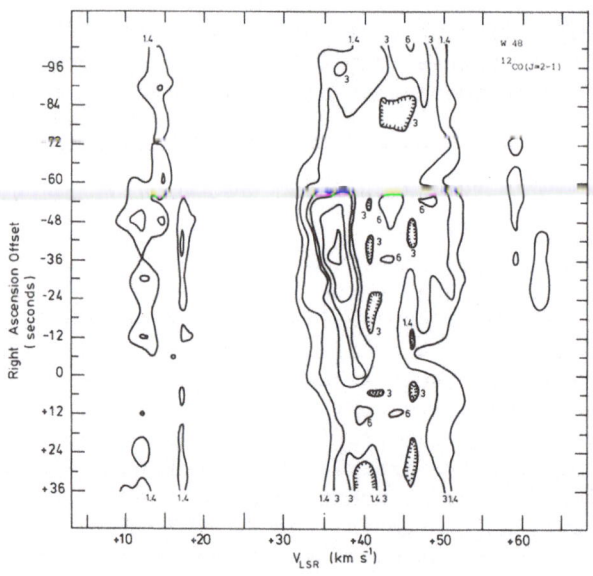

FIG.2 STRIPSCAN THROUGH W 48 AT DEC=+01°08'00"; PLOT OF RA-OFFSET VERSUS VELOCITY.
OFFSETS WITH RESPECT TO RA=18^{h}59^{m}15^s (1950.0)

References

Caswell, J.L., Robinson, B.J.: 1974,Austr. J. of Phys. 27,597
Cohen, R.S., Cong, H., Dame, T.M., Thaddeus, P.: 1980,Astroph. J. Lett.
     239,L53
Frogel, J.A., Persson, S.E.: 1974,Astroph. J. 192,351
Gardner, F.F., Whiteoak, J.B.: 1978,MNRAS 183,711
Goss, W.M., Caswell, J.L., Robinson, B.J.: 1971,Astron. Astroph. 14,481
Lichten, S.M.: 1982,Astroph. J. 253,593
Persson, S.E., Frogel, J.A., Aaronson, M.: 1976,Astroph. J. 208,753
Retallack, D.S.: 1980,Ph. D. Thesis,University of Sydney
Sherwood, W.A., Arnold, E.M., Schultz, G.V.: 1980,IAU Symposium 87
     "Interstellar Molecules" (ed. B.H. Andrew),pp.133
Wilson, T.L., Mezger, P.G., Gardner, F.F., Milne, D.K.: 1970,Astron.
     Astroph. 6,364
Wilson, W.J., Schwartz, P.R., Epstein, E.E., Johnson, W.A., Etcheverry,
     R.D., Mori, T.T., Berry, G.G., Dyson, H.B.: 1974,Astroph. J. 191,357

DISTRIBUTION OF CO IN THE SOUTHERN MILKY WAY

W.H. McCutcheon
Department of Physics, University of British Columbia
Vancouver, B.C.

B.J. Robinson, J.B. Whiteoak, and R.N. Manchester
Division of Radiophysics, CSIRO
Sydney, Australia

ABSTRACT
    A survey of J = 1-0 ^{12}CO emission in the ranges $294° \leqslant \ell \leqslant 358°$, $-0°.075 \leqslant b \leqslant 0°.075$ has been made with observations spaced at 3' arc intervals. Results pertaining to the large-scale properties of the Galaxy are presented.

INTRODUCTION

    CO emission is a tracer of cold, dense, molecular clouds and the results of large-scale surveys of this emission have yielded a picture of our galaxy substantially different from that based solely on 21 cm HI observations. There have been several CO surveys of the northern hemisphere (Burton and Gordon 1978, Solomon et al. 1979, Cohen et al. 1980, Liszt et al. 1981, Sanders 1981).

    We report here results of the first CO survey of the southern hemisphere. The goals of this survey are a comparison of the results with those from northern surveys and a determination of large-scale structure in the Galaxy.

OBSERVATIONS

    Observations were made with the 4-m telescope at the CSIRO Division of Radiophysics (Gardner et al. 1978) in 1980 and 1981. The beamwidth at 115 GHz is 2'.8 arc and the receiver single-sideband system temperature was 1000 K. The telescope was operated in a beam-switched mode with a switch rate of 0.5 Hz using a pneumatically-driven plane mirror. Data were obtained using a 512-channel acousto-optical spectrograph with a velocity coverage of 244 km s^{-1} and an effective resolution of 0.5 km s^{-1}.

    An observation at a given longitude consisted of integrating for 1 min at each of nine positions on a square grid of spacing 3' arc.

W. L. H. Shuter (ed.), Kinematics, Dynamics and Structure of the Milky Way, 165–170.
Copyright © 1983 by D. Reidel Publishing Company.

These observations were interspersed with observations at reference
positions, which were 10° above or below the plane, to remove instrumental baselines from the spectra. Recorded data were smoothed to an
effective velocity resolution of 1.5 km s^{-1} and angular resolution of
9' arc, values similar to those of Cohen et al. (1980). Further details
of equipment and observing procedure, and sample spectra, are given in
Robinson et al. (1982).

LARGE-SCALE DISTRIBUTION OF CO

The emissivity as a function of galactocentric radius is shown in
Figure 1. Spectral points were assigned to a galactocentric distance
and hence to a particular 500 pc wide annulus, using the Schmidt (1965)
galactic rotation curve in the analytical form given by Burton (1971).
This distribution is based on 30% of the total data, that is data from a
range of 1°.8 every 6° in longitude, in the interval 294° ≤ ℓ ≤ 350°.

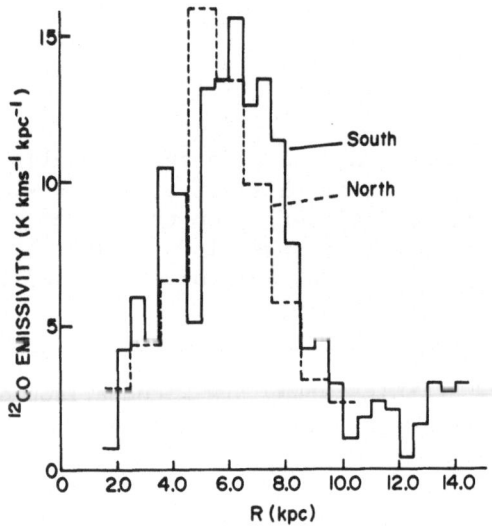

Figure 1. Distribution of CO emissivity with galactocentric radius.
The dashed line is the northern CO distribution (Sanders and Scoville,
private communication).

Figure 1 shows that the CO is distributed predominantly inside the
solar circle in the form of a molecular ring. Also shown in Figure 1
is the northern distribution. The two distributions are similar,
although the main peak for the south, at a radius of 6.5 kpc, is shifted
outward compared to that in the north by about 0.7 kpc. This outward
shift is also seen in the distribution of southern OH maser sources

(J.L. Caswell, private communication) which coincides with the main
peak for the south in Figure 1. This displacement could reflect a real
difference in the radial location of the gas in the two hemispheres.
However, the distributions are model-dependent; for example, differences
in the positions of the peaks could also result from differences in the
northern and southern rotation curves.

Figure 2 shows a longitude velocity diagram of the CO emission.
This diagram can be compared to that published by Cohen et al. (1980),
since the two data sets are both well sampled and have approximately the

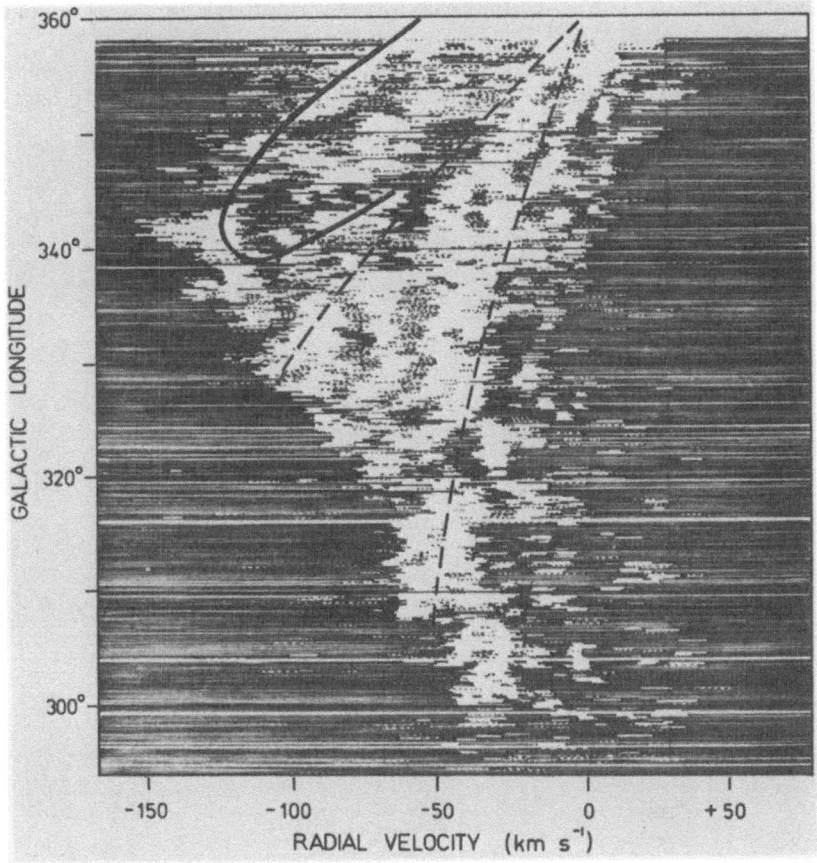

Figure 2. Distribution of CO emission along the galactic equator from
ℓ = 294° to ℓ = 358°. White represents significant CO emission. Broken
horizontal lines show regions of low emission. The upper and lower
dashed curves are the loci of points at galactocentric radii of 5 and
8 kpc respectively. The solid curve is the locus of points of part of
an expanding, rotating, 3.5 kpc ring.

same spatial and velocity resolutions. There is a strong similarity
between the two ℓ-v distributions, indicating an overall symmetry between
the northern and southern hemispheres.

The dashed lines enclose the emission that has contributed to the
main peak in Figure 1 between radii 5.0 and 8.0 kpc. The solid curve in
Figure 2 represents part of a 3.5 kpc expanding arm, or resonance ring,
detected in the 21 cm HI line (see Shane 1972) and traditionally called
the 3 kpc arm. HI emission occurs along the top part of this curve from
a velocity of -53 km s^{-1} to the terminal velocity. There is also strong
CO emission along this arm. The continuation of the arm past the
terminal velocity is shown only for the area where there is CO emission.
This is the region of the ℓ-v diagram which contributes to the secondary
peak in Figure 1, near 4 kpc.

The terminal velocities of all CO profiles are plotted in Figure 3.

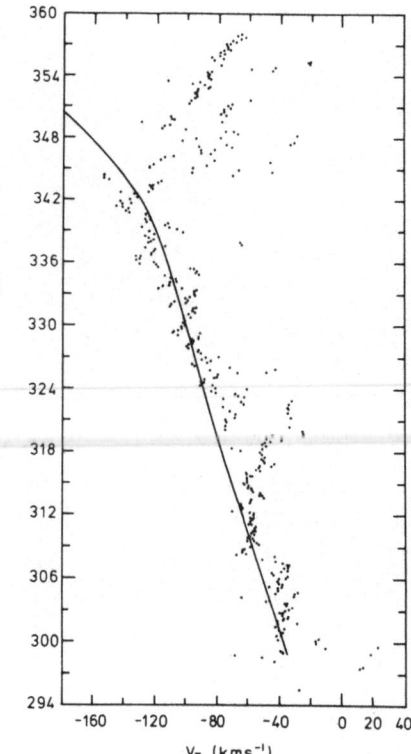

Figure 3. Variation with longitude of the terminal velocities measured
from the CO profiles. The solid line is the rotation curve from Burton
and Gordon (1978) based on northern CO data.

The terminal velocity was taken to be the velocity of the half maximum point of the extreme velocity feature. The rotation curve allows a comparison with the northern CO terminal velocities (Burton and Gordon 1978). There is good agreement between the absolute maximum velocities in the two hemispheres over most of the longitude range. The biggest difference involves data in the range $312° < \ell < 327°$, where the southern absolute velocities are significantly lower than those in the corresponding northern range. This hole is easily seen around $\ell = 320°$ in Figure 2. This longitude range is similar to that in which the southern HI terminal velocities are lower than those in the north (Kerr 1969).

The maximum negative velocities on a V_T-sin ℓ plot, excluding the range $-0.58 \geqslant \sin \ell \geqslant -0.73$, are fitted well by the straight line

$$V_T = -181(1+\sin \ell) - 19.6$$

with a possible error in the slope of about ± 10 km s^{-1}. For a flat rotation curve the solar velocity $\theta_0 = 181$ km s^{-1}. This is similar to the recent value determined by Shuter (1981) from an analysis of northern HI data.

SUMMARY

The southern galactic plane, in the ranges $294° \leqslant \ell \leqslant 358°$, $-0°.075 \leqslant b \leqslant 0°.075$, has been surveyed in the J = 1-0 line of ^{12}CO with a sampling interval of 3' arc.

The radial distribution is that of a molecular ring with the area under the emissivity curve comparable to that in the north.

A comparison of our ℓ-v diagram with that of Cohen et al. (1980), indicates that there is a large degree of symmetry between the two hemispheres. The southern portion of the HI 3.5 kpc expanding arm has a clearly defined CO counterpart in Figure 2.

The absolute maximum velocities in the two hemispheres are in good agreement except for the range $312° < \ell < 327°$, where the southern absolute velocities are significantly lower than those in the corresponding northern range.

ACKNOWLEDGEMENTS

We thank C.J. Rennie and R.E. Otrupcek, who helped considerably with the observations and data reduction, and the many members of the Division of Radiophysics who contributed to the success of this project.

W.H.M. acknowledges research grants from the Natural Sciences and Engineering Research Council of Canada.

REFERENCES

Burton, W.B.: 1971, Astron. Astrophys. 10, pp. 76-96.
Burton, W.B. and Gordon, M.A.: 1978, Astron. Astrophys. 63, pp. 7-27.
Cohen, R.S., Cong, H., Dame, T.M. and Thaddeus, P.: 1980, Astrophys. J.
    239, pp. L53-L56.
Gardner, F.F., Batchelor, R.A., McCulloch, M.G., Simons, L.W.J. and
    Whiteoak, J.B.: 1978, Proc. Astron. Soc. Aust. 3, pp. 264-266.
Kerr, F.J.: 1969, Annu. Rev. Astron. Astrophys. 7, pp. 39-66.
Liszt, H.S., Delin, X. and Burton, W.B.: 1981, Astrophys. J. 249,
    pp. 532-549.
Robinson, B.J., McCutcheon, W.H. and Whiteoak, J.B.: 1982, Int.J. Infrared
    Millimeter Waves. 3, pp. 63-76.
Sanders, D.B.: 1981, Paper presented at NRAO Workshop on Extragalactic
    Molecules, Green Bank.
Schmidt, M.: 1965, in A. Blaauw and M. Schmidt (eds.), *Stars and Stellar
    Systems*, Vol. 5, pp. 513-530, Univ. Chicago Press.
Shane, W.W.: 1972, Astron. Astrophys. 16, pp. 118-148.
Shuter, W.L.H.: 1981, Mon. Not. R. Astron. Soc., 194, pp. 851-861.
Solomon, P.M., Sanders, D.B. and Scoville, N.Z.: 1979, in W.B. Burton
    (ed.), *The Large-Scale Characteristics of the Galaxy* (Proc. IAU
    Symp. No. 84), pp. 35-52, Reidel, Dordrecht.

GALACTIC EMISSION OF OH AT 1720 MHz AS A TRACER OF SPIRAL ARMS

B. E. Turner
National Radio Astronomy Observatory

ABSTRACT

New data show, more clearly than ever, that the anomalous emission of galactic OH at 1720 MHz is unequalled as a tracer of large-scale structures which appear to be spiral arms. The properties of the 1720 MHz OH clouds associated with these arms are those of giant molecular clouds.

INTRODUCTION

In an earlier work (Turner 1982), based on the Green Bank survey of galactic OH (Turner 1979), OH was found to exhibit widespread, spatially extended, anomalous emission at 1720 MHz throughout the inner Galactic plane. The distribution of this emission in (ℓ,v) showed marked concentrations corresponding to the expected Sagittarius, Scutum, and local spiral arms, and an absence in the interarm regions. These "spiral arm" configurations appeared much more distinctly in the 1720 MHz OH line than in other tracers such as CO, H_2CO, and HII regions. Collisional excitation of OH at kinetic temperatures above ~15K but below ~40K seem responsible.

The present work is based on a larger data set, obtained in 1981 October with the NRAO 140-foot telescope. In the interval 337° < ℓ < 50° the new data, together with the Green Bank OH survey, comprises 91 fully mapped clouds radiating anomalously at 1720 MHz and 58 other positions which show the requisite OH excitation but which have not yet been fully mapped. In the interval 50° < ℓ < 270° there are an additional 35 regions of anomalous 1720 MHz OH, 9 of which have been fully mapped.

THE (ℓ,v) DISTRIBUTION OF 1720 MHZ OH

The (ℓ,v) distribution of anomalous 1720 MHz OH emission in the inner galactic plane is shown in Figure 1. The distribution is strikingly more patterned than is seen in other tracers such as CO

W. L. H. Shuter (ed.), Kinematics, Dynamics and Structure of the Milky Way, 171–178.

(Cohen et al. 1980), H$_2$CO (Downes et al. 1980) or HII regions (Lockman 1982). Similarly, neither OH absorption nor type I OH masers show patterns suggestive of spiral arms. The local arm, and Sag and Scutum arm loops show up particularly clearly in Figure 1, with well-defined interarm gaps between. A few points (at 10° < ℓ < 20°, v > 80 km/s) are associated with the inner or Norma arm, while the 3 kpc arm is revealed at negative ℓ and v. Two aspects of the 1720 MHz OH distribution depart somewhat from the traditional picture. First, the Sag arm loop appears to stop at v = 60 km/s rather than at the usual v = 70 km/s. Second, the apparent Scutum arm loop appears to stop at ℓ = 39° rather than at the usual ℓ = 30°.

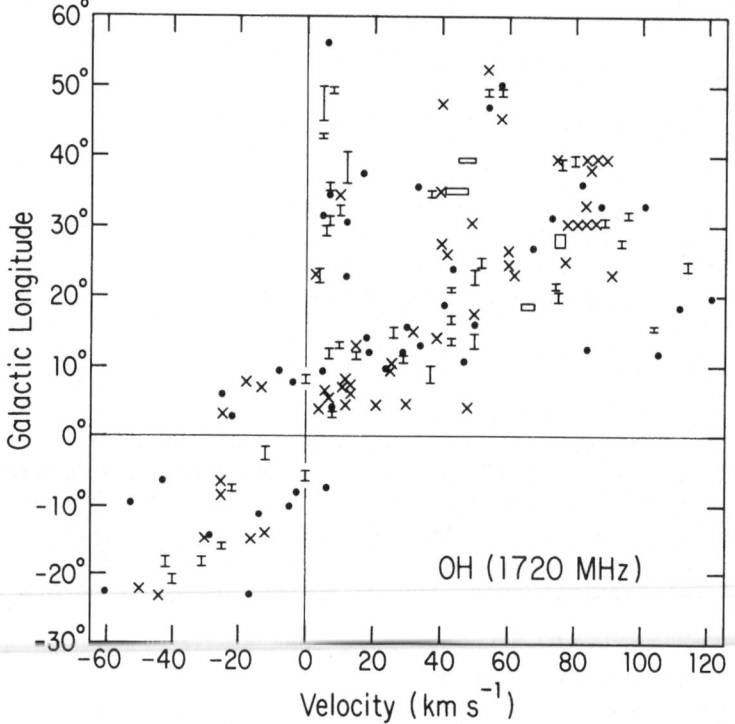

Fig. 1. The (ℓ,v) distribution of anomalous 1720 MHz OH emission in the inner galactic plane. Vertical bars indicate extension of the emission in longitude at a given velocity (typical linewidths are 2-3 km/s), while enclosed regions indicate extension in both ℓ and v. Crosses indicate clouds whose mapped extent is less than 2° in ℓ; dots represent emission not yet mapped.

A SPIRAL-ARM PATTERN FOR 1720 MHz OH

We adopt the rotation curve of Burton (1974) which involves purely circular rotation. The distance ambiguity for R < R$_0$ can be resolved more reliably than for other tracers by making use of the fact that the excitation temperature of the two main lines is essentially unaffected by the excitation that enhances the 1720 MHz transition, and therefore

remains at 6 - 12 K, typical of ordinary "thermal" OH (Turner 1982).
Thus the presence or absence of main line absorption at the velocity of
the 1720 MHz emission indicates whether the OH lies in front of or
behind HII regions which usually can be seen projected within the
extent of the OH clouds. Because the excitation temperature of the OH
main lines is higher than that of H_2CO ($\approx$ 1.5 K), this type of OH
avoids the problem common to H_2CO that absorption may occur against the
galactic background as well as against HII regions. Thus we have
revised a few of the distances for HII regions as given by Downes
et al. (1980) in the sense that some near-kinematic distances have been
revised to far-kinematic distances. (It is emphasized that there is no
physical association between OH and HII regions; the velocities are
entirely different, and there is no correlation between the two tracers
in (ℓ,v) space).

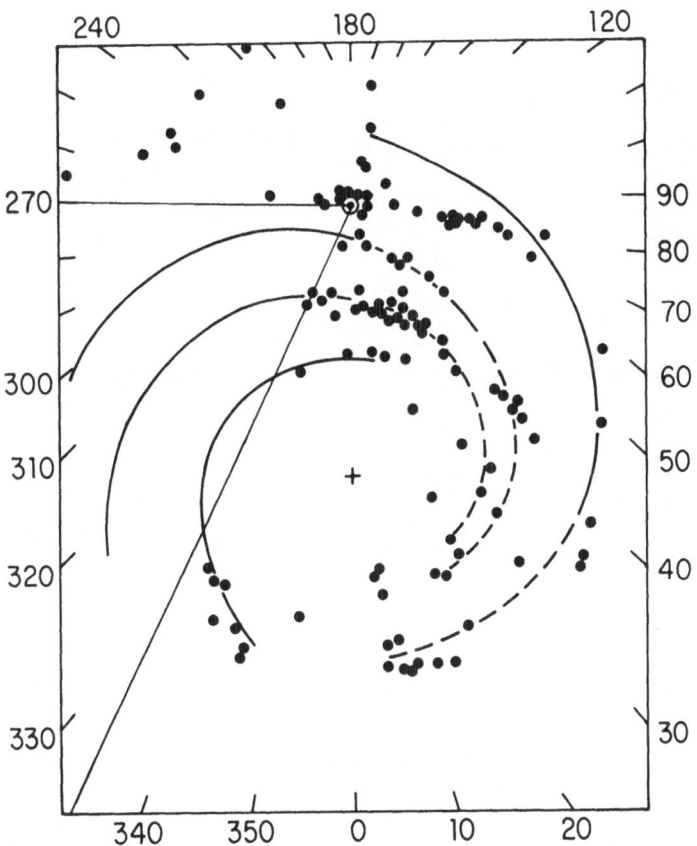

Fig. 2.   The spatial distribution of anomalous 1720 MHz OH clouds in
          the galaxy. Survey limits are ℓ = 337° and ℓ = 270°. Solid
          lines are the 4-arm pattern of Georgelin and Georgelin (1976);
          dashed lines are the extensions of Downes et al. (1980).

     Figure 2 gives the resulting spatial distribution of 1720 MHz OH
in the inner galactic plane. Solid continuous lines are those given by

Georgelin and Georgelin (1976) which represent their empirical fit of a
4-arm system to the distribution of HII regions studied optically.
Dashed lines represent extensions drawn by Downes et al. (1980) to
include additional HII regions studied at radio wavelengths. The dots
represent the spatially extended 1720 MHz OH clouds; it is seen that
they follow the 4-arm pattern well. The separation of the Sag and
Scutum arms at $337° < \ell < 30°$ is much more distinct than seen in HII
regions (Downes et al. 1980), as is the "local" arm or Orion arm spur,
which in 1720 MHz OH extends clearly to the Perseus arm in one
direction, and well past the sum in the other. On the far sides of the

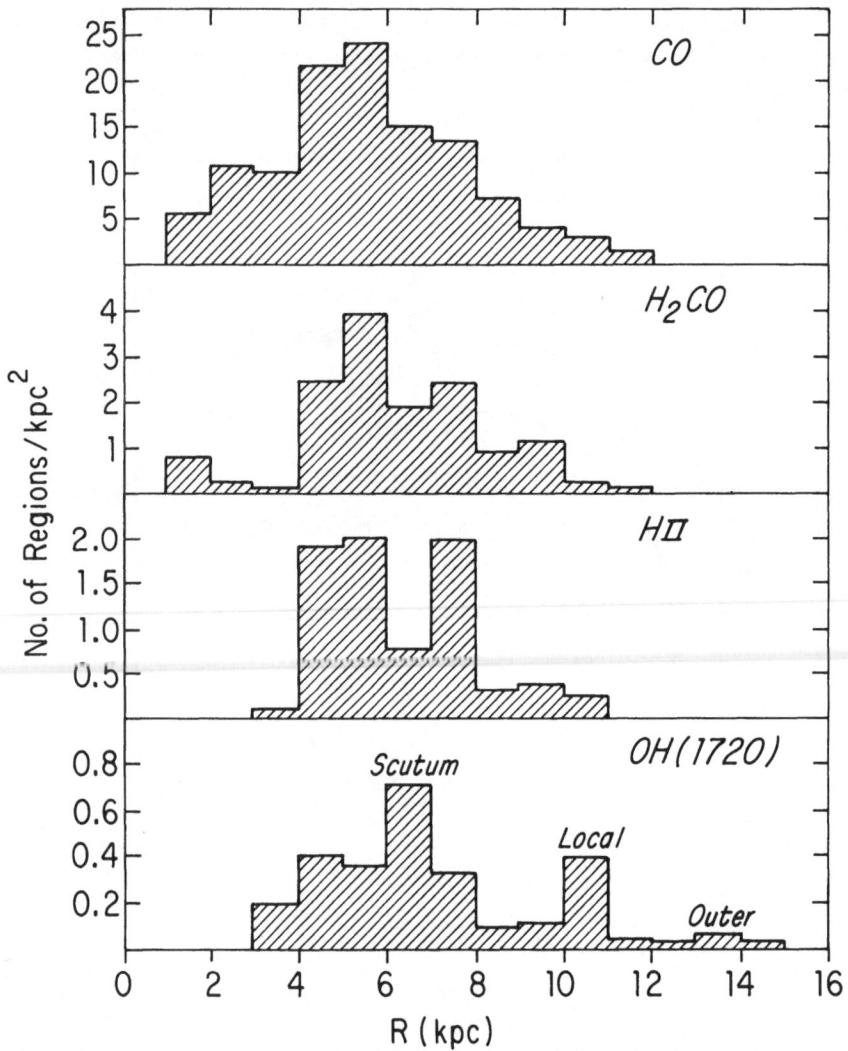

Fig. 3. The radial distribution of four tracers of galactic structure.
The CO, H_2CO, and HII region distributions are from Downes
et al. (1980). Vertical axis units for CO are integrated
intensity units K km s^{-1} kpc^{-2}.

inner and Perseus arms 1720 MHz OH also establishes several points
which do not show up in HII regions. The "anomalous" clump of points
centered at $\ell \approx 39°$, $v \approx 80$ km/s appears to lie on the Sag arm as
extended by Downes et al. (1980). While 1720 MHz OH traces the farther
side of the Perseus arm quite well, we note a dearth of points on this
arm in the region $100° < \ell < 140°$, which is well represented in CO
(Cohen et al. 1980). Stretching a point, we might claim a "gap" in the
1720 MHz OH distribution in the Sag arm similar to that suggested for
CO by Thaddeus (this volume).

Figure 3 gives the radial distributions of several tracers in the
inner galaxy. CO, H_2CO, and HII region distributions are from Downes
et al. (1980). All distributions are derived from circular rotation
curves. There are significant apparent differences between the
tracers, but these may not all be real. Many HII regions lie on the
Scutum and inner arms in the region $270° < \ell < 337°$ where there is no
1720 MHz OH data. This may account for the larger number of HII
regions in the interval $4 < R < 6$ kpc. The "excess" number of HII
regions at $7 < R < 8$ kpc seems to be the result of a very liberal
estimate by Downes et al. of the number of distinct sources in the W51
region. The importance of the Scutum, local, and outer (Perseus) arms
does seem to be much greater in 1720 MHz OH than in the other tracers.

While certain arm-like patterns in 1720 MHz OH appear to stand out
well (Sag, Scutum, and local arms near the sun), one cannot claim any
corroboration for the 4-arm pattern sketched by Georgelin and Georgelin
(1976). It seems clear that patterns other than those of Georgelin and
Georgelin could be drawn to represent equally well the 1720 MHz OH
distribution. And the use of circular rotation curves presents
well-known difficulties. Nevertheless, 1720 MHz OH appears to be a
superior tracer of some sort of regular structures, whose nature may
best be determined through approaches such as used by Bash and
colleagues (this volume), given that the 1720 MHz OH clouds indeed
appear to be Giant Molecular Clouds (GMCs).

THE NATURE OF THE 1720 MHz OH CLOUDS

Cloud sizes in the inner galaxy cannot be well defined for CO,
owing to severe blending in both ℓ and v, or for H_2CO, owing to
dependence on background continuum. In addition, there are problems in
removing the kinematic distance ambiguity for both of these tracers.
By contrast, we can directly map the angular sizes of the 1720 MHz OH
clouds in the inner galaxy, since they are well separated in velocity
one from the other. And, as mentioned above, the distance ambiguity is
more reliably removed by use of the main-line OH absorption.

Figure 4 gives the distribution of sizes for 88 of the 91 fully
mapped 1720 MHz OH clouds. The distribution is similar to that deduced
by Solomon and Sanders (1980) for GMCs as observed in CO. The
"typical" size of 30 - 40 pc is similar to the average size of 40 pc
determined by Solomon and Sanders. Our distribution is not, however,

determined for sizes <30 pc because we fail to sample all of the
smaller clouds, which suffer beam dilution unless closer than ~5 kpc,
in which case they occupy a large range in galactic latitude, not
sampled in this survey. Clouds indicated as "local" in Figure 4 are
those with $|v_{lsr}| <$ 10 km/s, which have been arbitrarily assigned a
distance of 1 kpc.

     For sizes >30 pc, the distribution of size (S) can be fit by a
power law $N(S) \propto S^{-n}$ with $n \approx$ 4/3. This corresponds to a mass
distribution law of $\phi(m) \propto m^{-2/9}$ if we assume that mass scales with
size cubed. Cloud collision models of GMC formation predict exponents
of −1.5 or less (Kwan 1979).

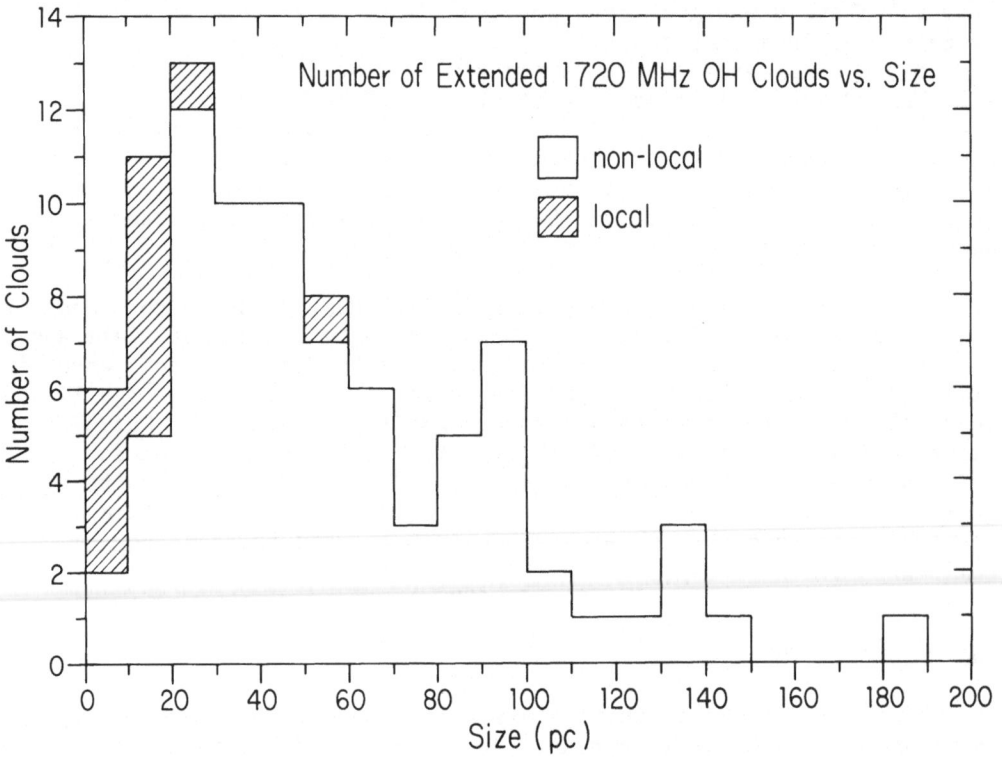

Fig. 4. The distribution of sizes for the anomalous 1720 MHz OH
         Clouds.

     A <u>general</u> excitation mechanism, such as is needed to explain the
widespread nature of the anomalous 1720 MHz emission, involves
collisions of OH with H_2 (or H) at low temperatures (15 < $T_k \lesssim$ 40 K)
and low densities (200 $\lesssim$ n $\lesssim$ 600 cm^{-3}). Pumping of the 1720 MHz
transition will not occur outside these ranges (Guibert <u>et al</u>. 1978;
Turner 1982). Thus the central question is whether the anomalous
1720 MHz OH clouds are monitoring a specific range of T_k, or a specific
range of n. Both ranges given above seem to characterize the overall

envelopes of GMCs. We suspect that T_k is the operational parameter, for the following reasons:

(1) There is no correlation of 1720 MHz OH clouds with radio HII regions, or with other regions of star formation. Thus there is no correlation with regions of higher density.

(2) The arm/interarm contrast ratio for the brightness of the 1720 MHz OH emission is > 8, as revealed by unsuccessful searches for 1720 MHz emission in the interarm gap regions. This ratio is greater than that believed to apply to CO in other galaxies, and apparent in the (ℓ,v) distribution of CO in our galaxy. If 1720 MHz excitation is according to theory, there should be a threshold at $T_k \simeq 15$ K, below which no excitation occurs.

(3) Rather little 1720 MHz OH emission is seen in the anticenter region, consistent with the evidence from CO that clouds are on average colder at larger galactocentric radii. There is also some evidence (Cohen et al. 1980) that the hotter CO clouds in the inner galaxy show (ℓ,v) patterns suggestive of spiral structure.

If higher temperature is indeed the physical parameter that confines the 1720 MHz OH clouds to spiral arms, what agent heats the spiral arms in the necessary global fashion? Mechanical energy from HII regions or snr's is ruled out because of the lack of correlation of these objects with 1720 MHz OH. Ultraviolet radiation from early-type stars is unlikely because of the large uv optical depths within arms. IR heating from star forming regions is too short-range. The most likely agent is heating by the spiral density wave shock, which also can serve to form the GMCs (Elmegreen 1979).

On this picture, we cannot rule out the existence of colder GMCs between spiral arms, which simply get warmed as they pass through the sdw shock. Tidal disruptions, if not star formation, would limit the sizes of GMC from growing too large with successive passages through the shock, probably in a way consistent with the results of Figure 4 (Stark and Blitz 1978).

CONCLUSIONS

(1) Anomalous 1720 MHz OH emission appears to provide the most reliable, if not the first, molecular tracer of "spiral-arm" structure in the inner part of our galaxy.

(2) The OH clouds thus related to the "spiral-arm" structures have the properties of GMCs.

REFERENCES

Burton, W. B. 1974, in Galactic and Extragalactic Radio Astronomy,
    ed. G. L. Verschuur and K. I. Kellermann, Springer-Verlag.
Cohen, R. S., Cong, H., Dame, T. M., and Thaddeus, P. 1980, Ap. J.
    (Letters) 239, L53.
Downes, D., Wilson, T. L., Bieging, J., and Wink, J. 1980, Astron.
    Astrophys. Suppl. 49, 379.
Elmegreen, B. 1979, Ap. J. 231, 372.
Georgelin, Y. M., and Georgelin, Y. P. 1976, Astron. Astrophys. 49, 57.
Guibert, J., Elitzur, M., and Rieu, N-Q 1978, Astron. Astrophys.
    66, 395.
Kwan, J. 1979, Ap. J. 229, 567.
Lockman, F. J. 1982, in preparation.
Solomon, P. M., and Sanders, D. B. 1980, in Giant Molecular Clouds in
    the Galaxy, ed. P. M. Solomon and M. G. Edmunds, Pergamon Press.
Stark, A. A., and Blitz, L. 1978, Ap. J. (Letters) 225, L15.
Turner, B. E. 1979, Astron. Astrophys. Suppl. 37, 1.
Turner, B. E. 1982, Ap. J. (Letters), 255, L33.

GALACTIC HII REGIONS, ANOMALOUS 1720 MHz OH CLOUDS AND SPIRAL
STRUCTURE IN THE GALAXY

Frank Bash and Gabriella Turek
Department of Astronomy, University of Texas
Austin, Texas

ABSTRACT

We have compared velocity-longitude diagrams of bright, galactic
HII regions and 1720 MHz OH clouds against diagrams predicted by the
ballistic particle model. Independent of our model, both velocity
longitude diagrams suggest the presence of spiral arms. In addition,
our ballistic particle model produces velocity-longitude diagrams
which resemble both observed ones. Our model assumes a two-arm
spiral pattern and that the objects in question are seen less than
50 million years after their launching, at the post-shock velocity,
from the TASS wave.

INTRODUCTION

The Ballistic Particle Model for the Galaxy was proposed by
Bash and Peters (1976) and refined by Bash (1979). This model suggests
that giant molecular clouds are born in the TASS wave, leave it at the
post-shock velocity, orbit ballistically in the Galaxy and live for
$\sim$ 40 million years. We assume that stars are born in the clouds about
30 million years after the cloud's birth and that the stars continue
in a ballistic orbit which can be computed for times well less than
the relaxation time. Since proposing our model, we have published a
set of papers, in the Ap. J., giving various, different pieces of
evidence that our model is correct. It assumes the Yuan (1979) spiral
pattern for our galaxy, with two spiral arms. The ballistic, giant
clouds and stars have non-circular velocity components of $\sim$ 20 km s^{-1}
and for ages since birth in the TASS wave of a few tens of millions
of years, they lie in a clear, spiral pattern.

Figure 1 shows a velocity longitude diagram of HII regions
observed in the radio recombination line H110α or H109α by Downes,
et al. (1980), Wilson, et al. (1970), or compiled by Lockman (1979).

W. L. H. Shuter (ed.), Kinematics, Dynamics and Structure of the Milky Way, 179–182.

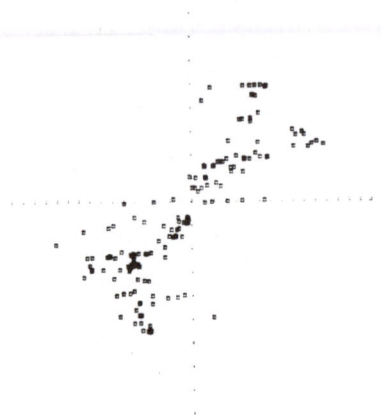

Fig. 1 - Velocity-longitude diagram
for HII regions observed in the recombination lines H109α and H110α.
The center of the diagram is at
velocity = 0 and longitude = 0 and
the velocity tick marks are separated
by 10 km s^{-1} and the longitude ones
by 10°.

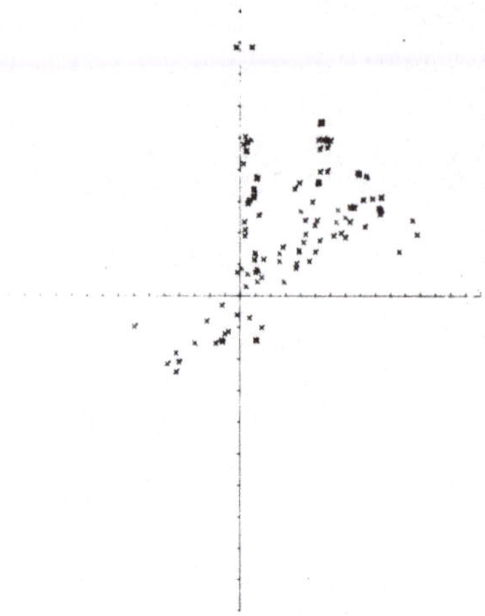

Fig. 2 - Velocity-longitude
diagram for OH clouds with
anomalous 1720 MHz emission.
The southern side cuts-off
due to the survey's limit and
the scale is exactly the same
as Fig. 1.

We desire to show only the most luminous HII regions, which trace
spiral arms, but since we don't know the distances to these HII
regions, we plot those whose continuum flux density is greater than
5 Jy. There must be some contamination from nearby, low luminosity
HII regions. Especially on the northern side, there seem to be two
clear arm features with an interarm hole, centered at $\ell \sim 40°$,
$V \sim 80$ km s^{-1}, in between.

Figure 2 shows a velocity-longitude diagram of the OH molecular
clouds with anomalous 1720 MHz emission and found by Turner (1982a,b).
As Turner point-out there seems to be evidence for the same arm
features as seen in the HII regions; however, the **hold**, referred to
above, seems to be smaller. Turner (1982a) argues that the objects
which fill the interarm hole in the OH diagram may be connected with
interarm HII regions and thus possibly may be neglected.

Figure 3 shows the ballistic particle model velocity-longitude
diagram on the same scale as Figures 1 and 2. Each model ballistic
particle's position is plotted each million years along its orbit.
The "herringbone pattern" is composed of individual particle tracks
and the pattern can be seen due to a limited number of particles in
the model. The age span in Fig. 3 is 10 million to 25 million years
after the particles were born in the TASS wave. At least on the
northern side, the model fits the HII regions in Fig. 1, quite well.

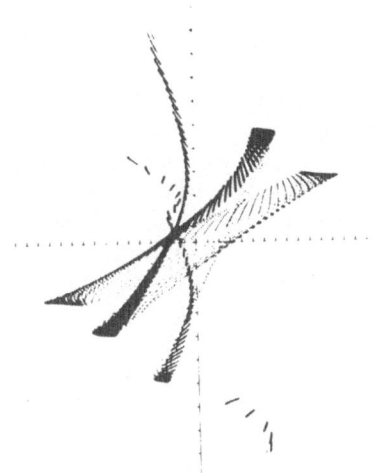

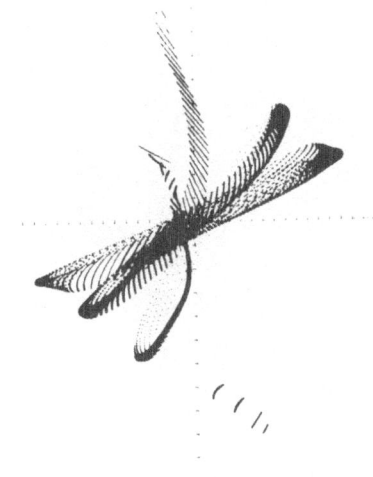

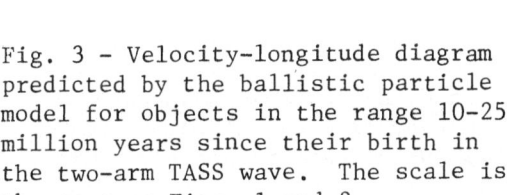

Fig. 3 - Velocity-longitude diagram predicted by the ballistic particle model for objects in the range 10-25 million years since their birth in the two-arm TASS wave. The scale is the same as Figs. 1 and 2.

Fig. 4 - The same as Fig. 3 but for ballistic particles in the age range 26-51 million years.

The model does not fit the OH clouds in Fig. 2 as well especially near $v \sim 80$ km s^{-1}, $\ell \sim 40°$.

Fig. 4 is based on exactly the same model of the Galaxy as Fig. 3 but the age range is 26 million to 51 million years after birth in the TASS wave. Fig. 4 perhaps fits the OH clouds better than Fig. 3 does but the Sagittarius arm is at too high a velocity to fit the HII regions in Fig. 1. Here we attempt to make the point that our ballistic particle model covers about the same amount of the velocity-longitude diagram as the HII region and OH cloud data do, that the model's age range which fits best compares well with our prediction of about 30 million years for the HII region phase and that the shape of the arms in the model's velocity-longitude diagram resemble the data. The model here is the simplest possible one where one age range, everywhere in the Galaxy, is compared to the data.

Leisawitz and Bash (1982) present a more elaborate model in which we compute the energy input into the cloud following Bash, Hausman and Papaloizou (1981). We assume that if the energy input is insufficient to replace the turbulent energy losses, the cloud collapses and forms stars (and HII regions). Fig. 5 shows our predicted velocity-longitude diagram for HII regions and for that model. It fits the HII region data in Fig. 1 extremely well.

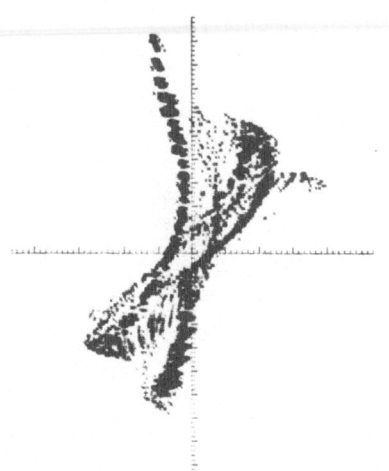

Fig. 5 – Velocity-longitude diagram for HII regions predicted by a more elaborate version of our ballistic particle model. The consequence of this model is that the HII region phase corresponds to different ages at different parts of the Galaxy. The scale is the same as Figs. 1-4 but with finer tick marks, 5 km s^{-1} in velocity and 2° in longitude.

We believe that these HII regions and OH clouds with anomalous 1720 MHz emission are consistent with a general model of the Galaxy in which these objects lie in a two-arm spiral pattern and also with a specific model, the ballistic particle model, and its predicted non-circular velocities and, Yuan (1969) spiral pattern.

We would like to thank Barry Turner for sending us his data on the OH clouds ahead of publication. This work was supported by NSF grant AST-8116403.

REFERENCES

Bash, F. N., and Peters, W. L. 1976, Astrophys. J., 205, 786.
Bash, F. N., 1979, Astrophys. J., 233, 524.
Bash, F. N., Hausman, M., and Papaloizou, J. 1981, Astrophys. J., 245, 92.
Downes, D., Wilson, T. L., Bieging, J., and Wink, J. 1980, Astron. Astrophys. Suppl. Ser., 40, 379.
Leisawitz, D., and Bash, F. N. 1982, Astrophys. J., to appear in Part 1, August, 1st Issue.
Lockman, F. J. 1979, Astrophys. J., 232, 761.
Turner, B. M. 1982a, Astrophys. J., 255, L33.
_____ 1982b, private communication.
Wilson, T. L., Mezger, P. G., Gardner, F. F., Milne, D. K. 1970, Astron. Astrophys., 6, 384.
Yuan, C. 1969, Astrophys. J., 158, 871.

THE LARGE SCALE DUST DISTRIBUTION IN THE INNER GALAXY

M. G. Hauser, E. Dwek, D. Gezari, R. Silverberg, T. Kelsall, and M. Stier
Laboratory for Extraterrestrial Physics
NASA/Goddard Space Flight Center, Greenbelt, MD   20771

and

L. Cheung
Department of Physics and Astronomy
University of Maryland
College Park, MD   20742

ABSTRACT

Initial results are presented from a new large-scale survey of the first quadrant of the galactic plane at wavelengths of 160, 260, and 300 μm. The submillimeter wavelength emission, interpreted as thermal radiation by dust grains, reveals an optically thin disk of angular width $\sim$ 0.9° (FWHM) with a mean dust temperature of 23 K and significant variation of the dust mass column density. Comparison of the dust column density with the gas column density inferred from CO survey data shows a striking spatial correlation. The mean luminosity per hydrogen atom is found to be 2.5×10^{-30} W/H, implying a radiant energy density in the vicinity of the dust an order of magnitude larger than in the solar neighborhood. The data favor dust in molecular clouds as the dominant submillimeter radiation source.

I. INTRODUCTION

In recent years there has been a substantial observational effort to map the far infrared and submillimeter emission of our Galaxy (see review by Okuda (1981) and references therein). These long wavelength surveys are of particular interest in the context of this workshop because they reveal the spatial and temperature distribution of interstellar dust throughout the Galaxy, and because they reveal the magnitude and distribution of a major component of galactic luminosity. We are in the process of surveying the galactic plane at submillimeter wavelengths using a newly developed 1.2-meter diameter balloon-borne telescope (Silverberg et al. 1979). Important characteristics of this survey include the use of three spectral bands,

W. L. H. Shuter (ed.), Kinematics, Dynamics and Structure of the Milky Way, 183–195.

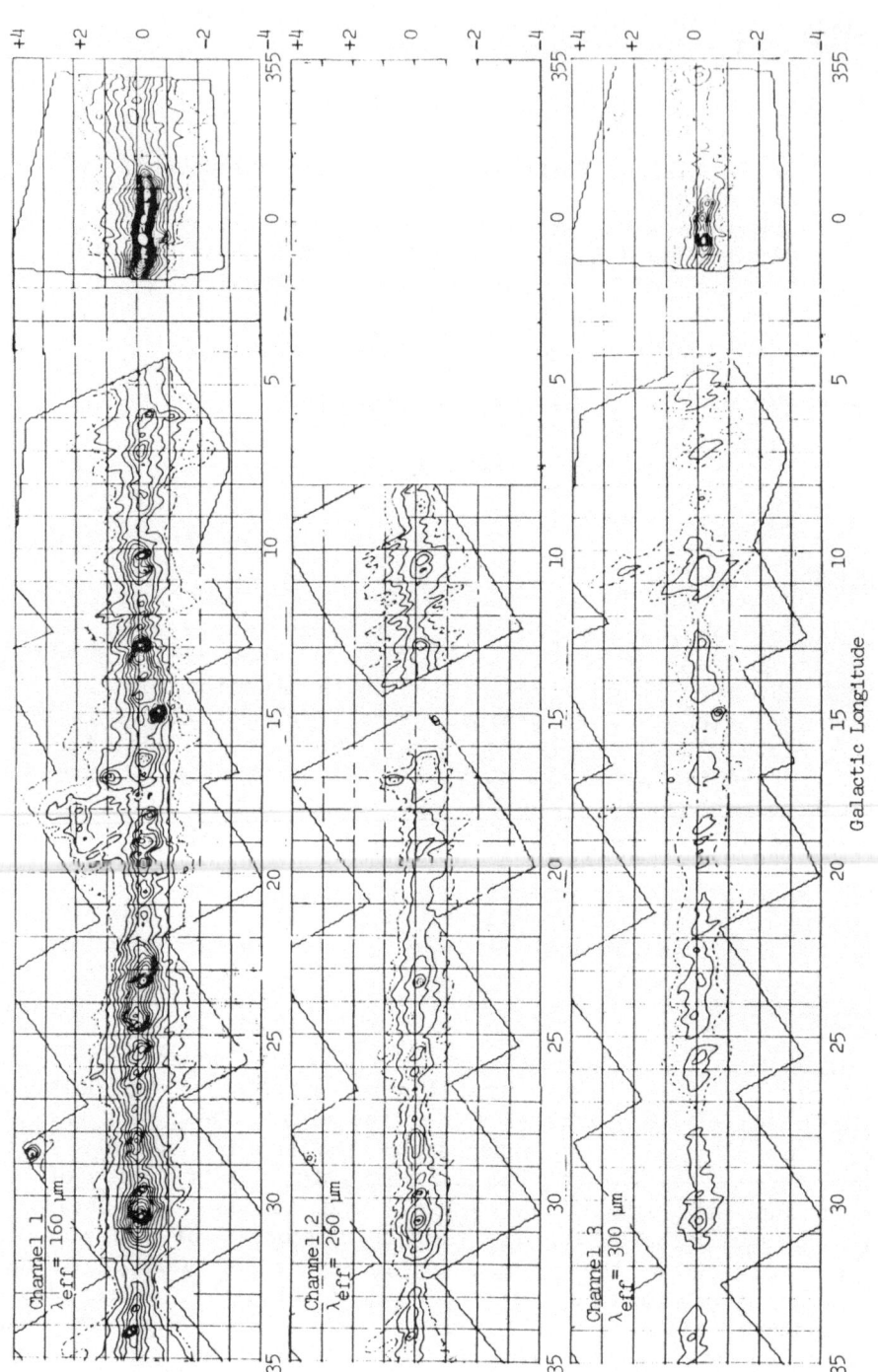

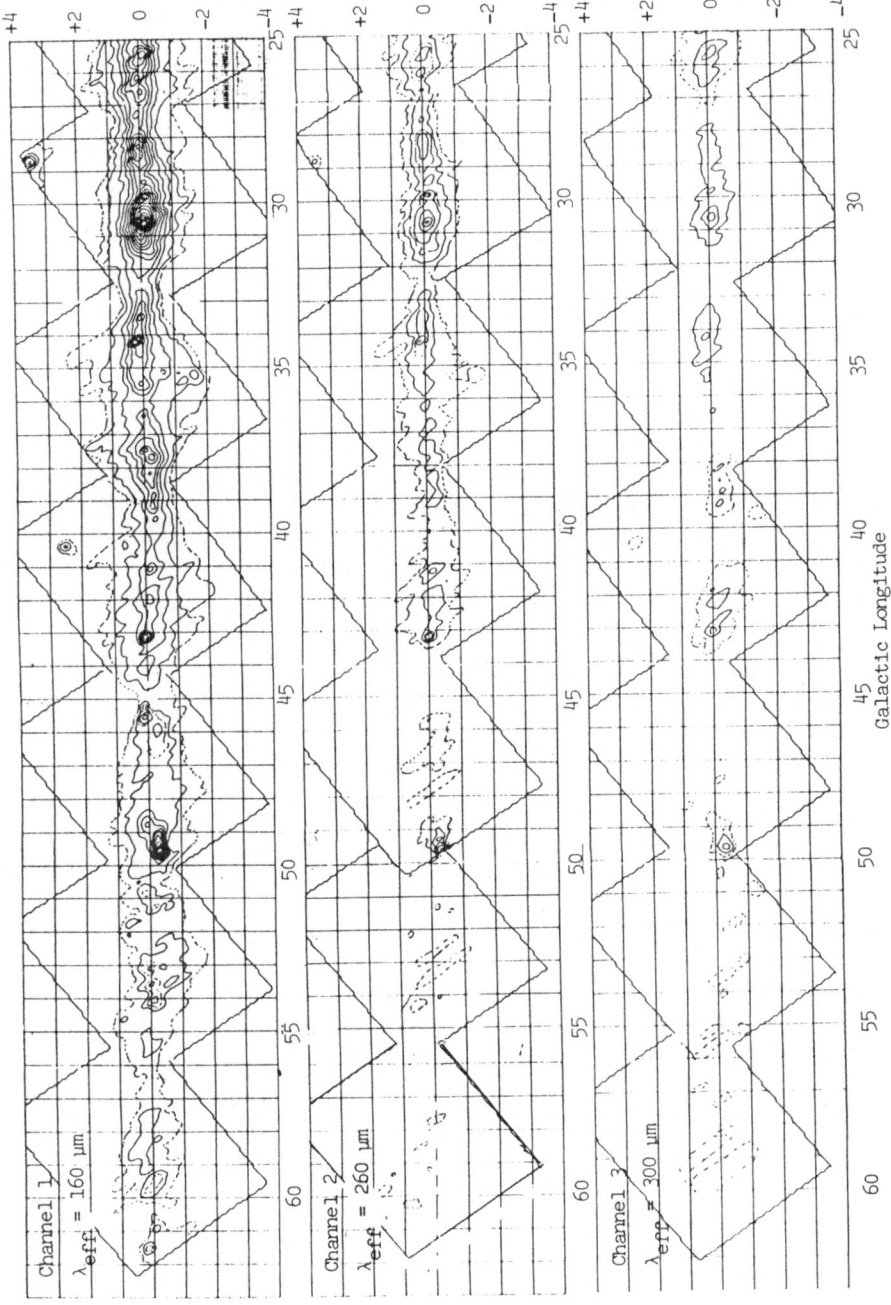

Figure 1. Flux density contour maps for the three survey bands. The solid contours begin at 20 data numbers (DN), and increase in steps of 20 DN. The dotted contour is at 10 DN. A DN corresponds to approximately 230, 330, and 270 Jy in a 10 arcmin x 10 arcmin beam in bands 1, 2, and 3 respectively.

wavelength coverage extending far into the submillimeter range, and high sensitivity. Results are presented here for the submillimeter wave surface brightness in the range $0 \leq \ell \leq 60°$, $|b| \leq 3°$, and for implied consequences such as the scale height of the dust distribution and longitudinal distributions of dust opacity, dust mass column density, dust temperature, and total infrared surface brightness. Comparison of these results with CO data then yield longitudinal distributions of the dust-to-gas mass ratio and the infrared luminosity per H atom. Brief comments are offered on the currently conflicting interpretations of the origin of the diffuse submillimeter emission.

II. OBSERVATIONAL RESULTS

The data presented here were primarily obtained during an 11.5 hour flight on November 15, 1979, during which scans of Jupiter provided the instrument calibration. The data in the vicinity of the Galactic Center, obtained in a second flight on August 15, 1980, have been discussed elsewhere (Stier et al. 1982) and will not be considered in detail here. Observational parameters of the instrument during the 1979 flight are summarized in Table 1.

Table 1. Submillimeter Survey Instrument Parameters – Flight 1

Telescope Configuration:	Cassegrain	
Primary Diameter:	1.2 m (48 inch)	
Primary Focal Ratio:	f/0.38	
Effective Focal Ratio:	f/4	
Field-of-view:	10 x 10 arcmin	

Chopper:	Waveform	square
	Frequency	10 Hz
	Throw	20 arcmin

Photometer Characteristics:	Band 1	Band 2	Band 3
50% Cut-on λ (μm)	106	238	270
Effective λ (μm)*	162	260	302
Effective bandwidth (Hz)*	1.3×10^{12}	6.4×10^{11}	5.3×10^{11}
NEFD (Jy $Hz^{-1/2}$)	350	270	190
NER (W $cm^{-2} sr^{-1} Hz^{-1/2}$)	5.9×10^{-11}	2.3×10^{-11}	1.6×10^{-11}

*for T = 20 K, n = 2.

Figure 1 shows maps of the submillimeter surface brightness in the three survey bands. These maps reveal continuous diffuse submillimeter emission from the inner Galaxy, as well as emission associated with prominent discrete sources. Most of the discrete sources can be identified with known HII regions. Systematic characterization of the discrete sources is in process (Kelsall et al. 1982) so that their contribution can be separated from the diffuse emission. As the integral distributions discussed here are not influenced in a significant way by the discrete sources, no attempt to separate discrete source contributions has been included in what follows.

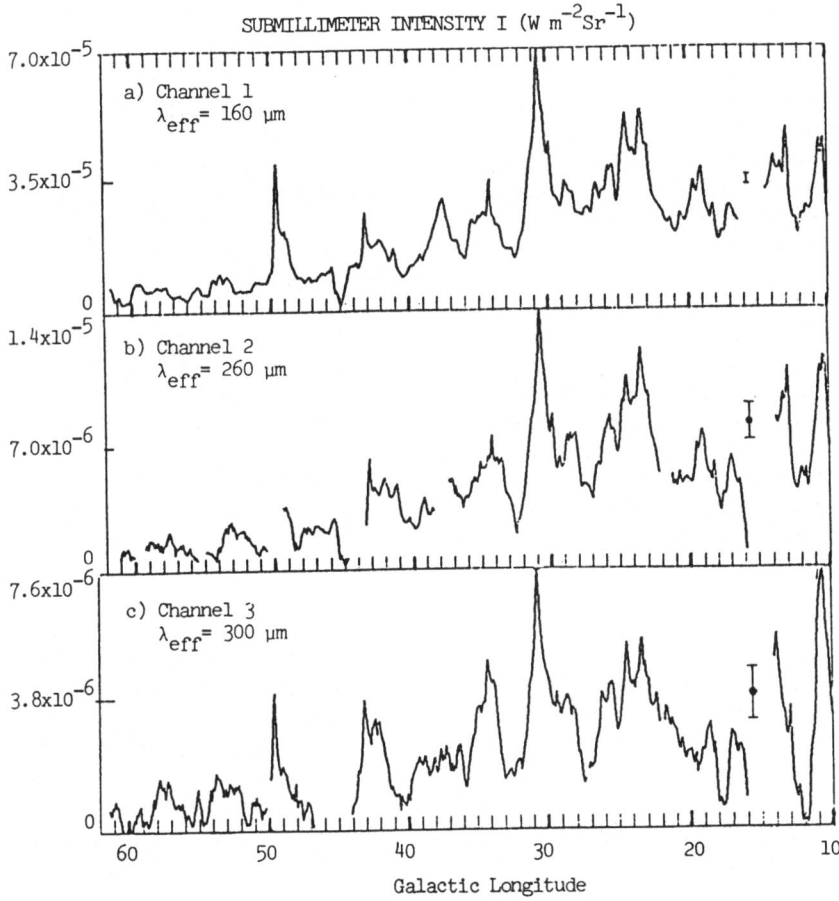

Figure 2. Galactic longitude profiles of submillimeter wavelength surface brightness I in the three survey bands averaged over the galactic latitude range $-1/2° \leq b \leq 1/2°$. A typical random error estimate for the $\ell = 10 - 43°$ range is indicated.

III. DERIVED RESULTS

The radiation detected in this survey is generally interpreted in terms of thermal emission from dust grains, the only controversy residing in the nature and location of those grains and the sources which illuminate them. At these wavelengths, the grain spectral emission (absorption) efficiency Q_ν can be characterized as having a power law dependence on frequency, $Q_\nu \propto \nu^n$. The survey was designed with three spectral bands so that both the spectral index n and the dust temperature T could be determined directly from the data. A technical problem in filter fabrication caused the effective wavelengths of bands 2 and 3 to be more similar than intended, making determination of n difficult. We adopt the typical value n = 2 where needed in what follows, and note the sensitivity of our results to this assumption.

The latitudinal distribution of the diffuse surface brightness in band 2 integrated over 2° wide strips in longitude was examined at several longitudes selected to avoid prominent discrete sources. The brightness variation in this channel largely reflects column density variations (since T, determined from the band 1/band 2 ratio, does not vary significantly with latitude). The width of this distribution of $\Delta b \sim 0.9^\circ$ (FWHM) implies, for a typical source distance of 7 kpc, a scale height of ~ 80 pc. No significant variation of this width with longitude can be discerned in the present data.

Surface brightness longitudinal profiles for the three survey bands are presented in Figure 2. In order to smooth random fluctuations and diminish the impact of discrete sources, these data, and consequently all derived quantities, have been averaged over the latitude range $-1/2^\circ \leq b \leq +1/2^\circ$. The dramatic contrast evident in these profiles can be better understood if these data are used to derive profiles of dust temperature (T_d), optical depth (τ_d), dust mass column density (M_d), and total far infrared surface brightness (I_{tot}). These are presented in Figures 3 and 4.

The temperature distribution shown in Figure 3(a) is derived from the ratio of intensities in bands 1 and 2. The temperature is seen to show no dramatic trend over this longitude range. The large fluctuations for $\ell \geq 45^\circ$ are largely observational artifacts (the signal-to-noise ratio in channel 2 is fairly low here, and the derived temperature varies roughly as the square of I_1/I_2). The mean temperature over the range $10^\circ \leq \ell \leq 43^\circ$ is 23 K, with an rms variation of 1.4 K. Use of an emissivity spectral index n = 1.5, as suggested by the grain model of Mezger, Mathis, and Panagia (1982), would raise the implied mean temperature to 27 K.

The derived temperature profile has been combined with the observed intensity profile of band 2 to yield the optical depth profile at an effective wavelength of 260 μm shown in Figure 3(b). This profile clearly shows that the large contrast observed in the

submillimeter emission profiles is primarily attributable to column
density variations. Note also that, averaged over the 10-arcmin beam
of this experiment, the Galaxy is optically thin at these wavelengths:
the mean opacity from $10^{\circ} \leq \ell \leq 43^{\circ}$ is $\tau_d = 6.2 \times 10^{-3}$, and the maximum
value is about 10^{-2}. This illustrates clearly one reason why sub-
millimeter radiation provides a useful probe of galactic content.

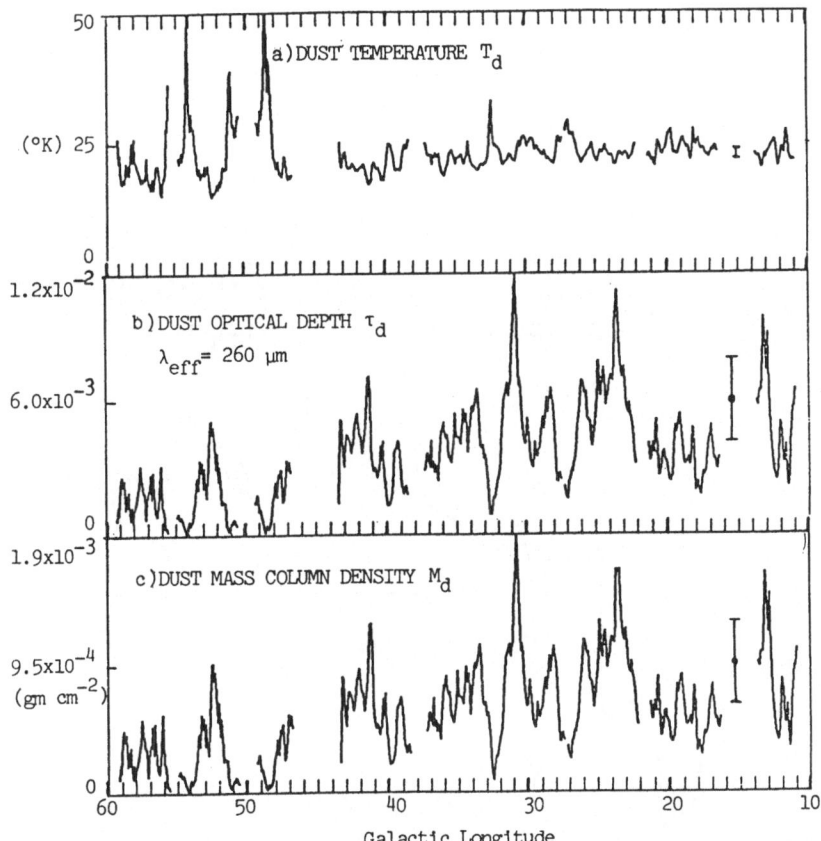

Figure 3. Submillimeter source properties derived from the surface
brightness profiles. A typical estimated random error is shown in
each case for the $\ell = 10^{\circ}$ to 43° range. The large spikes for $\ell > 43^{\circ}$
result from very low brightness values at some longitudes in band 2
and correspondingly have large uncertainties. Mean values presented
are weighted (by the random errors) means for the $\ell = 10 - 43^{\circ}$ range
in each case. (a) Dust temperature profile inferred from I_1 / I_2
assuming n = 2. The mean temperature is 23 K. (b) Optical depth profile at $\lambda_{eff} = 260$ μm. The mean optical depth is 6×10^{-3}. (c) Dust
mass column density profile. The mean value is 1×10^{-3} g cm^{-2}.

The optical depth profile can be converted to a dust column density profile using the dust mass absorption coefficient K_ν, where $\tau_\nu = K_\nu M_d$. Unfortunately, this quantity is poorly known. We have used the value $K_\nu = 4.6(\nu/10^{12} \text{ Hz})^2$, which, as will be shown below, gives a mean gas-to-dust mass ratio in the inner Galaxy of ~ 100, consistent with cosmic abundance arguments. This value of K_ν is within a factor of 2 of that found by Aanestad (1975) for silicate grains with ice mantles if the grain material density is $\sim 1 \text{ g cm}^{-3}$. The derived dust mass column density profile is shown in Figure 3(c), revealing large variations in the integrated mass along different lines of sight within the inner Galaxy.

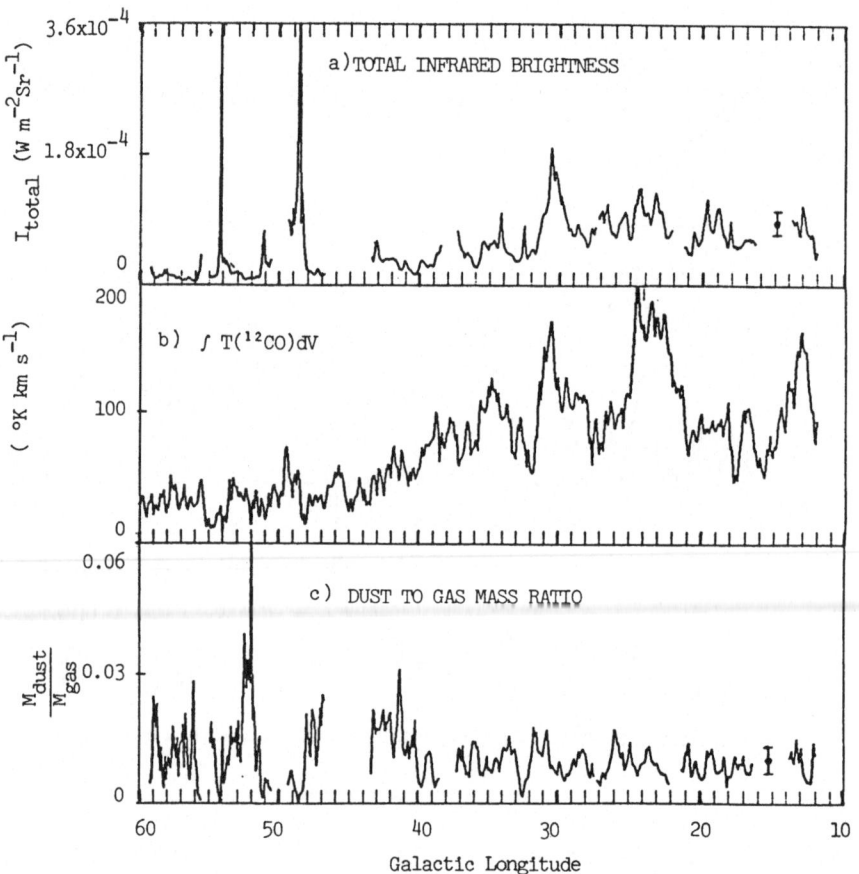

Figure 4. (a) Longitude profile of total submillimeter surface brightness ($n = 2$). The mean brightness (see remarks for Fig. 3) is $9 \times 10^{-5} \text{ W m}^{-2}\text{sr}^{-1}$. (b) Profile of the ^{12}CO intensity of Cohen et al. (1980) integrated over velocity and averaged over the latitude range $-1/2° \leq b \leq +1/2°$. The mean value in the $\ell = 10 - 43°$ range is $1.2 \times 10^2 \text{ K km sec}^{-1}$, implying a hydrogen column density of $4 \times 10^{22} \text{ cm}^{-2}$. (c) Profile of the dust-to-gas mass ratio inferred from the profiles in Figures 3c and 4b. The mean value is 1×10^{-2} (see remarks for Figure 3).

The spectrally integrated far infrared surface brightness I_{tot} (Figure 4(a)) is derived from the brightness in band 1 and the derived temperature assuming the source has a spectrum given by $\nu^2 B_{\nu}(T)$, where $B_{\nu}(T)$ is the Planck function (the inferred total brightness is increased by only 20% if n = 1.5 is assumed). For the source temperatures typical here, band 1 measures approximately 40% of the total flux emitted by the cold dust. Measurements of the diffuse galactic plane brightness at shorter infrared wavelengths (Price 1981; Hayakawa et al. 1979) indicate that the submillimeter emission is the dominant part of the total luminosity of the plane.

IV. COMPARISON WITH GAS MEASUREMENTS

To interpret the model-independent consequences of these data further, we have compared our results with the molecular gas distribution found in the CO survey of Cohen et al. (1980). This CO survey was made with a 7.5 arcmin beam, very similar to ours, and is almost fully sampled in the region of interest, making direct comparison particularly meaningful. Figure 4(b) shows the longitude profile of the velocity-integrated ^{12}CO intensity averaged over the latitude range $-1/2° < b < +1/2°$.

Taking the integrated CO intensity to be proportional to the total mass column density of gas in the line of sight with the calibration

$$N_H \ (cm^{-2}) = 3.2 \times 10^{20} \int T(^{12}CO) \ dv \ (K \ km \ sec^{-1})$$

(Solomon, Scoville, and Sanders 1979; Savage and Mathis 1979; and Dickman 1978) yields the dust-to-gas mass ratio profile in Figure 4(c). Examination of this profile and comparison of the dust mass and CO profiles directly (Figures 3c and 4b) show that a strong correlation exists between the dust and gas distributions. This is very suggestive of a close physical association between the molecular gas and the dust responsible for the submillimeter emission, though general concentration of interstellar material into spiral arms will tend to produce such a correlation whether or not the components are physically associated. However, Figure 4(c) shows that there is very little dispersion in the dust-to-gas mass ratio at the resolution of these surveys (rms variations are only 25% of the mean). There is also no apparent trend in the mean ratio over the longitude range covered here, and a very similar value is found in the 1°x2° region around the galactic center (Stier et al. 1982). Finally, the mean gas-to-dust mass ratio over the longitude range $\ell = 10°$ to $43°$ is $\sim$ 100, consistent with expectations from cosmic abundance arguments; i.e., virtually all possible interstellar dust is required to produce the observed emission. As discussed above, this is not a firm conclusion because of considerable uncertainty in the dust mass absorption coefficient. Overall, the evidence here seems to favor a picture of close physical association of molecular gas and the dust responsible for the submillimeter emission.

Further insight can be gained by considering the luminosity distribution. Figure 5 shows the infrared luminosity per hydrogen atom, obtained from the ratio of total submillimeter surface brightness (Figure 4a) to the gas column density (Figure 4b). The mean value of 2.5×10^{-30} W/H-atom in the range $\ell = 10^{\circ}$ to 43° is very close to the value of 2.1×10^{-30} W/H-atom found by Ryter and Puget (1977) in a study of nine molecular clouds, again suggesting close association of the dust responsible for the submillimeter emission with molecular clouds. As noted by these authors, this L_{TR}/H value is more than an order of magnitude larger than that which would arise from dust exposed to a stellar radiation density comparable to that in the solar vicinity (0.5 eV cm^{-3}); i.e., the dust radiating in the submillimeter is located in a radiation field typically more than an order of magnitude more intense than that in the solar neighborhood. It is also evident from Figure 5 that L_{TR}/H shows significant variation, implying variation in the mean energy density along the line of sight of as much as a factor of 2 for $\ell \lesssim 32^{\circ}$, with a suggestion of a modest decrease at large longitudes.

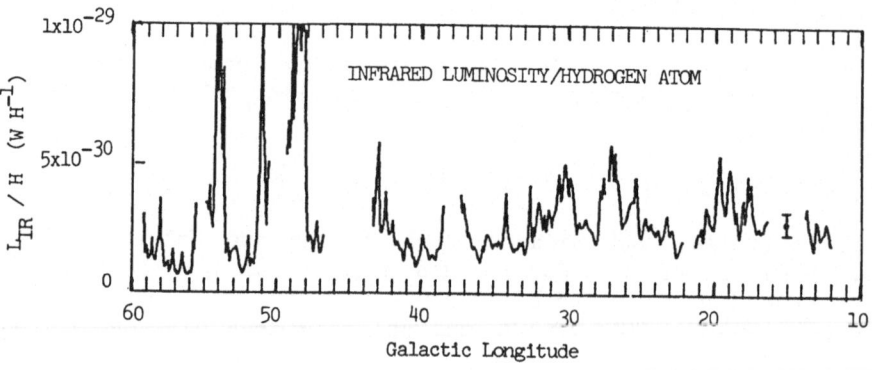

Figure 5. Longitude profile of the mean luminosity per hydrogen atom in the line of sight. The mean value is 2.5×10^{-30} W/H (see remarks for Figure 3).

V. ORIGIN OF THE DIFFUSE EMISSION

Two distinctly different models for the origin of this diffuse submillimeter emission from the galactic plane have been discussed in the literature. On the one hand, Mezger and his colleagues (see Mezger, Mathis, and Panagia (1982) and references therein, hereafter MMP) have argued that the dominant emission arises from a small fraction of the interstellar dust located in extended, low density (ELD) HII regions heated by the OB stars responsible for the ionization. On the other hand, Fazio, Stecker, and collaborators (Fazio and Stecker 1976; Cheung et al. 1982) argue that the emission is dominated by dust in molecular clouds heated by the interstellar radiation field.

The latter investigators (Cheung et al. 1982) developed a model for the submillimeter emission in which the dust and gas are thermally and dynamically coupled; i.e., the dust temperature is related by a constant factor to that of the gas, and the dust-to-gas mass ratio is constant throughout the galactic plane. Using the small-beam CO survey data of Solomon and Sanders (1981) to determine the gas temperature and column density, they then derived the expected longitude profile for submillimeter wavelength dust emission. The result of this modelling was compared with the surface brightness profile in band 1. Though the modelled and observed band 1 profiles showed considerable similarity, they were not as strongly correlated as the observed dust mass and integrated CO profiles (Figures 3c and 4b), suggesting some deficiency in the model assumptions. This deficiency is most probably related to the assumption that the gas-to-dust temperature ratio is constant. The premise that the observed submillimeter emission arises predominantly from dust in the molecular clouds would seem to be strongly supported by the constancy of the observed dust-to-gas mass ratio. However, if the dust mass absorption coefficient K_ν is substantially larger than assumed here and by Cheung et al. (e.g., a value closer to $K_\nu = 40(\nu/10^{12}\ \text{Hz})^2$ is indicated by the MMP analysis), this model would fail by a large margin to account for the high average brightness of the submillimeter emission. Perhaps more seriously, this would imply that only a small fraction of the cosmically available dust material would be needed to produce the observed emission, weakening the case that this dust must be in the molecular clouds. Clearly, a better determination of K_ν at submillimeter wavelengths is essential to settling this issue.

The MMP interpretation is based upon a determination of the dust absorption cross section per H atom using astrophysical observations of a variety of galactic sources and a two component (graphite and silicates) dust model (Mathis, Rumpl, and Nordsieck 1977). As indicated above, this analysis implies a K_ν value substantially larger than that used here. In addition, their analysis leads to a high temperature (T $\sim$ 40 K) for the graphite component when exposed to the interstellar radiation field in the ELD HII regions. This picture allows them to account for the observed submillimeter emission with a dust column density two orders of magnitude smaller than that shown in Figure 3(c).

To what extent do these data allow discrimination between the models? The strong dust/CO correlation and close correspondence between the observed L_{IR}/H and that found in molecular clouds would seem to offer at least strong circumstantial evidence in favor of the dominant emitting dust being located in molecular clouds. It is interesting to note that Gispert et al. (1982) reach a similar conclusion based on comparison of their far infrared maps with the galactic radio continuum emission. More definitive measurements of K_ν would be very helpful in assessing these models. A further critical discriminant would be a careful measurement of the spectrum of the diffuse galactic emission in the 40-200 μm range. If the warm dust required in the MMP model were a dominant source, the spectrum would be distinctly broader than that implied by the 23 K dust found in our single component analysis. Available far infrared/submillimeter data do not definitively determine this spectrum, but future balloon-borne observations and the IRAS survey data should make this possible in the near future.

VI. CONCLUSIONS

Analysis of the large scale properties of the submillimeter wavelength maps of the galactic plane in the range $\ell = 0^{\circ}$ to $+60^{\circ}$ leads to the following conclusions:

(1) The Galaxy is an optically thin disk at submillimeter wavelengths, showing continuous diffuse emission with a typical width of $\Delta b \sim 0.9^{\circ}$ (FWHM).

(2) The mean derived dust temperature in the plane is 23 K (assuming thermal emission from dust with an emissivity spectral index $n = 2$) with no significant large scale trends.

(3) The dust optical depth (mean value 6×10^{-3}) and mass column density (mean value $1 \times 10^{-3} g\ cm^{-2}$) show large variations with longitude, but little large scale gradient (averaged over the latitude range $|b| \le 1/2^{\circ}$).

(4) The spectrally integrated submillimeter surface brightness shows a prominent peak at $\ell = 30^{\circ}$, due at least in part to the W43 region, and a mean value of $9 \times 10^{-5}\ W\ m^{-2} sr^{-1}$, with a general decrease of diffuse brightness at $\ell \ge 30^{\circ}$.

(5) The dust mass column density and gas mass column density are strongly correlated, with a nearly constant ratio of gas mass to dust mass of 100 (assuming $K_{\nu} = 4.6\ (\nu/10^{12}\ Hz)^{2}$). This ratio is almost identical to the average ratio in the $1^{\circ} \times 2^{\circ}$ region around the galactic center.

(6) The mean luminosity per H atom over this longitude range is 2.5×10^{-30} W/H, essentially identical to the mean value found for individual molecular clouds and implying a radiant energy density in the vicinity of the dust more than an order of magnitude larger than the stellar radiation density in the solar vicinity.

(7) The present evidence favors an interpretation in which the emitting dust is associated with molecular clouds.

(8) Additional measurements, especially of K_{ν} and the spectrum of I_{ν} from 40 μm to 200 μm, are highly desirable to clarify the origin of the large scale submillimeter emission of the Galaxy.

REFERENCES

Aanestad, P.A. 1975, Ap. J. 200, 30.
Cheung, L. H., Fazio, G. G., Stecker, F. W., Sanders, D. B., and Solomon, P. M. 1982 (preprint).

Cohen, R. S., Cong, H., Dame, T. M., and Thaddeus, P. 1980, Ap. J. 239, L53.

Dickman, R. L. 1978, Ap. J. Supp. 37, 407.

Fazio, G. G. and Stecker, F. W. 1976, Ap. J. 207, L49.

Gispert, R., Puget, J. L., and Serra, G. 1982, Astron. Astrophys. 106, 293.

Hayakawa, S., Matsumoto, T. Murakomi, H., Uyama, K., Yamagami, T., Thomas, J. A. 1979, Nature 279, 510.

Kelsall, T., Hauser, M. G., Silverberg, R. F., Dwek, E., Stier, M. T., and Gezari, D. Y., 1982, B.A.A.S. 13, 809.

Mathis, J. S., Rumpl, W., and Nordsieck, K. H. 1977, Ap. J. 217, 425.

Mezger, P. G., Mathis, J. S., and Panagia, N. 1982, Astron. Astrophys. 105, 372.

Okuda, H., 1981, in Infrared Astronomy, IAU Symposium 96, C. G. Wynn-Williams and D. P. Cruikshank, eds., p. 247.

Price, S. D. 1981, A. J. 86, 193.

Ryter, C. E. and Puget, J. L. 1977, Ap. J. 215, 775.

Savage, B. D. and Mathis, J. S. 1979, Ann. Rev. Astron. Astrophys. 17, 73.

Silverberg, R. F., Hauser, M. G., Mather, J. C., Gezari, D. Y., Kelsall, T., and Cheung, L. H., 1979, Proc. SPIE 172, 149.

Solomon, P. M. and Sanders, D. B. 1981 (in preparation).

Solomon, P. M., Scoville, N. Z., and Sanders, D. B. 1979, Ap. J. 232, L89.

Stier, M. T., Dwek, E., Silverberg, R. F., Hauser, M. G., Cheung, L., Kelsall, T. and Gezari, D. Y. 1982, in American Institute of Physics Proceedings of the Workshop on the Galactic Center (in preparation).

COSMIC RAYS AND MAGNETIC FIELDS IN THE GALAXY

Robert L. Brown
National Radio Astronomy Observatory*

ABSTRACT

The Galactic distribution of cosmic rays and magnetic fields is accurately reflected in the Galactic non-thermal radio background. But models of the radio data only provide estimates of the Galactic variation of synchrotron emissivity, the product of cosmic ray intensity and the square of the magnetic field--one cannot derive either quantity uniquely from the radio data alone. Here we show how one can circumvent this problem by using knowledge of the distribution of thermal nucleons in the Galaxy to determine the Galactic magnetic field; once this is known the radio data can be used to map the Galactic cosmic rays. We stress the intimate relation that exists between interpretations of the radio background, the γ-ray background and the distribution of thermal material in the Galaxy.

INTRODUCTION

The Galactic distribution of the non-thermal interstellar medium-- cosmic rays and magnetic fields--has been an active field of inquiry for more than 25 years. The initial stimulus for work in this field came with the recognition that the Galactic non-thermal radio background was synchrotron radiation; hence one could attempt to interpret the radio observations in terms of the Galactic magnetic field and the Galactic distribution of cosmic ray electrons. The most comprehensive works in this regard are those by Webster (1974); French and Osborne (1976); Badhwar, Daniel and Stephens (1977); and Brindle, French and Osborne (1978). More recently, however, observations of the high energy, E > 100 MeV, gamma ray background have provided an additional probe of the Galactic distribution of cosmic rays--in this case of cosmic ray nuclei--that has greatly expanded our knowledge of energetic particles in the Galaxy. Moreover, by considering the non-thermal radio background in concert with the observed gamma ray background and the inferred Galactic distribution of thermal nucleons one can begin to disentangle the distributions of the magnetic field, cosmic rays and thermal gas from one another and consequently establish

197

W. L. H. Shuter (ed.), Kinematics, Dynamics and Structure of the Milky Way, 197–207.

the Galactic distribution of each of these components separately.
First attempts to this end are described by two excellent papers:
Paul, Cassé and Cesarsky (1976); and Higdon (1979).

Here we will focus on the non-thermal Galactic radio background
and we will use this radiation to motivate a discussion of the global
distribution of the non-thermal interstellar medium in the Galaxy.

LONGITUDE DISTRIBUTION OF GALACTIC SYNCHROTRON RADIATION

Observations

The Galactic radio background at any frequency is a superposition
of thermal and non-thermal radiation. But because the characteristic
spectral indices of the brightness of these two components are -2.0 and
-2.7 respectively, the non-thermal component becomes increasingly
dominant at lower frequencies. Moreover, the thermal emission tends to
be largely confined to spatially localized regions whereas the
non-thermal emission is more pervasive; hence along most lines of sight
the non-thermal component is increasingly favored as the telescope
beamwidth increases. Since beamwidth increases as frequency decreases
these two effects separately and together compel one to go to lower
radio frequencies to study the non-thermal radio background. Of
course, one gives up angular resolution to do so, but when the
questions one asks involve the global properties of the emission in the
Galaxy this limitation on angular resolution is tolerable. As a
reasonable compromise between wavelength and resolution we will discuss
observations made at 327 MHz with the 140-foot (43-m) telescope which
provides a beamwidth of 88' at this frequency.

Figure 1 illustrates the run of observed antenna temperature with
Galactic longitude along the Galactic equator at 327 MHz; we can use
this figure to identify the principle characteristics of the Galactic
non-thermal radiation. The observational features most germane to a
discussion of the global Galactic properties of this radiation are:

(1) The temperature T of the Galactic background
     (or equivalently the specific intensity) is
     greatest at $\ell = 0$.

(2) T decreases with increasing displacement in
     longitude from $\ell = 0$.

(3) The gradient $dT/d\ell$ is very steep in the
     restricted longitude range $35° < \ell < 50°$.

(4) The brightness varies reasonably smoothly with
     longitude: it is not dominated by particularly
     intense, angularly small, emission regions.

(5) $T(\ell)$ along $b = 0$ is symmetric about $\ell = 0$.

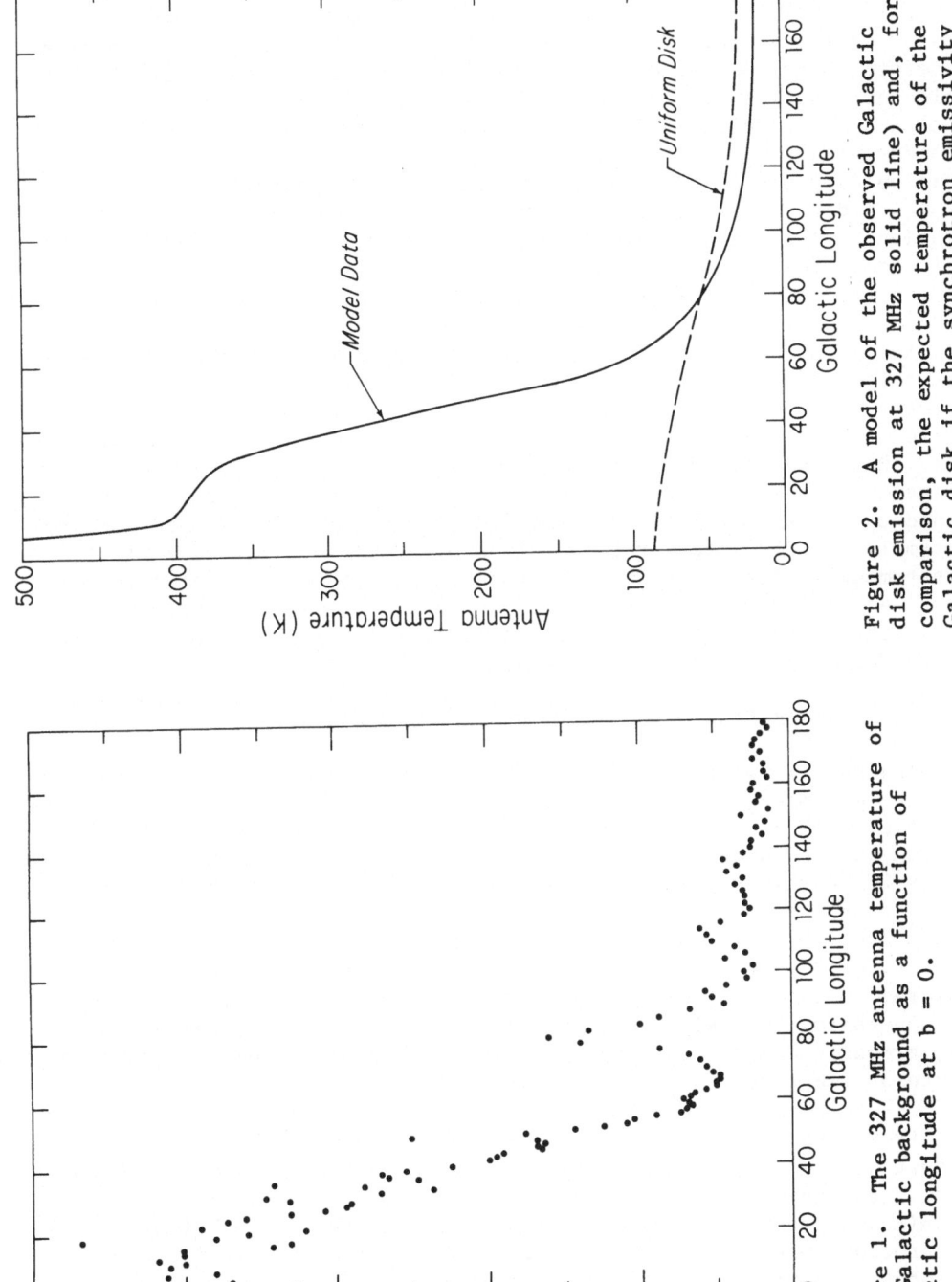

Figure 2. A model of the observed Galactic disk emission at 327 MHz solid line) and, for comparison, the expected temperature of the Galactic disk if the synchrotron emissivity was uniform everywhere is the disk (broken line).

Figure 1. The 327 MHz antenna temperature of the Galactic background as a function of Galactic longitude at b = 0.

Inferences

All the features enumerated above except (3) suggest that we
understand the Galactic non-thermal background as emission from a disk
symmetric about the Galactic center. The spiral arm contribution is a
small perturbation on the base disk emission: this result is
consistent with that found in other nearby spiral galaxies
(van der Kruit 1978). (Looking at Figure 1 the only feature that one
would be tempted to identify as a spiral arm is the region
75° < ℓ < 85°, that is, the Cygnus X region. However, most of the
emission in this direction is thermal bremsstrahlung, not synchrotron
radiation, from an association of HII regions; we are, thus, not
looking at a non-thermal spiral arm).

Although the Galactic non-thermal background is well described by
disk emission, the synchrotron emissivity in the disk cannot be
uniform. Figure 2 illustrates both a smooth fit to the data of
Figure 1 as well as the brightness expected from a homogeneous Galactic
disk. Here one can see that pronounced departures from such a
homogeneous model are evident everywhere interior to ℓ = 75°.
Consequently, we conclude that the synchrotron emissivity must increase
extremely rapidly interior to a radius of ~8 kpc. This is, in fact, a
restatement of the observational property (3) above, and the effect it
describes is not at all subtle. The abrupt increase in non-thermal
emissivity interior to the solar circle is the salient characteristic
of the Galactic non-thermal background.

LATITUDE DISTRIBUTION OF GALACTIC SYNCHROTRON RADIATION

The angular width of the Galactic plane as measured, for example
by telescope scans made along Galactic latitude b at constant longitude
show the width of the plane to be ~2° (FWHM) for longitudes ℓ < 40°
whereas in the outer Galaxy, ℓ > 90°, this same quantity may be 10° or
more. While this appears to show that the thickness of the Galactic
plane is greater at R = 10 kpc than at R < 4 kpc, such a result is a
little deceptive because the emission at ℓ < 40° arises, in the main,
at considerably greater distance from the sun than does the emission
measured in the outer Galaxy. To determine the linear scale-height of
the non-thermal emission as a function of galactocentric radius one
must properly compare the observations to a model of what is expected.
Figure 3 illustrates such a comparison.

In Figure 3 we plot the ratio of 327 MHz sky brightness at b = 2°
to that at b = 4° as a function of Galactic longitude. We then compare
these data with that expected from 3 models of the Galactic disk each
of which is chosen to have a specific value of the scale height for the
synchrotron emissivity. Here we have assumed that the variation in
synchrotron emissivity with z is described by a gaussian centered on
b = 0. From this figure it can be readily appreciated that no single
value of the scale height satisfies the observations. Rather, at

longitudes $\ell < 30°$ the scale height $z_{1/2}$--defined here as height above

the plane to half intensity--is on the order of 375 pc whereas in the outer Galaxy, $\ell > 120°$, $z_{1/2} = 850$ pc. Thus we conclude that the scale

height of the synchrotron emissivity increases monotonically with galactocentric radius; $z_{1/2}$ at R = 10 kpc is approximately twice that

at R < 3 kpc. This result is in keeping with the conclusions of Brindle et al. (1978) and Higdon (1979): one should refer to these papers for a more thorough elaboration of this point.

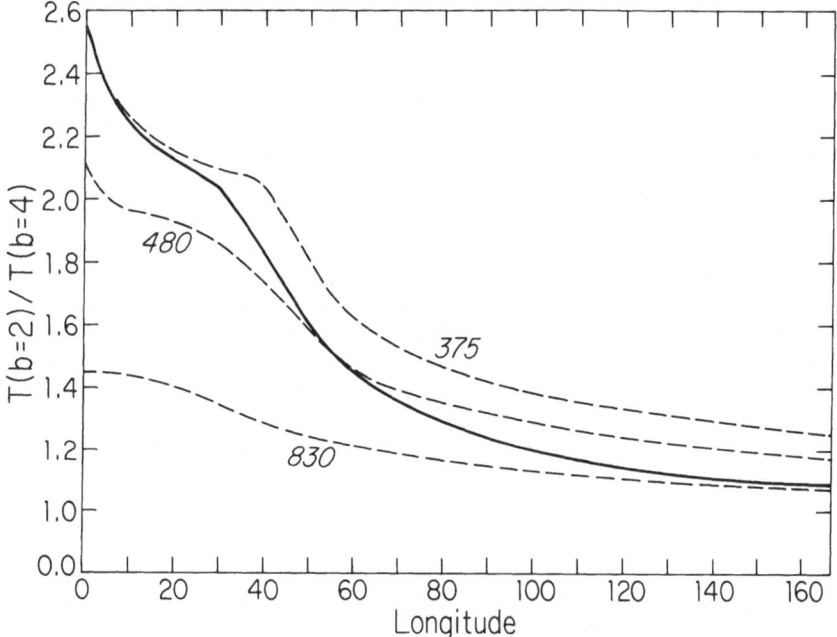

Figure 3. The ratio of the 327 MHz temperature at b = 2° to that at b = 4° as a function of Galactic longitude (solid line). Shown by dashed lines are models of this ratio that would obtain if the disk had a constant scale height--models with scale heights of 375, 480 and 830 pc are illustrated.

MODELS OF THE GALACTIC NON-THERMAL BACKGROUND

If we wish to determine the global Galactic distribution of cosmic ray electrons and magnetic fields we must model the Galactic background radiation. To first order this is quite simple because, as we have seen above, the non-thermal background is principally disk emission (not a superposition of spiral arm contributions) and the synchrotron volume emissivity is markedly greater interior to R = 8 kpc than it is at the sun (assumed here to be at 10 kpc) or in the outer Galaxy. Thus we know a priori the central feature of any viable model that we can construct: the magnetic field, or the cosmic ray intensity, or both, increases rapidly at $8 \gtrsim R \gtrsim 6.5$ kpc to a value considerably in excess of the mean value at the sun. To progress beyond this simple statement requires us to be able to establish either the cosmic ray intensity or the mean Galactic magnetic field as a function of radius independent of the non-thermal background radiation.

One way to do this is to note that whereas the cosmic ray distribution will reflect the (unknown) distribution of particle injection or acceleration sites, the magnetic field has no such ambiguity—it must be anchored in thermal matter. Thus, if we know how the magnetic field depends on gas density or gas quantity and we know the Galactic distribution of interstellar gas, we can describe the Galactic magnetic field.

Brown and Chang (1983) attempt to establish a relation between interstellar gas density and magnetic field strength by interpreting the correlation found between 21 cm column density and 327 MHz continuum brightness temperature. Specifically, they plot 21 cm column density against 327 MHz continuum antenna temperature in 5° longitude intervals with measurements made each 1° in latitude, b = -10° to 10°.

They find that these data can be fit by a simple linear function

$$\log N_H = \log \alpha + \beta \log T \qquad (1)$$

over the whole northern Galactic plane ($\ell = 10° -220°$), and moreover, the slope β is found to be remarkably constant: the mean value of β is 1.13 ± 0.11.

The correlation noted here between HI column density and non-thermal continuum temperature is not entirely unexpected. Indeed if the ratio of magnetic to gas pressure were everywhere constant in the Galaxy such a correlation would be a straightforward consequence. Locally, however, this quantity is not constant: we expect it to vary from ~1/3 in the interarm region to ~3 behind the Galactic shock (Mouschovias 1974). But if we restrict our attention to the average properties of a region large compared to a "Parker wavelength" (viz. ~1 kpc), then $\int B^2 dz \propto \langle \sigma \rangle$ (Mouschovias, Shu and Woodward 1974) and hence

$$T^{4/(\gamma+1)} \propto N_H \tag{2}$$

which provides the empirical correlation noted above.

We can now use this correlation to determine the scaling of magnetic field with gas density. The parameters of equation (1) can be expressed as

$$N_H = \langle n \rangle \ell_H = 1.823 \times 10^{18} \int_{-\infty}^{\infty} T_B \, dv \tag{3}$$

and

$$I = \frac{2kT}{\lambda^2} = c_5(\gamma) \, N_0 \, B_{\perp}^{(\gamma+1)/2} \, (\nu/2c_1)^{(1-\gamma)/2} \, \ell_S \tag{4}$$

where ℓ_H and ℓ_S are the path lengths through the HI and synchrotron emitting regions respectively, γ is the spectral index of cosmic ray electrons ($\gamma = 3.0$) and c_1 and c_5 are parameters tabulated by Pacholczyk (1970). Inserting these expressions in equation (1) we conclude that

$$\langle B \rangle \propto \langle n \rangle^{0.44 \pm 0.06} \tag{5}$$

(cf. Brown and Chang 1983). This proportionality depends only on the mean slope of the correlations such as noted earlier and on the power law index of the spectrum of relativistic electrons. The only other assumption that is implicit here is that cosmic ray electrons freely penetrate interstellar clouds.

Having found this proportionality between magnetic field and gas density we can now express the synchrotron emissivity $\varepsilon(R)$—which, as we have seen above, is a sensitive function of galactocentric radius— as

$$\varepsilon(R) \propto N_0(R) \, B^2(R) \propto N_0(R) \, n^{0.88}(R)$$

where N_0 is the normalization of the spectrum of cosmic ray electrons and $n(R)$ is the mean, large-scale, gas density. Since the synchrotron emissivity, and its variation with radius, $\varepsilon(R)$, is a quantity which we presumably know from the fits to the non-thermal background described previously, we can now estimate either $N_0(R)$ or $n(R)$ and derive the other quantity uniquely. Let us illustrate both approaches.

Uniform Cosmic Ray Spectrum

Let us first assume that the spectrum of cosmic ray electrons as measured, and demodulated, at the sun applies everywhere in the Galaxy: in this case $N_0(R) = N_0$. With such an assumption the entire variation of synchrotron emissivity with radius is a direct reflection of the variation of gas density with radius. We can, therefore, use $\varepsilon(R)$ to determine $n(R)$; the result is shown in Figure 4. The curve labeled "synchrotron" in this figure shows the mean gas density required to reproduce the derived variation of synchrotron emissivity $\varepsilon(R)$. This curve is in surprisingly good agreement with the radial distribution of nucleons in the Galaxy as derived by Gordon and Burton (1976) over the range R = 2-12 kpc. Interior to 2 kpc the gas density goes up to values consistent with those shown here, but the form of the variation with R in the inner 2 kpc is not well established (Liszt and Burton 1980).

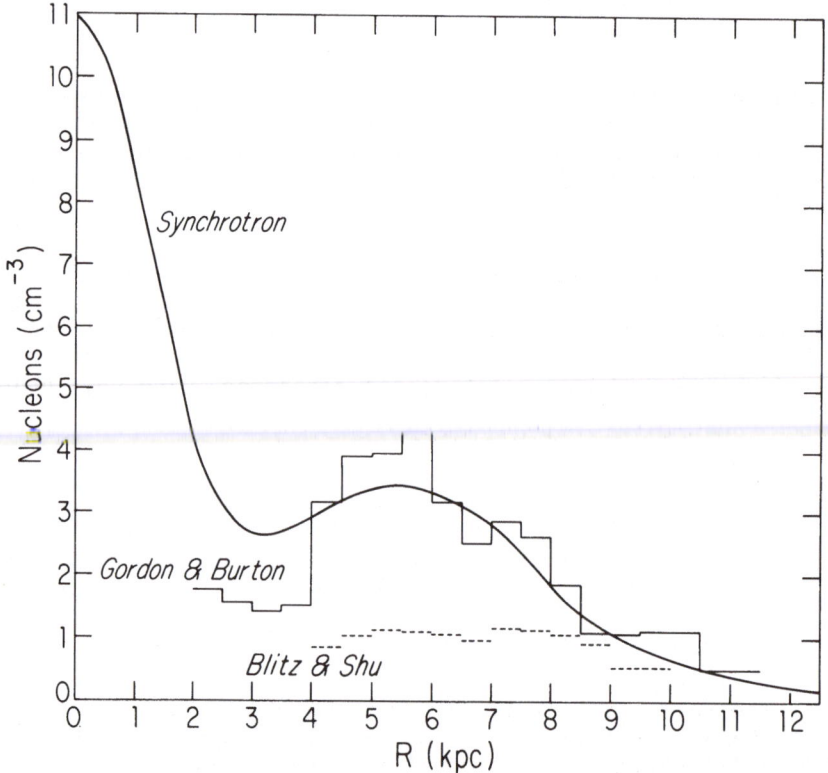

Figure 4. The radial distribution of nucleons in the Galaxy. The solid curve labeled "synchrotron" shows the distribution required if the flux of cosmic ray electrons is everywhere uniform in the Galaxy. Also shown are the distributions suggested by Gordon and Burton (1976) and by Blitz and Shu (1980).

Radial Distribution of Nucleons is Known

The alternative approach that we can use is to adopt a model of
the radial distribution of nucleons; we can then use the variation in
synchrotron emissivity to establish the radial variation in cosmic ray
intensity. For the purpose of illustration let us adopt Gordon and
Burton's (1976) radial distribution of nucleons as modified by Blitz
and Shu (1980) to incorporate the conversion $N(H_2/N(^{13}CO) = 4 \times 10^5$ and
to account for the Galactic metallicity gradient. The resultant
distribution n(R) is shown by a broken line in Figure 5.

Since n(R) as determined by Blitz and Shu varies by only a factor
~2 between 6 and 10 kpc whereas the gradient in synchrotron emissivity
over this same range of radius is much steeper, the difference in
synchrotron emissivity must be provided by a compensating increase in
cosmic ray intensity. The form of the Galactic variation of cosmic ray
intensity needed here is shown in Figure 5: the intensity of cosmic
ray electrons at R = 5-6 kpc is ~1.75 times greater than the intensity
at the sun.

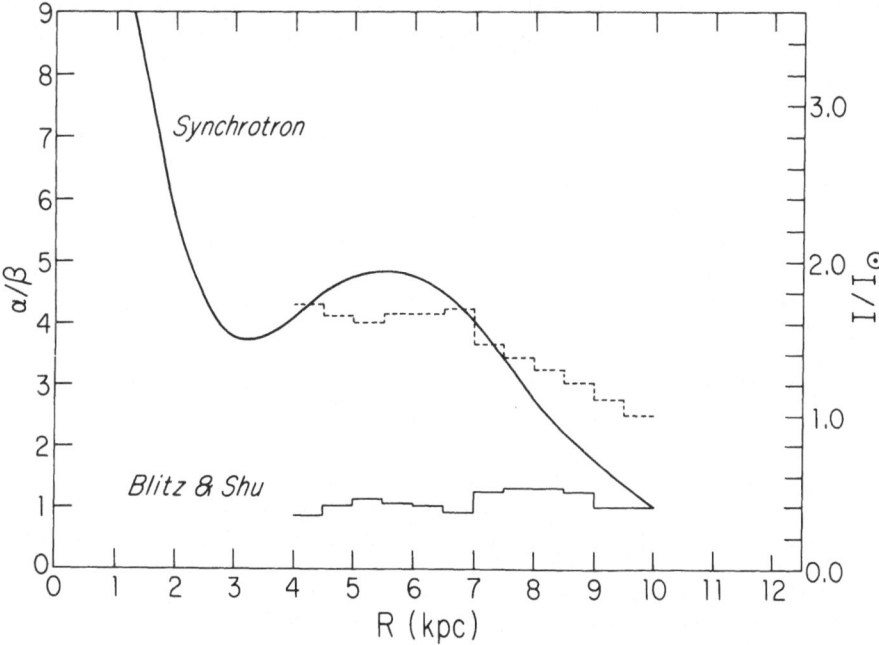

Figure 5. The ratio of magnetic to cosmic ray pressure, α/β, in the
disk of the Galaxy as a function of galactocentric radius. The curve
labeled "synchrotron" obtains if the cosmic ray electrons are
everywhere uniform in the Galaxy. Also shown is the ratio α/β that
would obtain if the Blitz and Shu (1980) distribution of nucleons in
the Galaxy is adopted. However, in this latter case, the intensity of
cosmic ray electrons would also vary with R; the required variation is
shown by a dashed line (referred to the ordinate on the right side).

Finally, it is instructive to compare the two alternatives outlined above--uniform cosmic ray flux and the Blitz and Shu (1980) radial nucleon distribution--in light of considerations regarding the stability of the Galactic disk. Several years ago Parker (1970) and others noted that in the vicinity of the sun the cosmic ray pressure P was comparable with the pressure of the interstellar magnetic field $B^2/8\pi$ and the local turbulent kinetic pressure of the interstellar gas. This equivalence assured that the distribution of gas and field in the disk was in large-scale hydrostatic equilibrium. If we now know the radial distribution of cosmic rays and magnetic fields--we have provided two such estimates above--we can ask whether the equilibrium established locally applies elsewhere in the Galaxy as well. To address that question we plot in Figure 5 the ratio of magnetic to cosmic ray pressure α/β (in Parker's notation) for each of the two alternatives that have been discussed. Here it can be seen that the assumption of a uniform cosmic ray distribution throughout the Galaxy also implies that α/β varies from 1, the value at the position of the sun, to more than 4.5 at R = 5.5 kpc (and to even higher values in the inner 2 kpc). On the other hand, with the Blitz and Shu radial distribution of nucleons, and the consequent radial gradient in cosmic ray intensity, the ratio α/β retains a remarkably constant value throughout the disk of the Galaxy, R = 4-10 kpc.

CONCLUSIONS: FUTURE DIRECTIONS

The non-thermal Galactic background accurately reflects the distribution of cosmic rays and magnetic fields in the Galaxy. But now neither of these constituents can be considered as a free parameter. As a result of recent and continuing radio spectroscopic surveys of the interstellar gas distribution as well as γ-ray surveys of the Galactic disk, both the mean radial gas density (and hence the mean magnetic field) and the Galactic cosmic ray distribution are severely constrained. And it is precisely these constraints that provide the challenge and the opportunities for the future. Specificially, the challenge is to construct a single model of the Galactic non-thermal radiation, the Galactic gamma ray emission and the CO and HI observations that is self-consistent. The observations are sufficiently good to suggest that this should be possible now. With such a model, and the understanding that it provides, we will be in a position to discuss, with reasonable confidence, questions involving the equilibrium and stability of the Galactic disk as well as questions pertaining to the production and acceleration of energetic particles and the convection and diffusion of particles and fields perpendicular to the Galactic plane.

*The National Radio Astronomy Observatory is operated by Associated Universities, Inc., under contract with the National Science Foundation.

REFERENCES

Badhwar, G. D., Daniel, R. R., and Stephens, S. A. 1977,
      Ap. Space Sci., 49, 133.
Blitz, L., and Shu, F. H. 1980, Ap. J., 238, 148.
Brindle, C., French, D. K., and Osborne, J. L. 1978, MNRAS, 184, 283.
Brown, R. L., and Chang, C. 1983, Ap. J. (in press).
French, D. K., and Osborne, J. L. 1976, MNRAS, 177, 569.
Gordon, M. A., and Burton, W. B. 1976, Ap. J., 208, 346.
Higdon, J. C. 1979, Ap. J., 232, 113.
Liszt, H. S., and Burton, W. B. 1980, Ap. J., 236, 779.
Mouschovias, T. Ch. 1974, Ph.D. Dissertation, U. Calif., Berkeley.
Mouschovias, T. Ch., Shu, F. H., and Woodward, P. 1974, Astron. Ap.,
      33, 73.
Pacholczyk, A. G. 1970, Radio Astrophysics (San Francisco: Freeman).
Parker, E. N. 1970, in Interstellar Gas Dynamics, ed. H. J. Habing
      (Dordrecht: Reidel), p. 168.
Paul, J., Cassé, M., and Cesarsky, C. J. 1976, Ap. J., 207, 62.
van der Kruit, P. C. 1978, in Structure and Properties of Nearby
      Galaxies, ed. E. M. Berkhuijsen and R. Wielebinski (Dordrecht:
      Reidel), p. 33.
Webster, A. S. 1975, MNRAS, 171, 243.

THE DISTRIBUTION OF STARS IN THE GALAXY

John N. Bahcall and Raymond M. Soneira
Institute for Advanced Study, Princeton, N. J. 08540

The basic ideas underlying the use of star counts to determine Galactic parameters are described. The techniques are illustrated by applications to the differential number-magnitude counts in SA 57, the frequency-color diagrams in SA 57 and SA 68, the paucity of intermediate population stars in the range $+5 \leq M_V \leq +8$, and the M/L ratio for the unseen matter in the halo.

Star counts used to be an important technique for learning about the Galaxy (see the wonderful description in Bok 1937), but fell into disuse for forty years or so. Why revive the technique now? The reasons are simple and cogent. Automated methods of acquiring and processing star counts and improved photometric techniques make feasible observational programs that were impossible only a few years ago (see e.g., the talks by Freeman and Sandage in this volume). Also the availability of computer models of the Galaxy enables us to study easily the effects of varying different galactic parameters so that we can isolate and determine the quantities that are most important in a given direction and in specified color and apparent magnitude ranges.

In this talk, we discuss first the basic ideas that are used in predicting star counts. Then we review some of the successes of the work so far and indicate future observational programs that will be important in increasing our knowledge of the stellar content of the Galaxy. Finally, we indicate the already achieved, and the achievable, limits on the (M/L)-ratio for the so far unseen material that makes up the massive halo. Most of the work we concentrate on here has been described in a series of papers by Bahcall and Soneira (1980, 1981a,b) (hereafter Papers I-III).

The basic departure from the classic work, (see for example, Seares 1924, Bok 1937, Oort 1938, and references therein, and also the work of the Basle group Becker 1965, Becker and Steppe 1977) on star counts is that we assume the large scale geometry of the stellar components, using the photometric observations of other galaxies as a guide to the determination of the density laws for the Galaxy. In previous work, the star

W. L. H. Shuter (ed.), Kinematics, Dynamics and Structure of the Milky Way, 209–216.

counts were used to try to determine the general structure of the mass distribution. In our opinion, this overall distribution is best determined from the observations of other galaxies, exploiting the measurements of star counts in the Galaxy to determine the scale parameters and the normalization factors.

We have found it adequate to assume an exponential disk for the population I stars, a de Vaucouleurs (1959) or a Hubble law (1930) for the population II spheroid stars (relative normalization spheroid/disk ~ 0.001 at the solar position), a nuclear component (confined to the inner kiloparsec), and a massive (unseen) halo. Luminosity functions and scale heights are taken from the existing data on stars in the solar vicinity. We have, of course, tried many different variants of these ingredients (such as an exponential disk with or without a truncation radius, an exponential disk with a hole in the middle, a flattened spheroid, etc.), but we will limit ourselves here to the few variations that make a difference for observations that are feasible now. The ingredients described above are combined in the basic equation of stellar statistics (see below) in order to predict the number of stars in a given direction and in a specified apparent magnitude range.

How could the above prescription be wrong? This is an important question because it is by comparison between observations and models that we confirm or improve our knowledge of the stellar content of the Galaxy.

We could be wrong for the following reasons. (1). The density laws for the absolute magnitude range accessible to star counts in the Galaxy (usually ~ $4 \lesssim M_V \lesssim 12$) might be different from the distribution of light in external galaxies where other absolute magnitude ranges are most important (typically half the light is from stars brighter than $M_V = +2$). Half the surface brightness at the solar position is contributed by stars with apparent magnitudes $m_V \leq 7.5$ mag. (2). The luminosity functions could vary from place to place in the Galaxy (or be dominated by stars that have not been fully included in the available determinations of the local luminosity function). (3). The scale heights could vary from place to place. (4). The numerical values of the most important parameters we use could be incorrect. We adopt here as standard values the disk scale length h = 3.5 kpc, R_0 = 8 kpc, μ_0(Disk) = 75 $M_\odot$ pc^{-2}, and $\rho_{Spherical}/\rho_{Disk}$ ~ 0.001 at the Sun, and then determine the sensitivity of the model predictions to each of these assumptions. (5). The color-magnitude diagrams might be different from what we expect (because, e.g., of a metallicity gradient). (6). There may be unexpected components - something new.

The basic equation by which these ingredients are combined is an expression for the number of stars per magnitude interval:

$$A(m,\ell,b)d\Omega \; dm = dm \; d\Omega \sum_{i=1}^{4} \int_0^\infty dR \; R^2 \; \rho_i(\vec{r}_i,M)\Phi_i(M) , \qquad (1)$$

where ρ_i is the density function of the ith stellar component, $d\Omega \, dR \, R^2$ the volume element about the Sun, $\Phi(M)$ the locally determined stellar luminosity function, and A the number of stars per apparent magnitude interval per steradian in a specified direction.

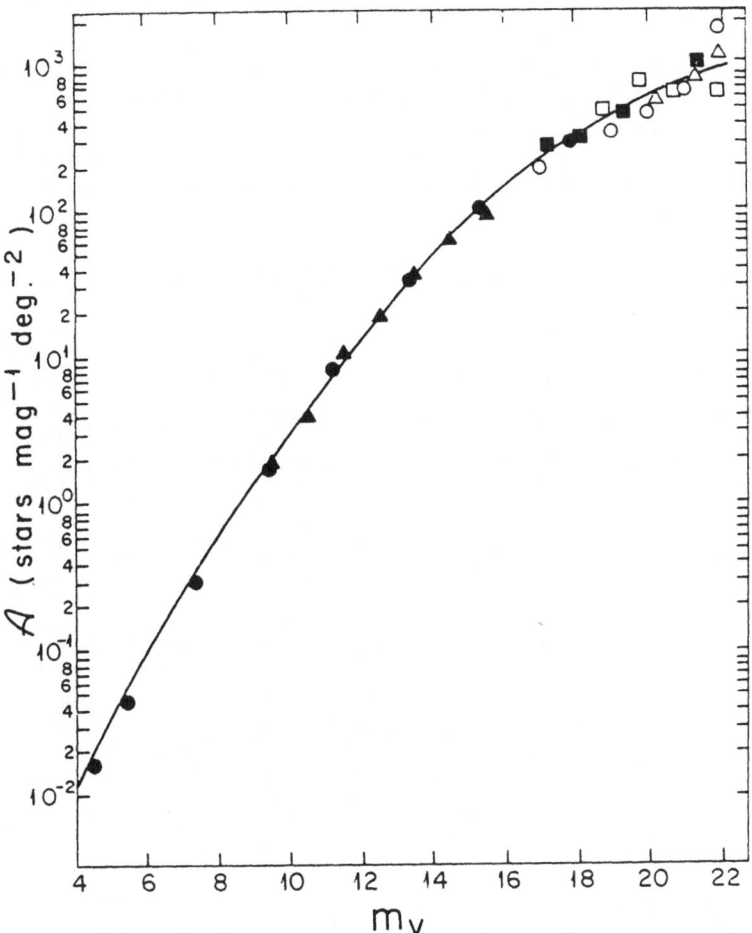

Figure 1. Differential star counts A per magnitude per square degree for the galactic pole. The solid curve is the counts predicted by the standard Galaxy model of Paper I assuming no obscuration at the pole and including a separate contribution from giants. Data from Seares et al. (1925) as reduced to the Visual band in Paper I are plotted as filled circles, data from Weistrop (1972, 1980) are plotted as filled triangles, data from Brown (1979) corrected to the pole (see Paper I) are plotted as filled squares, data from Kron (1978) are plotted as open triangles, data from Tyson and Jarvis (1979) are plotted as open circles and data from Peterson et al. (1979) as open squares. For further details, see Papers I and III.

We will not discuss here the various models we have used for describing the obscuration since we will limit ourselves in this discussion to fields above $|b^{II}| = 30°$ in which all the obscuration laws lead to only small corrections to the predictions (see Sandage 1972, de Vaucouleurs, de Vaucouleurs, and Corwin 1976, Burstein and Heiles 1978). The local parameters - luminosity functions and scale heights - are described in Papers I - III.

The comparison with the predicted total star counts is SA 57 (near the North Galactic Pole) with the observations is shown in Figure 1. The predicted and observed differential counts are in satisfactory agreement over the five orders of magnitude variation in A between $m_V = 5$ and $m_V = 22^m$; comparable data over such a wide magnitude range does not yet exist in other fields.

We predict frequency-color diagrams (see Figure 2 below) by including on the right hand side of equation (1) only those absolute magnitudes corresponding to the specified color bin.

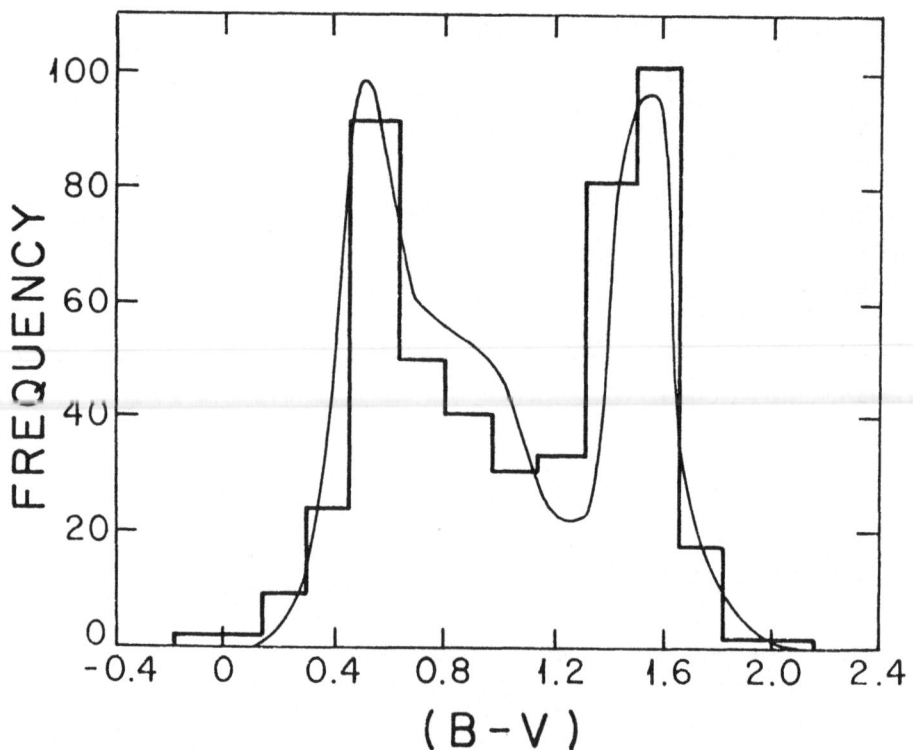

Figure 2. The distribution of (B-V) colors in the direction of the galactic pole (SA 57) for stars of apparent visual magnitude between 19.75 and 22.0 (adapted from Figure 8b of Paper I). Data from Kron (1978) is plotted as a histogram. The smooth curve is the distribution predicted by the standard Galaxy model of Paper I.

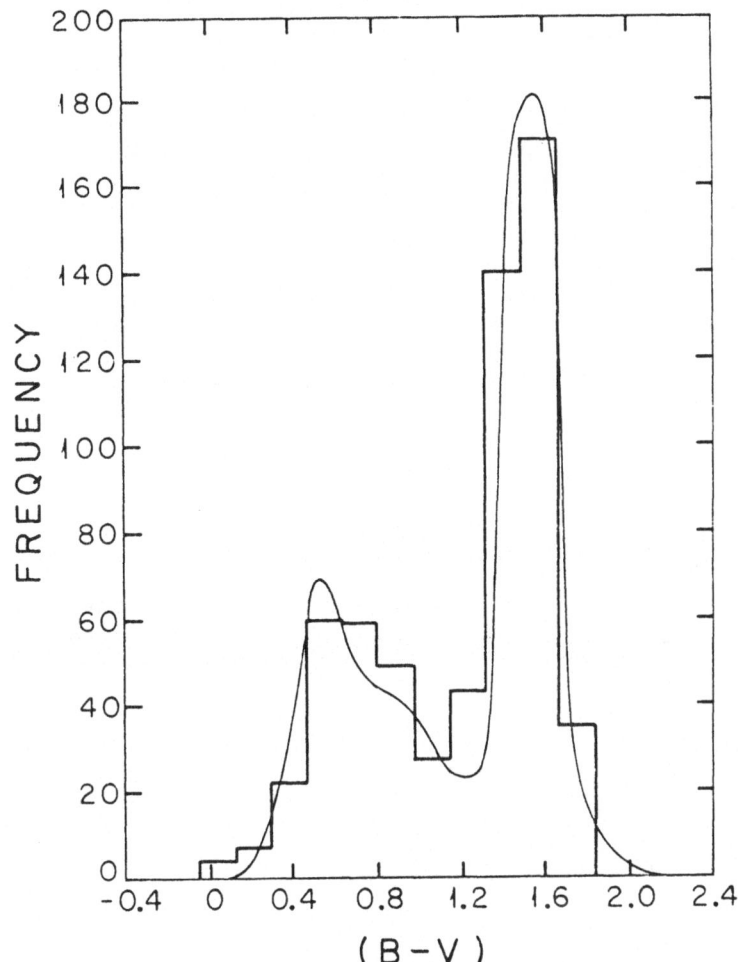

Figure 3. The Distribution of (B-V) Colors in the Direction of SA 68 ($\ell = 111°$, $b = -46°$) [adapted from Figure 8d of Paper I]. Data is from Kron (1978) and the smooth curve from the standard Galaxy model of Paper I. The small amount of reddening is calculated following Sandage (1972).

The comparison of the observed (Kron 1978, 1980) and the calculated frequency-color diagrams for SA 57 and SA 68 are shown in Figures 2 and 3. In the magnitude range that Kron's data refers to ($19.75^m \leq m_V \leq 22.0^m$) and for these fields, the disk and the spheroid appear as two distinct peaks in both the observed and the predicted counts. But notice that the disk occupies the red peak (see Paper I for an explanation). In different directions and in different apparent magnitude ranges, the predicted frequency-color diagrams are very different, with the bi-modal distribution disappearing entirely at bright or very dim apparent magnitudes or in the direction of the galactic anti-center. These predictions are of course subject to direct observational tests (see Paper I or

II); these tests will determine if we have understood correctly the relative importance of population I and II stars and their geometry and luminosity functions.

Now we wish to discuss two consequences of adopting the Galactic model for the star counts described above. These are: (1). the estimation of Galactic parameters; and (2). the determination of limits on the (M/L)-ratio for the unseen matter in a massive halo.

Some of the galactic parameters that are determined by the above-described model for the stellar content of the Galaxy are summarized in Table I. A description of how these parameters have been determined is given in Paper I and in Bahcall, Schmidt, and Soneira (1982); the latter paper deals with the determination of the characteristics of the spheroid.

TABLE 1. Some Computed Properties of the Galaxy Model
[Adapted from Table 4 of Paper I.]

	Luminous Disk	Luminous Spheroid
Solar Neighborhood ($M_V \leq 16.5$)		
Luminosity density ($L_\odot pc^{-3}$)	0.062	$(2 \text{ to } 13) \times 10^{-5}$ [a]
Mass density of main sequence stars ($M_\odot pc^{-3}$)	0.040	$(4 \text{ to } 14) \times 10^{-5}$ [a]
Mass density of white dwarfs	0.005	6×10^{-6}
Number density (stars pc^{-3})	0.11	$(1 \text{ to } 9) \times 10^{-4}$ [a]
Total		
Luminosity (visual $L_\odot$)	1.2×10^{10}	$(0.5 \text{ to } 3) \times 10^9$ [a]
Mass of visible stars ($M_V \leq 16.5$)	2.0×10^{10}	$(0.9 \text{ to } 3.2) \times 10^9$ [a]
Mass ($M_\odot$) [b]	5.6×10^{10} [b]	3×10^9 [b]
Number of Stars ($M_V \leq 16.5$)	6×10^{10}	$(0.2 \text{ to } 2) \times 10^{10}$ [a]
M_V (no obscuration)	-20.4	$-16.9 \text{ to } -19.0$ [a]

[a] Estimated in Bahcall, Schmidt, and Soneira (1982).

[b] These mass values are derived from the mass model given in Section VI of Bahcall, Schmidt, and Soneira (1982). Local mass density of 0.15 $M_\odot$ pc^{-3} assumed. Interstellar matter of local density 0.045 $M_\odot pc^{-3}$ is assumed to be distributed like the visible stars in the disk but with a scale height of 125 pc. Remaining (dark) matter is assumed to be distributed like the visible stars in both disk and spheroid.

One can place limits on the relative abundance of intermediate population stars, which have been discussed somewhat in this conference, by making use of the relative paucity of stars in the valley of the frequency-color diagrams that are shown in Figure 2 which would be populated by stars of intermediate scale heights. For $0.65 \leq (B-V) \leq 1.3$, i.e., $5 \leq M_V \leq 8$, there are relatively few stars. One concludes from this fact that the density in the solar vicinity of intermediate population stars, n_{Int}, in this absolute magnitude (or color) range is not large. For a fat disk with a scale height of 3 kpc or a flattened spheroid with an eccentricity of 0.5, one finds (Bahcall, Schmidt, and Soneira, 1982) $n_{Int} < 0.003 n_{Disk}$, or $n_{Int} \leq 2 n_{spheroid}$. Less strong limits are obtained for stars outside this range.

Next we want to indicate how star counts can be used to place a limit on - or lead to an observation of - the mass to light ratio of the objects that make up the matter that is so-far unobserved in massive halos (which we know from dynamical considerations must be present). Recall that from Figure 1 the calculations made with the standard Galaxy model give good agreement at the North Galactic Pole ($\sim$ SA 57) between predicted and observed star counts for $5^m \leq m_V \leq 22^m$. If halo stars had been observed, they would have produced a feature in the counts that increased as $10.^{0.6m}$ (since the halo stars must be faint and relatively abundant and nearby), whereas the observed counts increase significantly less steeply, $\sim 10^{0.1m}$ near $m_V = +22$. Thus we can set a limit on the absolute magnitude of stellar objects that would be abundant enough to lead to a flat rotation curve for the Galaxy (similar to what is observed for other galaxies of about the same Hubble type). One finds

$$(M/L)_{V,Halo} \geq 650 \text{ solar units} \left[\frac{\rho_{Halo}(R_o)}{0.01 M_\odot pc^{-3}} \right]^{2/3}, \tag{2}$$

for main sequence stars (for which equation (2) is equivalent to the statement that $M_V \geq 14^m$, i.e., later than M5). In deriving the equivalent of equation (2) in Paper I, we assumed that the halo stars must be redder than $(B-V) > 1.1^m$ in order not to interfere with the agreement between model and the observed frequency-color diagrams that are shown here in Figure 2. The limit $(M/L)_{V,Halo} \geq 650$ solar units applies for an assumed halo density of $0.01 M_\odot pc^{-3}$.

If the objects that make up the massive halo burn hydrogen ($M \gtrsim 0.085 M_\odot$), then they will be detectable by observations that could be carried out now with available techniques. Detailed calculations that are described in Paper II show that observations in the I-band down to $m_I = 22$ mag should reveal the stellar components of a massive halo if the stars are anywhere on the hydrogen burning main sequence. This result is valid for local halo densities in the broad range from 0.0025 to 0.020 $M_\odot pc^{-3}$ (or, equivalently, rotation velocities at large distances from the Galactic center of from 175 to 350 km s^{-1}). Stars that burn hydrogen will have an (M/L)-ratio in the visual band of

$$(M/L)_V \leq 4 \times 10^3 \text{ solar units} \quad . \tag{3}$$

The equal sign is achieved for a star of mass $0.085M_\odot$ ($M_I = 11.5$ mag). If the massive halo is composed of stars that have not yet reached the main sequence, then $(M/L)_V > 10^9$ and they will not be detectable by ordinary techniques (see Paper II).

This work was supported in part by the National Science Foundation under contract no. NSF-7919884 and by NASA under contract no. NAS8-32902.

REFERENCES

Bahcall, J.N. and Soneira, R.M.:1980, Ap. J. Suppl. 44, p. 73 (Paper I).
Bahcall, J.N. and Soneira, R.M.:1981, Ap. J. Suppl. 47, p. 357 (Paper II).
Bahcall, J.N. and Soneira, R.M.:1981, Ap. J. 246, p. 122 (Paper III).
Bahcall, J.N., Schmidt, M., and Soneira, R.M.:1982, Ap. J. (to be published).
Becker, W.:1965, Z. Astrophys. 62, p. 54.
Becker, W. and Steppe, H.:1977, Astron. and Astrophys. Suppl. 28, p. 377.
Bok, B.J.:1937, *The Distribution of Stars in Space*, (Chicago: University of Chicago Press).
Brown, G.S.:1979, A. J. 84, p. 1647.
Burstein, D. and Heiles, C.:1978, Ap. J. 225, p. 40.
de Vaucouleurs, G.:1959, in *Handbuch der Physik*, Vol. 53, ed. S. Flügge (Berlin: Springer-Verlag), p. 311.
de Vaucouleurs, G., de Vaucouleurs, A., and Corwin, H.G.:1976, *Second Reference Catalogue of Bright Galaxies* (Austin: University of Texas Press).
Hubble, E.:1930, Ap. J. 71, p. 231.
Kron, R.G.:1978, PhD. Thesis, University of California, Berkeley.
Kron, R.G.:1980, in *Two Dimensional Photometry* (ESO Workshop), ed. by P. Crane and K. Kjar.
Oort, J.H.:1938, Bull. Astr. Inst. Netherlands 8, p. 233.
Peterson, B.A., Ellis, R.S., Kibblewhite, E.J., Bridgeland, M.T., Hooley, T., and Horne, D.:1979, Ap. J. (Letters) 233, L109.
Sandage, A.:1972, Ap. J. 178, p. 1.
Seares, F.H.:1924, Ap. J. 59, p. 11.
Seares, F.H., van Rhijn, P.J., Joyner, M.C., and Richmond, M.L.:1925, Ap. J. 62, p. 320.
Tyson, J.A. and Jarvis, J.F.:1979, Ap. J. (Letters) 230, L153.
Weistrop, D.:1972, A. J. 77, p. 849.
Weistrop, D.:1980, private communication.

OPTICAL SPIRAL STRUCTURE BETWEEN l = 30° AND 70°

Douglas Forbes
University of Victoria and University of Nebraska

In the past two decades, a great deal of effort has gone into attempts to map the spiral structure of the Galaxy by optical means. This labour has been rewarded with some success, particularly in the southern Milky Way (e.g. Jackson 1976) and in the anticentre direction (Moffat, FitzGerald, and Jackson 1979), yet there remain some regions of the Galaxy where our knowledge of the distribution of optical spiral tracers is far from complete. Perhaps the best example of this is the region between l = 30° and 70°, as can be appreciated from any of a number of recent maps of optical tracers (Walborn 1973; Vogt and Moffat 1975; Crampton and Georgelin 1975; Humphreys 1979), all of which show a distinct lack of objects within this longitude range.

Inspection of a photograph of this section of the Milky Way suggests that it is important to consider to what extent interstellar absorption determines the observed minima in the number of optical tracers – minima that are all the more striking when one considers that portions of several major spiral features should be seen within this longitude range. Studies of absorption (FitzGerald 1968; Lucke 1978; Neckel and Klare 1980), although restricted in longitude and distance coverage, have indicated very strong absorption (typically 2-3 mag kpc^{-1}) setting in within ~ 500 pc of the Sun, apparently throughout the region. Such high values of obscuration might well be expected to contribute to the observed lack of optical tracers, and may also be partly to blame for the scarcity of modern systematic studies of optical spiral structure in this region (Sherwood 1974).

The combination of seemingly ubiquitous absorption and observational neglect has introduced substantial bias into maps of the overall optical spiral structure; in a number of cases the l = 30° - 70° "gap" has been interpreted as an inter-arm region (Becker and Fenkart 1970; Humphreys 1979), with the implication that the outer edge of the well-known Sagittarius arm should be seen tangentially near l = 30° - 40°. There are, however, a number of problems with such a model: it is in conflict with radio observations of intensity maxima in the continuum flux and in neutral hydrogen (Burton and Shane 1970) which indicate tangency at l = 50°,

217

W. L. H. Shuter (ed.), Kinematics, Dynamics and Structure of the Milky Way, 217–222.

and it serves to locate the streaming motions observed in H I at l = 50°
(Burton 1966) (which presumably arise from the presence of a spiral arm)
in an inter-arm region.   In an effort to resolve some of these problems,
new UBV and MK observations were made of OB stars in H II regions, young
open clusters and associations, and in the field (Forbes 1982).   In particular these observations were aimed at trying to trace the supposedly
major Sagittarius arm optically from l = 28° through tangency (as defined
by radio data) at l = 50° - 55°.

A detailed study of the distribution of absorbing matter based on
these and earlier data shows that absorption at l = 28° - 70° is mainly
local (within ~500 pc of the Sun), with essentially clear regions beyond,
extending from 1 to 5 kpc from the Sun, beyond which the data become unreliable.   Within 2° of the plane obscuration is greatest between l = 30°
and 60° and least near l = 28° and l = 60° - 70°.   This corresponds closely
with the strong discontinuity in the number of optical tracers observed
between 30° and 60°, and supports the contention that optical maps to date
have been strongly influenced by absorption in this part of the Galaxy.
Nevertheless, the relatively faint apparent magnitude limit of the present data, together with the apparently local nature of the dust complexes
and the discovery of a number of "windows" of lower obscuration, appear
to have significantly reduced the effects of absorption on the resulting
distribution of optical tracers.

Figure 1 shows the distribution of optical tracers at l = 28° - 70°
from this study.   Also included are radio sources between l = 28° - 55°
from the survey by Downes et al. (1978) for which the kinematic distance
ambiguity could be resolved.   An attempt was made to weight the objects
according to their importance as spiral tracers.   There are several points
worth noting:
     1) Tracers are now seen optically as distant as 6 kpc from the Sun -
an extent much more in line with similar data for other portions of the
Galaxy.
     2) Some of the bias due to strong absorption is still apparent in
the lanes or "fingers" extending radially from the Sun near l = 28° and
l = 60° - 70°, but a significant number of optical tracers are also seen
in the heavily obscured zone between l = 30° - 60°; a good example is the
newly discovered OB association with a probable 51-day Cepheid member at
l = 55°, d = 3.5 kpc (Forbes 1982b).
     3) The overall impression given by the data in Figure 1 is one of
"clumpiness", rather than one of continuous spiral features.   There is a
strong tendency for all types of tracers to be arranged in clumps; often
several types of tracers will define a given clump.   That these clumps
are seen in clear regions at l = 28° and l = 60° - 70°, as well as in the
intervening area, suggests they are not the result of selective obscuration.

To see how the distribution of optical tracers at l = 30° - 70° fits
into the overall pattern of the rest of the Galaxy, Figure 1 has been
combined with similar data for other longitudes (Georgelin and Georgelin
1976; Crampton and Georgelin 1975; Tammann 1970; Humphreys 1979).   The

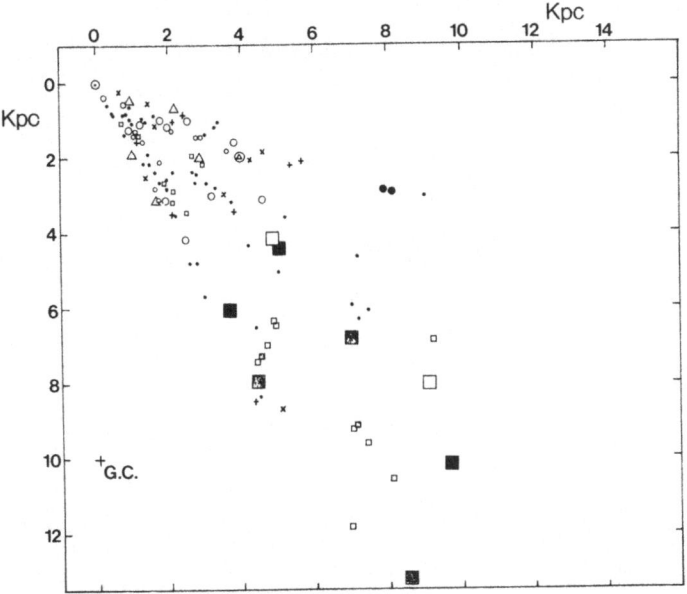

Figure 1 – The space distribution of optical and radio spiral tracers at
l = 28° – 70°. The size and density of each symbol is used to indicate the
importance of each object as a spiral tracer.

resulting distribution is shown as Figure 2. As in Figure 1 the symbols
have been drawn to indicate the importance of each object, but no discrimination has been made between distances derived optically and those
derived kinematically. In general, objects more than 4-5 kpc from the
Sun were located kinematically, using the rotation curve of Georgelin and
Georgelin (1976).

Figure 2 may be viewed several ways, if one wishes to try to trace
spiral features through longitudes 30° to 70°. Superposing the spiral
pattern of Georgelin and Georgelin (1976) on the data in Figure 2, as
shown in Figure 3, suggests that the new data for this longitude range
are not inconsistent with this pattern, especially if a reasonable arm
width of ~800 pc is assumed. In such an interpretation the Local (Orion)
feature appears to remain separate from both the Sagittarius and the
Perseus arms as far as 5 kpc from the Sun along l = 60° – 70°. The Sagittarius arm appears to become rather broad near l = 30°, which may be related to the branching of this arm in the Centaurus direction as reported
by Jackson (1976).

The data in Figure 3 do not however make a very strong case for a
continuation of the Sagittarius arm from the vicinity of the M16 - M17
complex (l = 20°, d = 2 kpc) through higher longitudes to the area of the
W 51 complex (l = 49°, d = 6.5 kpc). If the Sagittarius-Carina arm is, as

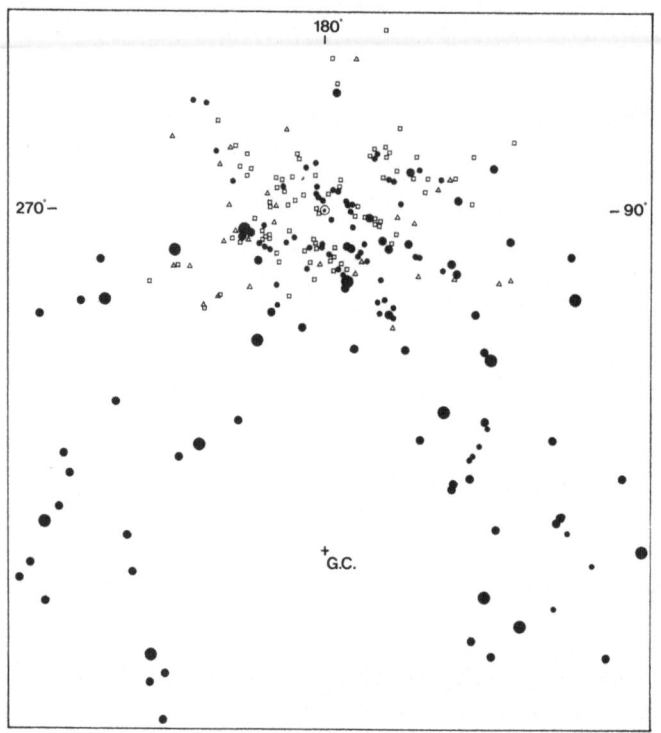

Figure 2 - The space distribution of spiral tracers over the entire
Galaxy. In the region 1 = 28° - 55° the distribution of radio H II regions
has been taken from Downes et al. (1978); elsewhere the data are from
sources given in text and from the present study.

is believed, a major "grand design" arm (Georgelin and Georgelin 1976;
Humphreys 1979), it is disconcerting that the Sagittarius portion appears
to be devoid of both optical and radio tracers between 1 = 30° - 50° - a
length of nearly 5 kpc along the arm.   The close correspondance between
histograms showing pronounced minima in the numbers of optical tracers
(Sherwood 1974) and radio H II regions (Lockman 1978; Wilson 1980) strongly
suggests that this gap is not an artifact of interstellar absorption but
represents a real lack of objects; if the Sagittarius-Carina arm is a
"string of beads" of bright spiral tracers, then something has broken this
part of the string and allowed most of the beads to be lost!

It is also troubling that a number of tests of linear density-wave
theory made in this study, based of the radial velocities of 75 H II
regions and OB stars with well-determined optical distances, showed no
evidence for the sort of streaming motions or velocity differences in the
stars and gas as have been observed in the Carina portion of this arm
(Humphreys and Kerr 1974), which are believed to be manifestations of a
galactic density wave.

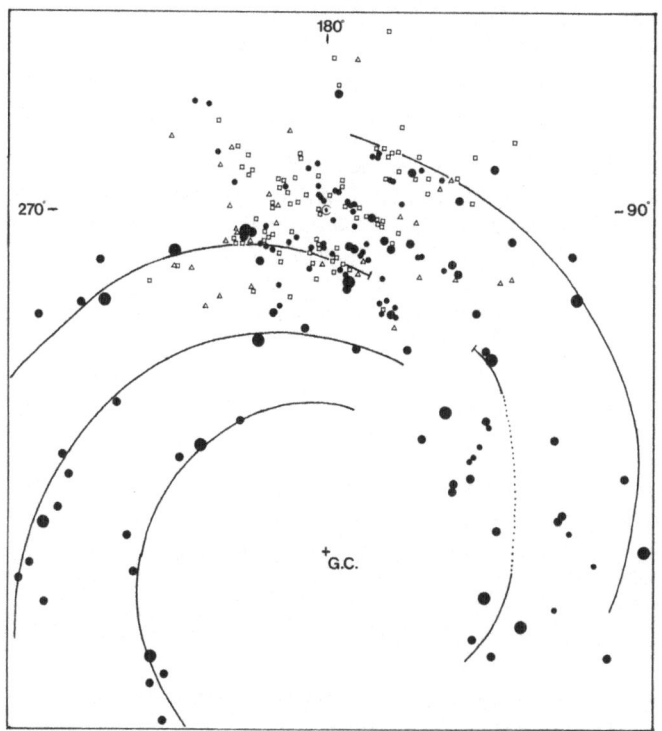

Figure 3 - The same as Figure 2 but with the spiral pattern of Georgelin
and Georgelin (1976) superposed. The "gap" in the Sagittarius arm is
indicated.

        The large and apparently real gap and the lack of density-wave kin-
ematics in this portion of the Sagittarius arm, as well as the observed
clumpiness in the distribution of both optical and radio tracers and the
presence of spiral fragments (like the Local feature) throughout the
observed Galaxy suggest that attempts to fit a global spiral pattern to
these bits and pieces have probably been overinterpretations of the
available data. The spiral structure of the Galaxy might better be
viewed as an irregular distribution of clumps and fragments of young objects that might only form a global pattern if seen from intergalactic
distances. In this respect one is reminded of the stochastic self-propagating star formation (SSPSF) models of Gerola and Seiden (1978; Seiden
and Gerola 1979), which (at least qualitatively) can account for the
irregular structure and lack of density-wave kinematics observed at l =
30° - 70°, either by acting alone or in concert with an underlying old-
population density wave (Kaufman 1979; Elmegreen 1979; Bash 1979).
What is needed at this time are deeper radio and optical surveys of this
and other portions of the Milky Way to confirm the irregular nature of
the spiral structure, a great many more high-quality radial velocities of
young objects for kinematic studies, and, if possible, some specific

predictions from both non-linear density-wave and SSPSF-type theories
with which to compare such observational evidence.

REFERENCES

Bash, F.N.: 1979, in I.A.U. Symposium No. 84, p. 165.
Becker, W. and Fenkart, R.B.: 1970, in I.A.U. Symposium No. 38, p. 205.
Burton, W.B.: 1966, Bull. Astron. Inst. Netherl., 18, 247.
Burton, W.B. and Shane, W.W.: 1970, in I.A.U. Symposium No. 38, p. 415.
Crampton, D. and Georgelin, Y.M.: 1975, Astron. Astrophys. 40, 317.
Downes, D., Wilson, T.L., Bieging, J., and Wink, J.: 1978, Astron.
      Astrophys. Suppl. 40, 379.
Elmegreen, B.G.: 1979, Ap. J. 231, 372.
FitzGerald, M.P.: 1968, A.J. 73, 983.
Forbes, D.: 1982, Ph.D. dissertation, Univ. of Victoria.
Forbes, D.: 1982b, A.J. (in press).
Georgelin, Y.M. and Georgelin, Y.P.: 1976, Astron. Astrophys. 49, 57.
Gerola, H. and Seiden, P.E.: 1978, Ap. J. 223, 129.
Humphreys, R.M. and Kerr, F.J.: 1974, Ap. J. 194, 301.
Humphreys, R.M.: 1976, Ap. J. 206, 114.
Humphreys, R.M.: 1979, in I.A.U. Symposium No. 84, p. 93.
Jackson, P.D.: 1976, Ph.D. dissertation, Univ. of Maryland.
Kaufman, M.: 1979, Ap. J. 232, 317.
Lockman, F.J.: 1979, Ph.D. dissertation, Univ. of Massachusetts.
Lucke, P.B.: 1978, Astron. Astrophys. 64, 367.
Moffat, A.F.J., FitzGerald, M.P., and Jackson, P.D.: 1979, Astron. Astrophy
      Suppl. 38, 197.
Neckel, Th. and Klare, G.: 1980, Astron. Astrophys. Suppl. 42, 251.
Seiden, P.E. and Gerola, H.: 1979, Ap. J. 233, 56.
Sherwood, W.A.: 1974, Pub. Royal Obs. Edin. 9, 85.
Tammann, G.A.: 1970, in I.A.U. Symposium No. 38, p. 236.
Vogt, N. and Moffat, A.F.J.: 1975, Astron. Astrophys. Suppl. 20, 85.
Walburn, N.R.: 1973, A.J. 78, 1067.
Wilson, T.L.: 1980, in Radio Recombination Lines, ed. P. Shaver, Dordrecht,
      Reidel, p. 205.

COS-B GAMMA-RAY MEASUREMENTS AND THE LARGE-SCALE DISTRIBUTION OF INTERSTELLAR MATTER

H.A.Mayer-Hasselwander
Max-Planck Institut für Physik und Astrophysik
Institut für Extraterrestrische Physik, Garching, Germany
On behalf of the 'Caravane' Collaboration

1. INTRODUCTION

Recently a complete survey of gamma radiation from the galactic disc within about 20 degrees from the equator in three energy bands above 70MeV has become available (Mayer-Hasselwander et al., 1982).

For the local galactic region the analyses of Lebrun et al., (1982) using this database as well as earlier work using the gamma-ray data from the SAS-2 satellite (Fichtel et al., 1978, Lebrun and Paul, 1979, Strong and Wolfendale, 1981, Strong et al.,1982) have demonstrated the proportionality between the observed gamma-ray intensity and the column density of the 'local' interstellar medium (ISM) and have deduced a value for the gamma-ray emissivity per atom. For these analyses the gamma-ray data at latitudes $|b|>10^{\circ}$ were used, thus restricting the investigated region to within $\simeq 1$ kpc from the sun (for z-scaleheight 120 pc) where good arguments exist for the cosmic-ray (CR) density being constant.

Since gamma-rays now have been proven to be a valuable tracer for the local ISM, it is interesting to explore their tracing capabilities for more distant regions of the Galaxy, which are observed at latitudes $|b|<10^{\circ}$. Clearly, compared to the local region the problem is much more complex as various parameters like CR density and gas density and their z-scale heights and also radiation fields affect the amount of gamma-radiation generated at a given site. Also a quantitatively unknown contribution from compact sources exists. All these parameters are expected to vary over galactic scales and most of them are poorly known.

The main advantage of the gamma rays as a tracer is their negligible absorption by the ISM which allows them to reach us even from the other side of the Galaxy. On the other hand gamma rays as a tracer encounter handicaps which basically are the lack of a direct distance indicator and the limited angular resolution of about 1.2 degrees HWHM at energies above 300MeV (respectively a few degrees at lower energies) together with the limited count statistics of presently available data.

W. L. H. Shuter (ed.), Kinematics, Dynamics and Structure of the Milky Way, 223–231.
Copyright © 1983 by D. Reidel Publishing Company.

For the local environment the proportionality between gamma rays and the ISM column density excludes a significant contribution from unresolved compact gamma emitting objects. On a large scale the contribution from an unresolved population of compact gamma-ray stars is expected (e.g. from pulsars (Harding and Stecker, 1981) where only 2 near objects are resolved so far), the amount of such a contribution is still largely unknown. So it is the presence of this quantitatively unknown contribution which introduces the dominating uncertainty into the conclusions.

Another systematic problem could be due to inverse-compton gamma-ray emission from CR-electron interactions with radiation fields, not being coupled to the gas-distribution. But the contribution from this process (Kniffen and Fichtel, 1981) should contribute only a small fraction to the observed emission especially at low galactic latitudes.

The intention of this paper is to describe what the available gamma-ray data (after a first analysis) are telling us on the properties of the ISM and of the CR's at various sites in our Galaxy.

2. THE GAMMA-RAY DATA

A detailed presentation and discussion of the gamma-ray survey is given in Mayer-Hasselwander et al. (1982). Two specific results are outlined here:

-      The Intrinsic Thickness of the Gamma-Ray Disc

The observed angular latitude extent exceeds the angular resolution of the experiment at all longitudes. This made it possible to derive the true angular extent of the gamma-ray emitting disc, a valuable parameter against which a gamma-emission model can be checked. This is shown in Figure 1. Inside the solar circle a narrow component attributed to more distant emission (2 to 6 kpc) appears to be superimposed on the very wide component corresponding to emission from more local structures (1-2 kpc). Both components appear also distinctly offset from the equatorial plane indicating the gamma-ray disc being warped on a large scale.

-      The Gamma-Ray Colour of the Milky Way

As was established in the previous paragraph the high-energy gamma rays observed at low galactic latitudes are a mixture of radiation originating in remote regions - whose counterpart in the latitude profile is the narrow component - superimposed on the contribution of photons produced in less distant regions. Early attempts were made to derive the energy spectrum of the narrow component (Paul et al., 1978; Hartman et al., 1979) in those regions of the galactic disc where such a component could be easily distinguished. However, this approach cannot be used in most of the regions pertaining to the second and third galactic quadrants ($90^{\circ} > l > 270^{\circ}$). A conclusion of the earlier spectral analyses of the galactic gamma-ray emission was that no particular feature could be seen in the measured spectra. In particular the expected π° bump - the trace of the proton-proton collision

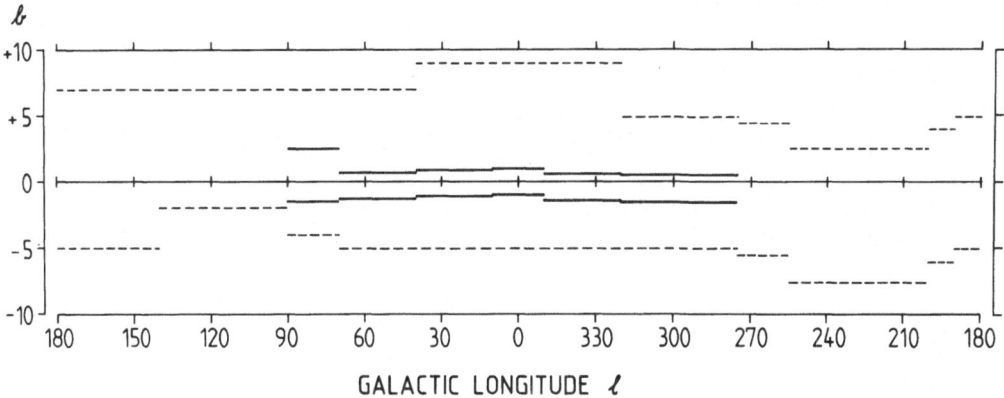

Fig. 1: The intrinsic latitude extent (FWHM) of the narrow component of
the galactic gamma radiation derived by an unfolding procedure (solid line)
and the latitude extent (FWHM) estimated for the wide component (dashed
line).

process - did not appear. This was interpreted in terms of significant contributions of either bremsstrahlung and/or inverse Compton radiation of
CR electrons, especially below 150MeV (Fichtel et al., 1976; Cesarsky et
al., 1978; Strong et al., 1978; Lebrun et al., 1982, Schlickeiser, 1982)

A straightforward illustration of the fact that the π^0-decay type of emission is not the only production process at work in the Galaxy is given in
Figure 2 where the colour index R=F(70-150MeV)/F(150MeV-5GeV) of the
observed radiation in different Milky-Way cells ($-10^0<b<+10^0$) exceeds significantly that of a pure π^0 process. The contribution of point sources has
been subtracted for this analysis only in the 3 longitude intervals
(180^0-190^0, 190^0-200^0, 255^0-275^0) where the emission of isolated bright
sources dominates the underlying galactic background. The validity of the
derived spectral ratio R strongly depends on the correct subtraction of the
instrumental and isotropic celestial background intensity. An error in the
absolute value of the subtracted background would affect regions of weak
disc emission and regions of intense disc emission in a different manner.
Since the analysis in Mayer-Haßelwander et al. (1982) inclusion of additional data led to an improved estimate of the background.

With the revised background spectral ratios as shown in Figure 2 are obtained. In contrast to the earlier result a systematic variation of the spectral ratio with longitude is now indicated: it appears that the spectrum at
longitudes $50^0>l>280^0$ is softer than at the other longitudes. This result
suggests that either the ratio of relevant CR electrons (>100MeV) to CR
protons (>1GeV) is not independent of galactocentric radius or of sites with
low or with very high concentration of dense clouds. Alternatively the
softer spectrum in the inner part of the Galaxy could be explained by a
significant contribution from an unresolved compact gamma-ray source
population, the average spectrum of which clearly is unknown.

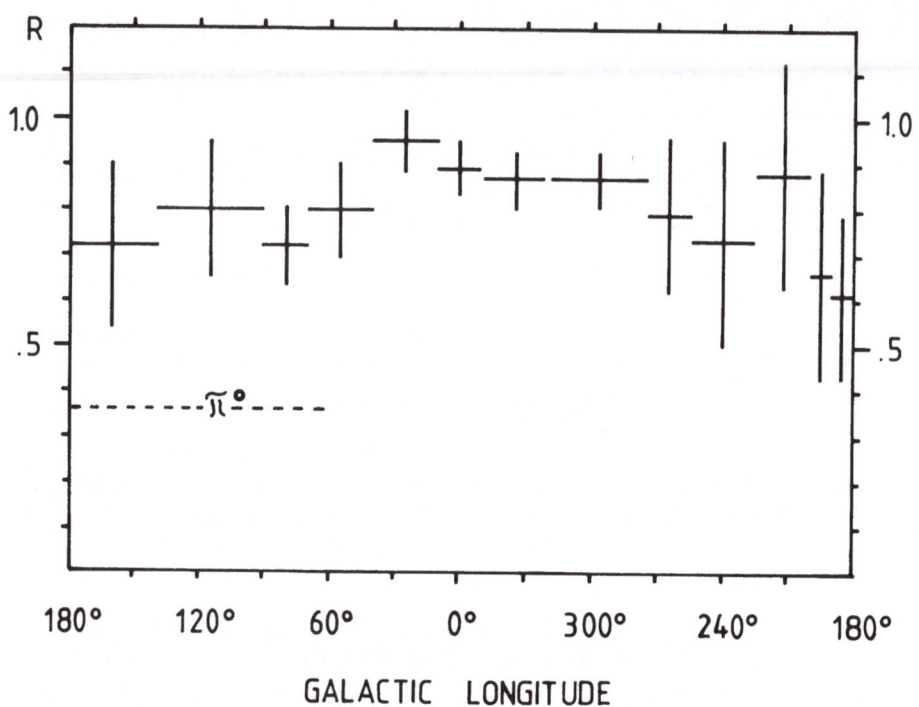

Fig. 2: Colour index R along galactic longitude for the gamma emission of the disc within $|b|<10°$. R is the ratio of the intensity in the range 70-150 MeV to the intensity in the range 150MeV - 5GeV.

3. THE LARGE-SCALE STRUCTURE OF THE GALACTIC GAMMA-RAY EMISSIVITY

The three-dimensional distribution of gamma-ray emissivity in the Galaxy can be inferred from the observed longitude and latitude profiles if one assumes that most of the radiation does not originate in a few small-scale structures. It is also necessary to impose certain assumptions on the large-scale geometrical structure of the Galaxy before any unfolding procedure, attempting to unravel the spatially varying source function from the observed line-of-sight integrals, can give meaningful results.

For this analysis data in the energy range from 150MeV to 5GeV have been used; longitude profiles were derived for the latitude range $|b|<7°$. In order to avoid systematic errors due to the limited angular resolution, further assumptions on the thickness of the emitting disc or, more generally, the dependence of the emissivity on the distance to the galactic plane have to be made. For the inner Galaxy the thickness of the gamma-ray emitting disc is commonly assumed to be constant, whereas an increase of thickness outside the solar circle can be considered (Caraveo and Paul, 1979). The scale height of the model gamma-ray disc has to be chosen to be compatible with the resolved intrinsic width of the latitude profiles as derived in section 2.

The aim of the unfolding technique is to convert the longitude profile of
the observed emission into a radial variation of the emissivity (ε) versus
galactocentric distance without resorting to physical models of source processes and their distribution in the Galaxy. Two forms of the underlying
geometrical pattern of ε have been used: firstly it was assumed that the
emissivity is arranged in galactocentric rings which when viewed from our
position in the Galaxy correspond to the bins of the longitude profile
(Caraveo and Paul, 1979 and references therein). As a second step, particularly in view of the evidence for spiral arms in the longitude profiles, a
spiral pattern can be imposed on the emissivity (Kanbach and Beuermann,
1979). The pitch angle was set to 13° which is consistent with the consen-
sus on the galactic structure as known from radio data.

The principal result of both methods applied to the COS-B measurements is
the radial distribution of gamma-ray emissivity averaged over 0.5kpc rings
as shown in Figure 3. In this figure the result for the spiral-pattern case
is given but the main features of this distribution are equally reproduced
in the method implying cylindrical symmetry. A scale height of 130pc was
chosen in agreement with the derived intrinsic latitude extent of the
gamma radiation from the inner Galaxy. The emissivity ε_0 at 10kpc was
taken to be 2.1×10^{-25} photons ($>$100MeV) cm^{-3} s^{-1} in agreement with the
value derived for the solar neighbourhood by Lebrun et al. (1981) assuming
a density of 1 H-atom cm^{-3}. The first ($l < 90^\circ$) and fourth ($l > 270^\circ$) galactic
quadrants have been treated separately. Towards the outer Galaxy, where
no unfolding can be performed, the scale length of the emissivity decrease
is directly derived from the observed average anticenter intensity and a
value of $\simeq$2kpc is obtained.

A ring of enhanced emissivity is evident in the 3-6 kpc region of the
Galaxy particularly in the fourth quadrant. The gamma-ray emission in this
galactic ring is 2-4 times stronger than in the solar region. From this result it is straightforward to compute the total gamma-ray luminosity of
the Galaxy. The value obtained is 2×10^{42} photon (70MeV-5GeV) s^{-1} equivalent to 10^{39} erg s^{-1} if one assumes an E^{-2} spectral dependence. Previous
analysis performed on the SAS-2 data have yielded values for the galactic
luminosity in agreement with the present result (Bignami et al., 1975;
Strong and Worrall, 1976; Caraveo and Paul, 1979).

An important consideration on the large-scale structure of our Galaxy and
its gas content stems from the comparison of figure 3 with the similar
distribution of atomic and molecular hydrogen as derived from radio measurements for the first quadrant. Both the radio-derived plots and the
gamma-ray ones are in agreement as far as the presence of a single broad
peak at R$\simeq$5kpc. The ratio of the total projected gas densities at the peak
to that at the sun is about 6 according to Solomon and Sanders (1980) and
about 3 according to Gordon and Burton (1976). The more recent work by
Liszt et al. (1980) favours the Solomon and Sanders result. On the other
hand, the emissivity ratio derived from the gamma-ray data is about 3.

Assuming that the gamma-ray emissivity is only due to interactions of
cosmic rays with the interstellar gas, ε traces the product of the densities

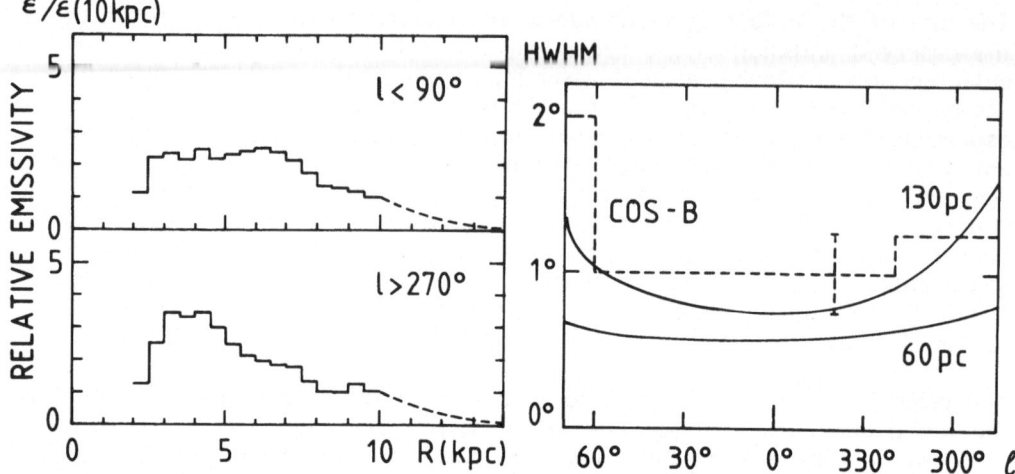

Fig. 3: Radial distribution of gamma-ray emissivity ε in the Galaxy as derived in the unfolding procedure. The normalisation value is $\varepsilon_0(10 \text{ kpc}) = 2.1 \times 10^{-25}$ photon (>100 MeV) cm^{-3} s^{-1} which agrees with the value for the solar neighbourhood for a gas density of 1 H-atom cm^{-3}. The first and fourth quadrant were treated separately.

Fig. 4: Latitude extent of the galactic gamma radiation observed (dashed line) and expected (solid lines) for two different scale-heights used in the unfolding procedure.

of the two components (CR flux and gas). On the assumption of a uniform cosmic-ray density it is seen that the total gas increase as traced by the COS-B gamma rays cannot be more than a factor of 3 between 10 and 5kpc. This value must be regarded as an upper limit also because the gamma-ray data do contain a significant component due to undetected sources (see Hermsen, 1980 and references therein). This upper limit is in contrast (by a good factor of 2) with the Solomon and Sanders distribution. However, even neglecting as unphysical the hypothesis of a negative galactocentric gradient of the cosmic rays, one could postulate that a significant fraction of the gas is not seen by the gamma-ray producing cosmic rays because they cannot penetrate dense clouds. Three arguments stand against this:

- The theoretical works of Skilling an Strong (1976) and of Cesarsky and Völk (1978) show that only very low-energy protons (<50MeV) and electrons fail to penetrate completely a dense cloud;
- in at least one case, that of the Orion cloud complex, COS-B data (Caraveo et al., 1980) show that the totality of the gas is involved in the production of gamma rays;
- most important, if by some mechanism a fraction of the cosmic-ray spectrum, presumably the lower-energy part of it, would be excluded from the gamma-ray producing process with the gas, one should

observe harder spectra from regions of dense clouds, contrary to the result presented in section 2.

One is thus forced to conclude that the amount of gas derived by Solomon and Sanders (1980) for the inner Galaxy cannot be reconciled with the maximum tolerable amount of gas compatible with the gamma-ray data.

Recently, Blitz and Shu (1980) also concluded that the Solomon and Sanders results are an overestimate. Arguing on the basis of the theoretical vs. observed value of the ratio $N(H_2):(^{13}CO)$ as well as on the postulated metallicity gradient in our Galaxy, Blitz and Shu (1980) decreased the peak in the Solomon and Sanders and Gordon and Burton radial H_2 abundance distribution, thus rendering it compatible with the present gamma-ray data. The Blitz and Shu analysis thus supported by the result given here has the additional advantage of reducing the lifetime of molecular clouds to a few 10^7 years, a value more easily acceptable than the several 10^8 years proposed by Solomon and Sanders (1980) and Scoville and Hersh (1979), since such long-lived molecular clouds would have to survive several passages through the spiral arms. A similar conclusion has already been reached by Cohen et al. (1980), on the basis of their much more complete CO data base. Actually in a very recent analysis of the Stony Brook - Massachusetts CO-survey by Sanders (1981) a maximum increase by only a factor 3 in the peak at 6kpc relative to the local surface density is indicated for the total gas. So at present the CO-derived values tend to get in agreement with the upper limit indicated by gamma-ray data under the assumption of constant CR density.

As has been mentioned before the question of the scale height of the gamma-ray disc can now be addressed in more detail since the COS-B measurements provide a value for the intrinsic latitude extent of the gamma-ray Galaxy. The unfolding was performed with two trial scale heights of 60pc and 130pc and line-of-sight integrals were computed for directions out of the plane of the Galaxy. In Figure 4 the resulting model latitude profiles in terms of HWHM are compared with the derived intrinsic width. Even with the uncertainties of the experimental results and the simplifying assumptions of the model a trend is visible: dominant gamma-ray emission with a scale height of much less than 100pc is not supported by the data. If the gas in the inner Galaxy were dominated by molecular hydrogen one would expect the scale height of this gaseous disc to be approximately 60pc (Cohen and Thaddeus, 1977). Therefore this result can be regarded as independent evidence for a rather small contribution of molecular gas in the inner Galaxy at the lower end of the bandwidth of values deducable from CO data.

4. PROSPECTS

High energy gamma-rays appear to be an interesting tracer for the ISM not only for the 'local' galactic area but also for the Galaxy on a larger scale. At present the unresolved contribution from compact gamma-ray sources and the incompleteness of the radio surveys of H I and H_2 in addition to uncertainties due to background and statistics limit the conclusions which can be drawn. More complete surveys and refined analysis procedures are required to make the next steps.

The COS-B mission has ended in April 1982 after nearly 7 years of successful operation. So the database can be brought into a final form now; best possible background corrections and refined analysis procedures will be applied and will allow to exploit the data to their full potential. On this basis significant progress in the near future is expected by the detailed comparative analysis of the now complete gamma-ray database in relation to the new H I survey for the southern hemisphere (A.W. Strong, priv. communication) and of the extended CO-surveys for both hemispheres which are partially completed or in preparation (P.Thaddeus, priv. communication).

For example the details of the latitude distributions hopefully will allow to disentangle the parameters CR density and H_2-density as a function of galactocentric radius. For various sites of the Galaxy a reliable calibration of the $N(H_2)/N(CO)$ ratio then should be possible.

REFERENCES

Bignami, G.F., Fichtel, C.E., Kniffen, D.J., Thompson, D.J.: 1975, Ap.J.
    199, 54
Bignami, G.F., Morfill, G.E.: 1980, Astron. and Astrophys. 87, 85
Blitz, L., Shu, F.H.: 1980, Ap.J. 238, 148
Caraveo, P.A., Paul, J.A.: 1979, Astron. and Astrophys. 75, 340
Caraveo, P.A., Bennett, K., Bignami, G.F., Hermsen, W., Kanbach, G.,
    Lebrun, F., Masnou, J.L., Mayer-Hasselwander, H.A., Paul, J.A., Sacco,
    B., Scarsi, L., Strong, A.W., Swanewnburg, B.N., Wills, R.D.: 1980,
    Astron. and Astrophys. 91, L3
Cesarsky, C.J., Völk, H.J.: 1978, Astron. and Astrophys. 70, 367
Cesarsky, C.J., Paul, J.A., Shukla, P.G.: 1978, Astrophys. Space Science 59,
    73
Cohen, R.S., Cong, H., Dame, T.M., Thaddeus, P.: 1980, Ap.J. 239, L53
Cohen, R.S., Thaddeus, P.: 1977, Ap.J. 217,L155
Fichtel, C.E., Kniffen, D.A., Thompson, D.J., Bignami, G.F., Cheung, C.Y.:
    1976, Ap.J. 208, 211
Fichtel,C.E., Simpson,G.A.,Thompson,D.J.: 1978, Ap.J. 222,833
Gordon, M.A., Burton, W.B.: 1976, Ap.J. 208, 346
Harding, A.K. and Stecker, F.W.: 1981, Nature 290,316
Hartman, R.C., Kniffen, D.A., Thompson, D.J., Fichtel, C.E., Ögelman,
    H.B., Tümer, T., Özel, M.E.: 1979, Ap.J. 230, 597
Hermsen, W.: 1980, Ph.D. Thesis, University of Leiden
Kanbach, G;, Beuermann, K.: 1979, Proc. 16th I.C.R.C. (Kyoto), 1, 75
Kniffen, D.A., Fichtel, C.E.: 1981, Ap.J. 250,389
Lebrun, F., Paul, J.A.: 1979, Astron. and Astrophys. 79,153
Lebrun, F., Bignami, G.F., Buccheri, R., Caraveo, P.A., Hermsen, W.,
    Kanbach, G., Mayer-Hasselwander, H.A., Paul, J.A., Strong, A.W.,
    Wills, R.D.: 1982, Astron. and Astrophys. 107, 390

Liszt, H.S., Xiang, D., Burton W.B.: 1980, preprint No. 80-43, University of
    Minnesota, Dept. of Astronomy
Mayer-Hasselwander, H.A., Bennett, K., Bignami, G.F., Buccheri, R.,
    Caraveo, P.A., Hermsen, W., Kanbach, G., Lebrun, F., Lichti, G.G.,
    Masnou, J.L., Paul, J.A., Pinkau, K., Sacco, B., Scarsi, L., Swanenburg,
    B.N., and Wills, R.D.: 1982, Astron. and Astrophys. 105,164
Paul, J.A., Bennett, K., Bignami, G.F., Buccheri, R., Caraveo, P., Hermsen,
    W., Kanbach, G., Mayer-Hasselwander, H.A., Scarsi, L., Swanenburg,
    B.N., Wills, R.D.: 1978, Astron. and Astrophys. 68, L31
Sanders, D.B.: 'The distribution of molecular clouds in the Galaxy', NRAO
    workshop on Extragalactic Molecules, Green Bank, Nov. 1981, preprint
Schlickeiser, R.: 1982, Astron. and Astrophys. 106,L5
Scoville, N.Z., Hersh, K.: 1979, Ap.J. 299, 578
Solomon, P.M., Sanders, D.B.: 1980 in "Giant Molecular Clouds in the
    Galaxy", eds. P.M. Solomon and E.G. Edmunds, Pergamon Press, P. 41
Skilling, J., Strong, A.W.: 1976, Astron. and Astrophys. 53, 253
Strong, A.W., Bignami, G.F., Bloemen, J.B.G.M., Buccheri, R., Caraveo,
    P.A., Hermsen, W., Kanbach, G., Lebrun, F., Mayer.Hasselwander, H.A.,
    Paul, J.A., and Wills, R.D.: 1982, Astron. and Astrophys., in press
Strong, A.W., Worrall, D.M.: 1976, J. Phys. A. Math. Gen. 9, 823
Strong, A.W., Wolfendale, A.W., Bennett, K., Wills, R.D.: 1978, M.N.R.A.S.
    182, 751
Strong, A.W., Wolfendale, A.W.: 1981, Phil.Trans.Roy.Soc.Lond.A. 301,541
Strong, A.W., et al.: 1982, submitted to Astron. and Astrophys.

THE ROTATION CURVE OF THE GALAXY

G. R. Knapp
Princeton University Observatory, Princeton, NJ08544, U.S.A.

I. INTRODUCTION

The determination of the rotation curve of the Galaxy, $\Theta(R)$, where Θ is the circular velocity at galactocentric distance R, involves the synthesis of a large body of complementary (and sometimes contradictory) data. The synthesis has to be guided to some extent by theoretical considerations such as studying the distribution of mass in the Galaxy, i.e. by what one "expects" to find. For many years, Galactic structure studies have been guided by the seminal work of Schmidt (1965).

For practical reasons, the determination of $\Theta(R)$ breaks down into four subtopics:
(a) Determination of the value of R_O, the distance between the Sun and Galactic center, and Θ_O, the circular velocity at this distance. The radial velocity of an object in the Galactic plane in circular motion is

$$V_r = (\Theta(R)R_O/R - \Theta_O) \sin\ell \qquad (1)$$

Thus, for kinematic studies of Galactic structure, the actual value of R_O is unimportant because all kinematic distances are scaled to it. This scaling provides a method for determining R_O, as will be discussed below.
(b) The rotation curve between 1 kpc $<R <R_O$. In this region, the rotation curve can be determined from the velocity extrema of HI (or CO) profiles observed at or near $b = 0^{\circ}$; the rotation curve is directly obtained from these tangent-point velocities V_T via

$$V_T = \Theta(R) - \Theta_O R/R_O \qquad (2)$$

In this region of the Galaxy, kinematic distances are also independent of the value of Θ_O, since a combination of (1) and (2) gives

$$V_r = V_T(R)R_O \sin\ell/R$$

W. L. H. Shuter (ed.), Kinematics, Dynamics and Structure of the Milky Way, 233–247.

(i.e. the distances are found from a comparison of two relative velocities).
(c) The rotation curve for $R > R_0$. Here the rotation curve is determined using eq. (1), and a group of objects is needed for which V_r and
d, the heliocentric distance, are known.
(d) $R < 1$ kpc. The kinematics of material in the inner Galaxy are very
complex (see the review by Oort 1977), and much of the gas is not in
circular motion or in the Galactic plane. It may be intrinsically
impossible to obtain a reliable rotation curve in this region. This is
particularly unfortunate because much higher linear resolution is
available for the center of our Galaxy than for any other, and there
are many important questions related to the mass distribution in the
inner regions of the Galactic spheroid.

The determination of the rotation curve is further
complicated by many other aspects of kinematics in a real Galaxy, such as:
- the possibility of a non-circular residual motion of the Local
Standard of Rest (e.g. Shuter 1982).
- perturbation of the local velocity field by the local stellar system
Gould's Belt (Lindblad, this conference)
- non-axisymmetry in the galactic disk (warps, bars, flares, etc.)
- differences in the kinematics of different disk components
- non-circular motions (streaming and cloud-cloud turbulence). Local
non-circular motions may strongly affect the determination of the
rotation constants A and B. Non-circular motions are of quite large
amplitude ($\sim 10\%$) relative to the value of the rotation velocity.

II. THE GALACTIC ROTATION CONSTANTS R_0 AND Θ_0

(a) R_0: Values of R_0 between 7 and 11 kpc have been suggested in recent
years, with the weight of evidence favouring (though not conclusively)
somewhat lower values than the I.A.U. standard of 10 kpc. There are
at present basically three approaches to determining R_0, all more or
less indirect.

(1) Determination of the centroid of the halo population: Shapley's
original demonstration that the Sun lies several kpc from the Galactic
center was based on the distance to the center of the globular cluster
population. Recent discussions of the globular cluster distribution
have suggested $R_0 = 8.5 \pm 1.6$ kpc (Harris 1976) and $R_0 = 6.8 \pm 0.8$ kpc
(Frenk and White 1982). The distances to the RR lyrae stars in the
directions of low obscuration towards the Galctic center give $R_0 =$
8.7 kpc (Oort and Plaut 1975). These determinations all rest on the
measurement of the absolute magnitudes of RR Lyrae stars and of the
effects of metallicity on the absolute magnitudes. The RR Lyrae data
have been combined with data on Mira variables in the same fields to
give $R_0 = 9.0 \pm 0.6$ kpc by Glass and Feast (1982).

(2) Distances to OB stars near $b = 0°$: Crampton et al. (1976) find $R_0 \sim$
8 kpc while Balona and Feast (1974) find $7.7 < R_0 < 10.9$ kpc. These
determinations may be strongly influenced by streaming motions in the
OB star population.

(3) Comparison of kinematic and spectroscopic distances: This method is rather similar to (2) above, and uses spectroscopic distances to the exciting stars of HII regions compared with the kinematic distances to the HII regions, whose velocities are measured by spectroscopy of the hot or cold gas associated with the region. The method has recently been applied to observations in the fourth quadrant by Quiroga (1980), who finds $R_O = 8.4 \pm 1$ kpc from HII region data.

A similar analysis can be carried out for the first quadrant using the data recently acquired by Blitz, Fich and Stark (1982). These data consist of spectroscopic distances to the HII regions with the radial velocities being measured by CO observations of the associated molecular cloud. Kinematic distances may be found from (1) and

$$d(kin) = R_O cos\ell \pm (R^2 - R_O^2 sin^2\ell \;)^{1/2} \qquad (3)$$

with the "near" kinematic distance assumed. Values of $d(kin)$ were found using the Schmidt (1965) model, which assumes $R_O = 10$ kpc, and the results are shown in Figure 1. The relationships expected for

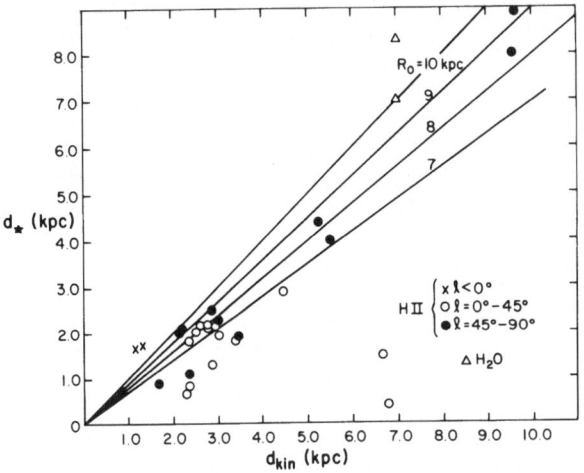

FIGURE 1. Plot of spectroscopic distances d(*) versus kinematic distances d(kin) for a sample of HII regions in the first Galactic quadrant; the latter are calculated assuming the Schmidt model ($R_O = 10$ kpc). Also plotted are the distances to two components of the W51 complex (Genzel et al. 1981b; Schneps et al. 1981).

various values of R_O are also drawn in. The error bars are large; for the spectroscopic distances, the random errors are $\sim \pm 30\%$, while for the kinematic distances the errors are $\sim \pm 2$ kpc. The two points with large kinematic distances but small spectroscopic distances may be due to the detection of high-velocity molecular gas along the line of sight

not associated with the HII region, and are ignored in the following
discussion. The fact that the points do not lie along the $R_O = 10$ kpc
line is due presumably to systematic error, either in the optical
distance scale or in the value of R_O; assuming the latter, we find that
$R_O = 8\pm2$ kpc, in good agreement with the analysis of fourth-quadrant data
by Quiroga (1978).

A new method for measuring the distances to HII regions, the moving-
cluster method applied to clusters of H_2O masers, has recently been
described by Genzel et al. (1981a) and is in principle applicable
throughout the Galaxy and free of systematic error. Distances have been
measured to two components of the W51 complex by this method (Genzel
et al. 1981b and Schneps et al. 1981). Unfortunately the W51 complex
is along the line of sight to the tangent point to the Sagittarius
spiral arm, and strong streaming motions are observed in the gas along
this arm. Assuming that both components are at the tangent point, the
kinematic distance is 7 ± 2 kpc ($R_O = 10$ kpc). These points are also
plotted in Figure 1. They give a value for R_O of 10.5 ± 4 kpc. The
determination of R_O by this method to 10% will probably require
measurements of about 20 sources.

The product AR_O may be found from the slope of the tangent-
point curve V_T versus $\sin l$ near R_O ($l = 90°$). HI data for the first
and fourth quadrants give $AR_O = 110\pm3$ km/s (Gunn et al. 1979), while
CO data give 103 ± 5 km/s (McCutcheon, this conference). Interestingly
enough, the tangent point data for both the HI and CO do not suggest
nearly such strong local streaming motions as do the OB star data
(papers by Ovenden, Pryce and Shuter, this conference). Published
values of A allow $6.5 < R_O < 10$ kpc.

In conclusion; the value of R_O is quite uncertain. Probably
the best current value is 9 ± 2 kpc, but the real possibility exists
that there are systematic effects in the distance scales of both the
young and old stellar populations. A value of 8.5 kpc will be assumed
in the rest of this paper in deriving the rotation curve of the Galaxy.

(b) Θ_O: The only direct method for determining Θ_O, using the Oort
formula $\Theta_O = R_O(A-B)$ defines Θ_O to $150 < \Theta_O < 340$ km/s. Other methods
for estimating Θ_O involve the measurement of the motion of the LSR
with respect to some group of very different objects whose motions
can be assumed to be different enough from that of the LSR that they
average to zero or to some definable quantity. This method has been
applied to the globular cluster system by several authors, including
Oort (1965), Woltjer (1975), Hartwick and Sargent (1978), and Frenk
and White (1980). Values of Θ_O between 180 and 230 km/s are found.

The value of Θ_O has also been estimated using the HI emission
from the distant edges of the galactic disk by Knapp, Tremaine and
Gunn (1978), who find $200 < \Theta_O < 230$ km/s. This analysis depends on
the constancy of the scale length of the Galactic disk emission with
azimuth; while this assumption seems to hold fairly well between $l = 90°$

and 225°, recent data for the fourth quadrant (Kerr, Jackson and Kerr
1982) suggest that the Galaxy may be lopsided (see Baldwin et al. 1980),
with extent to greater radii in the fourth quadrant - a similar analysis
of these data suggests Θ_0 = 250 km/s (Jackson, this conference).

In similar vein, an analysis of the Magellanic stream data
(Mathewson et al. 1974) suggests $\Theta_0 \sim$220 km/s. On the other hand, the
velocity of the LSR with respect to the centroid of the Local Group
is $\sim$300 km/s (Lynden-Bell and Lin 1977; Yahil, Tammann and Sandage
1977). This last value will be further discussed later.

A very straightforward and powerful method has recently been
applied to the motions of the galactic spheroid. This method makes use
of the fact that, if motions of the spheroidal component are isotropic
and the circular velocity is not changing rapidly with radius, then Θ_0
= $\sqrt{3}\ \sigma$, where σ is the one-dimensional velocity dispersion of the
spheroid. This relationship holds for the globular cluster population
of M31 (Tremaine 1982) and for the bulge components of many spirals
(Knapp et al. 1982). Using data for the velocity dispersion of the
bulge of the Galaxy, Einasto et al. (1979) find Θ_0 = 225±10 km/s, while
Lynden-Bell and Frenk (1982) analyse globular cluster radial velocities
to find Θ_0 = 212±16 km/s.

Thus most of the current data give a value for Θ_0 of about
220 km/s, with a probable uncertainty of about 20 km/s. This value
will be adopted in the remainder of this paper.

How, then, to understand the large velocity of the LSR with
respect to members of the Local Group? It is likely that some large
contribution to this is made by the motion of the center of mass of the
Galaxy. The magnitude and direction of this component can be calculated
once the value of Θ_0 is known (or, more properly, assumed) - the
direction of motion on the sky is plotted for various values of Θ_0 in

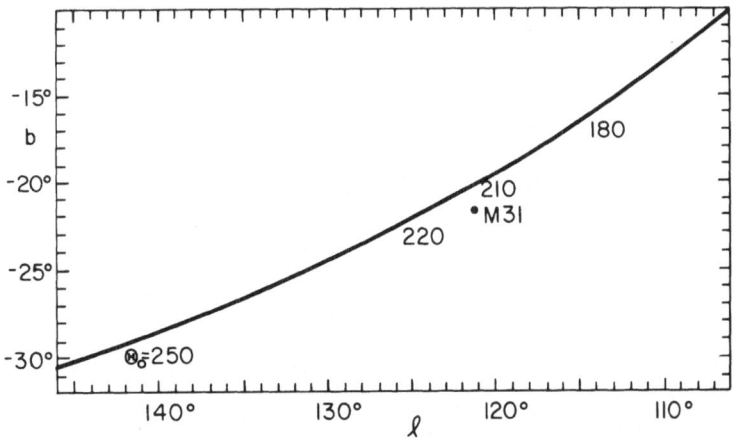

FIGURE 2. Plot of the apex of the motion of the center of mass of the
Galaxy on the sky for the solar motion of Yahil et al. (1977) and
several assumed values of Θ_0.

Figure 2. It can be seen that for any reasonable value of Θ_o, the Galaxy
and M31 are approaching each other on essentially radial orbits. More
thorough discussions of this important result are given by Gunn et al.
(1979), Einasto and Lynden-Bell (1982), and Lynden-Bell (this conference).
The total mass for the Galaxy inferred from the timing argument is $\sim 10^{12}$
$M_\odot$ (Θ_o = 220 km/s); the time to collision of the two systems is about
5 billion years.

III.   THE ROTATION CURVE BETWEEN 1 AND 8.5 KPC

        In principle, the measurement of the rotation curve in this
region is simple, since the tangent-point data are used in a straightforward derivation. In practise, the method is fraught with difficulties
created by the presence of non-circular motions, i.e. streaming
and turbulence. Evaluation of the rotation curve involves the assumption
that these two motions are separable from each other and the "underlying" circular velocity, and that the streaming motions average to
zero over large enough regions of the Galaxy. Analysis of the streaming
and turbulent motions is necessary for deciding where on a line profile
the tangent-point velocity should be read. The main requirement for
the definition is that it should ensure that the tangent point velocity
read from the profile at ℓ= 90° should be 0 km/s.

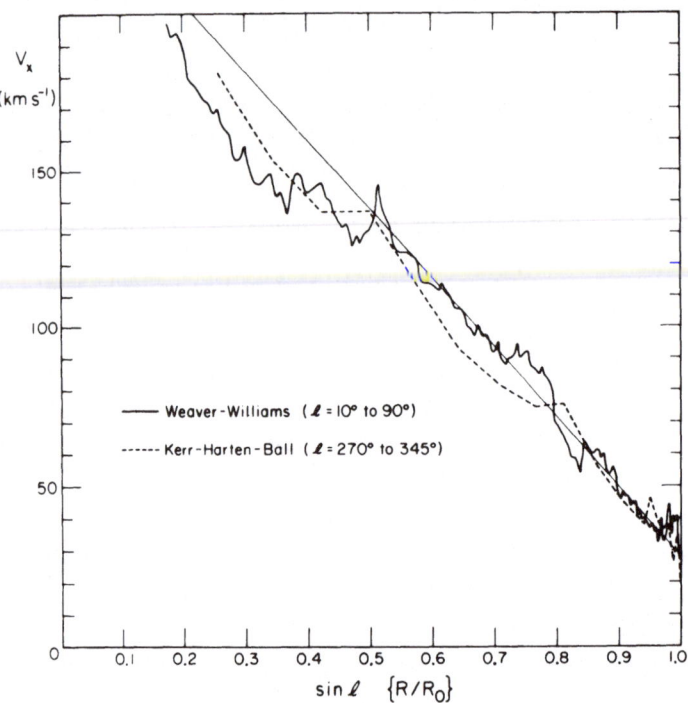

FIGURE 3. Plot of the velocity of HI at T_b = 1 K, versus sin ℓ.   Data
for the first and fourth quadrants are shown.

One attempt to do this is illustrated in Figure 3, from Gunn et al. (1979), where the velocity of a low brightness temperature contour was measured (using the data of Weaver and Williams 1974), and a constant velocity offset subtracted to give the tangent point velocity. The extreme-velocity gas tends to have a large latitude extent and to have velocity roughly constant with latitude (Lockman, this conference), and the emission may arise from a hotter, less cloudy component of the interstellar medium than does the more intense HI emission. As an aside, it is worth pointing out that the tangent-point analysis described by Gunn et al. (1979), the z-structure of the HI disk described by Lockman, and the measurement of the HI extent of the disk to large radii described in the next section all make use of the low-level HI emission at the velocity extrema of the profiles. Thus, somewhat paradoxically, information about small-scale galactic structure (spiral arms, etc.) comes from the intense (T $\sim$100 K) HI emission while information about the structure on very large scales (tens of kpc) is given by the low-level (T $\sim$1 K) emission, and adequate velocity coverage, baseline stability, and sensitivity to better than 1 K are required in Galactic HI surveys.

Immediately striking in Figure 3 is the straightness of the V_T versus $\sin\ell$ relationship between $\sin\ell$ = 0.5 and 1.0, apparent in the data from both the first and fourth quadrants. This shows that the rotation curve is linear over this range. The slope of the tangent-point curve is 220±4 km/s, so that, if Θ_O = 220 km/s, the rotation velocity is constant with radius over this region of the Galaxy. For R/R_O less than about 0.5, the rotation velocity falls slightly below this mean value. These trends are confirmed by the analysis of the globular cluster motions by Lynden-Bell and Frenk (1982).

Although the method of analysing the HI data described above seems to use the hotter HI, the kinematics of the HI and CO seem to be similar. In Figure 4, the first-quadrant tangent-point velocities for the HI and CO are compared, and the agreement is excellent. The CO velocities are obtained from a set of fully-sampled latitude-velocity maps at many longitudes (Knapp, Stark and Wilson 1982). The CO tangent-point velocities are found by averaging the latitude maps to make a composite profile at each longitude, and fitting a Gaussian profile to the last 50 km/s or so of the profile. The velocities of both the HI and CO tangent-point data agree well with those of Burton and Gordon (1978). In Figure 5 is a composite rotation curve for this region of the Galaxy found from these data and the adopted values of R_O and Θ_O. The rotation velocity drops below 220 km/s between about 1 and 4 kpc; the greater distance represents the inner edge of the molecular (and infrared) ring in the Galaxy. If, like most other galaxies with pronounced inner rings, our Galaxy also has a large (4 kpc radius) bar, the tangent-point velocities below R $\sim$4 kpc may no longer represent the true circular velocity. In the rotation curves compiled by Bosma (1981) there is some suggestion that rotation curves with this sort of shallow dip arise in galaxies with oval distortions in the isovelocity contours (due to a bar?). However, many more data are needed relating to this

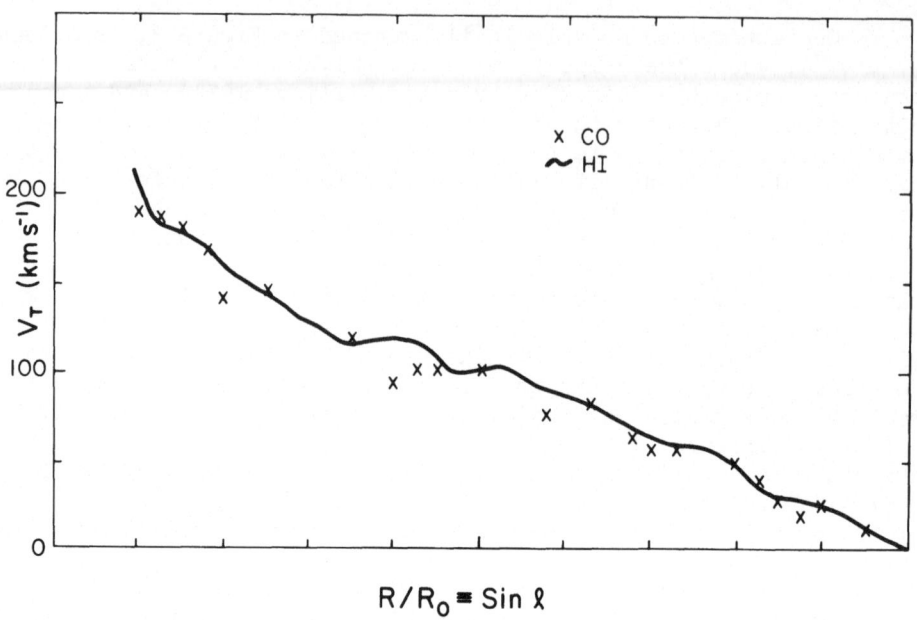

FIGURE 4. Comparison between the tangent-point velocities of CO (from Knapp, Stark and Wilson 1982), and of HI (Gunn et al. 1979) in the first quadrant.

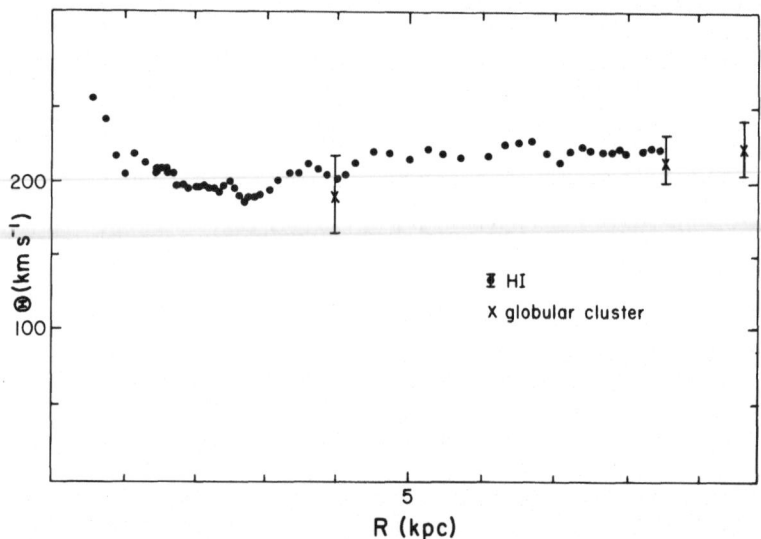

FIGURE 5. Rotation curve for the Galaxy between 1 and 8.5 kpc, from first quadrant data (Weaver and Williams 1974). The circular velocities found from globular cluster motions (Lynden-Bell and Frenk 1982) are also plotted.

question before any such conclusion can be drawn.

IV. THE ROTATION CURVE BEYOND R_O

Here, the rotation curve must be determined using a set of objects of known distance and radial velocity. Several recent advances have been made in finding the rotation curve in this region. Jackson, FitzGerald and Moffet (1979) have measured the spectroscopic distances to HII regions beyond the solar circle, and Blitz, Fich and Stark (1982) have measured the radial velocities of a large number of HII regions by observing the CO emission from their associated molecular clouds. These data extend the rotation curve to about 7 kpc beyond the solar distance. The scatter in the data is very high, and much of it is likely to be intrinsic to the sample of objects; the smallish radial velocities, for example, are likely to contain a large contribution

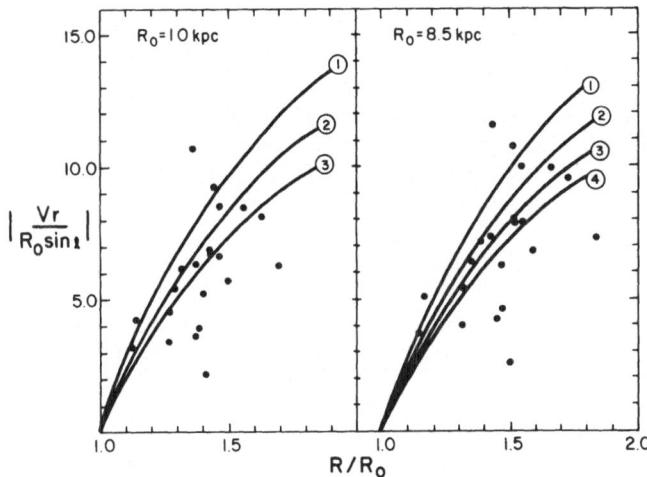

FIGURE 6. Plot of $\omega - \omega_o$ vs. R/R_O for HII regions beyond the solar circle (a) assuming R_O = 10 kpc, and (b) assuming R_O = 8.5 kpc.

from non-circular motions. In Figure 6 is a plot of $|V_r/R_O\sin\ell| = \omega - \omega_o$, versus R/R_O; the former quantity is derived from the radial velocities and the latter from distances; the sample used is that of FitzGerald, Jackson and Moffet (1979). The values of R/R_O are calculated for R_O = 10 and 8.5 kpc. Also shown in Figure 6 are the expected relationships for various rotation curves. In Figure 6a the relationships for (1) the Schmidt curve and (2) and (3) for values of Θ_o constant at 250 and 220 km/s respectively are shown. In Figure 6b, the relationships for (1) and (2) constant circular velocities of 250 and 220 km/s and (3) and (4) rotation velocity rising with radius as $\Theta = 220(R/R_O)^{0.1}$ and $220(R/R_O)^{0.2}$ are shown. The data are very noisy, but best agreement is provided with curves with values of Θ_o smaller than 250 km/s and rising slowly with radius. For curves (3) and (4), Figure 6b, the rotation velocity reaches 235 km/s and 253 km/s respectively at 17 kpc.

The data presented in Figure 6, and the more extensive compilation by Blitz et al. (1982) provide a worrisome disagreement with previous studies of local motions. The values of A and B obtained directly or from velocity ellipsoid data give $|A| > |B|$ in the solar neighbourhood (see the reviews by Schmidt 1965, Caldwell and Ostriker 1981 and Mihalas and Binney 1981), which means that the circular velocity is decreasing in the solar neighbourhood, in disagreement with Figure 6. This disagreement also appears in a somewhat different form in the discussion of the local stellar velocity field by Pryce, Ovenden and Shuter (this conference). Since the objects examined are the same, or at least overlap strongly in the various samples, one suspects that these discrepancies arise in the analyses used, and must be sorted out before we can understand local Galactic rotation.

Beyond the distances reached by these HII region samples, no further direct information on the behaviour of $\Theta(R)$ is available. Analyses of the globular cluster velocities suggests that $\Theta(R)$ is constant with a mean value of $\sim 220 \pm 15$ km/s to $R \sim 60$ kpc (Hartwick and Sargent 1978; Lynden-Bell, this conference). Analysis of the distant HI emission finds Θ constant with a similar value to about 45 kpc (Knapp et al. 1978), although the lopsided HI distribution in the Galaxy (Jackson, this conference) makes this analysis less useful.

The constancy of the rotation velocity to these large distances demonstrates, as all the world knows, the presence of an unseen component in the Galaxy whose mass grows with distance. Inferences about the distribution of this mass must be drawn indirectly from the scanty available data. The HI emission from large distances shows that the layer thickness increases approximately linearly with increasing distance. The HI scale height as a function of R, derived from the data of Knapp et al. (1978) is shown in Figure 7. Also shown are the expected relationships for various mass distributions. The mass implied by the rotation curve (Θ constant with R) is assumed distributed in a central spheroid (the bulge of the Galaxy), a disk (these two components representing the visible parts of the Galaxy), and a dark halo, assumed to have various degrees of flattening. The scale height is then calculated assuming that the velocity dispersion stays constant at $\sigma_z = 10$ km/s. Figure 7 shows that the flattening of the unseen matter is no greater than 2:1, and the data are consistent with a spherical distribution of matter (see Gunn 1980).

Beyond distances of ~ 50 kpc, the Galactic mass distribution is inferred from the motions of the distant satellites (e.g. the Magellanic stream analysis of Mathewson et al. 1974). Further yet, the total mass is derived from the gravitational interaction between the Galaxy and other members of the Local Group. A mass of $\geqslant 10^{12}$ $M_\odot$ is derived from the timing argument applied to the orbits of M31 and the Galaxy (Gunn et al. 1979; Einasto and Lynden-Bell 1982; Lynden-Bell, this conference). If the mass is distributed to make the circular

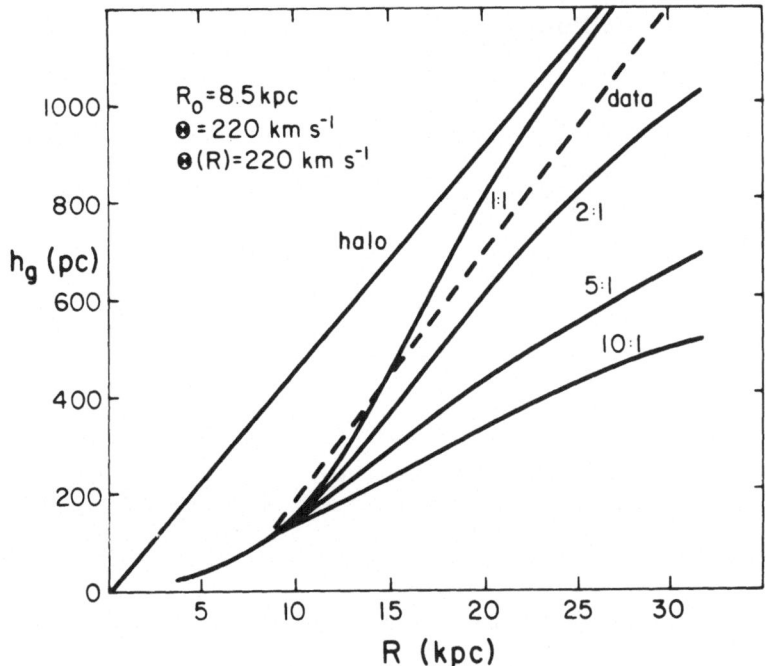

FIGURE 7.   Dependence of HI scale height h_g on radius R for the Galaxy.
The data are shown by a dashed line, and model distributions by
solid lines (Gunn 1980). (see text).

velocity constant at 220 km/s, the total radius of the Galaxy is ∿100
kpc, and the satellites (the globular clusters, the dwarf spheroidals
and the Magellanic clouds) are within the Galaxy.

V.   THE ROTATION CURVE AT R < 1 KPC

        In Section III, the possibility that the tangent-point velocity
no longer measures the circular velocity at R < 4 kpc was alluded to;
the situation grows much worse at R <1 kpc; it is entirely possible
that a direct measurement of the rotation curve here is out of reach.
Spectral-line maps of the inner Galaxy show that much of the gas is in
non-circular orbits, and at velocities forbidden by circular rotation
(e.g. Sanders et al. 1977; Bania 1977; Heiligmann 1982). The kinematics
and dynamics of the inner Galaxy are beyond the scope of this review,
and the ability of this reviewer; for the present purpose, the data
will be examined to see if any gas in circular motion can be found.
Figure 8 shows velocities measured from profile or map extreme velocities,
plotted as θ versus R, within 3 kpc. Within about 400 pc, the velocities

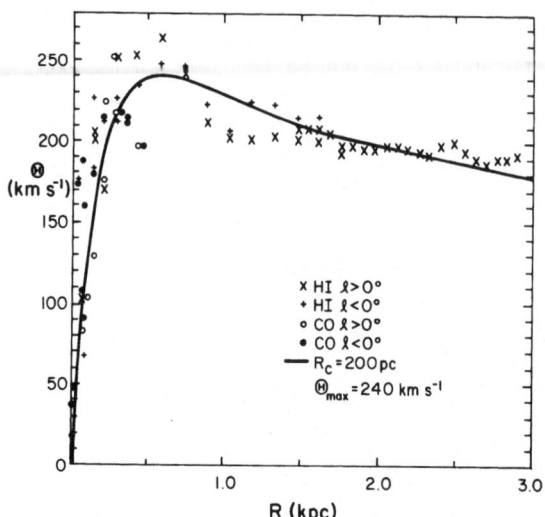

FIGURE 8. Rotation curve
for the inner 3 kpc of
the Galaxy (see text).

of all peaks not at forbidden velocities are shown. Also sketched in
Figure 8 is the distribution of circular velocity expected for a
spherical mass distribution, with

$$\rho(R) = \rho_0 \Big/ \left(1 + \left(\frac{R}{R_c}\right)^2\right)^{3/2} \qquad (4)$$

In Figure 8, the core radius is 200 pc and the peak velocity 240 km/s.
At R $\lesssim$ 1.2 kpc, the spheroid provides a reasonable fit to some of the
data, i.e. gas in circular motion appears to be present (this is the
"nuvlear disk" - see Oort 1977).

VI. SUMMARY, MASS MODEL, AND GENERAL REMARKS

The data discussed above suggest that the rotation velocity
in the Galaxy is roughly constant at about 220 km/s, with small and
large-scale fluctuations of about 10% of the rotation velocity, from
the inner few 100 pc of the Galaxy to distances of up to 100 kpc. The
approximate constancy of Θ with R is taken as evidence for the existence
of a massive unseen halo with $\rho \sim R^{-2}$. Detailed mass modelling of the
Galaxy is described by Ostriker at this symposium; here, only a few
general remarks are made. The Galaxy is known to have its mass
distributed among three main components, which may be simply described
by the central R^{-3} spheroid discussed above, the disk, and the unseen
halo.

The rotation curves expected for each of these components
separately are sketched in Figure 9. The surface density distribution
of the disk is

$$\Sigma = \Sigma_0 \exp\left(-\lambda \cdot \frac{R - R_0}{R_0}\right) \qquad (5)$$

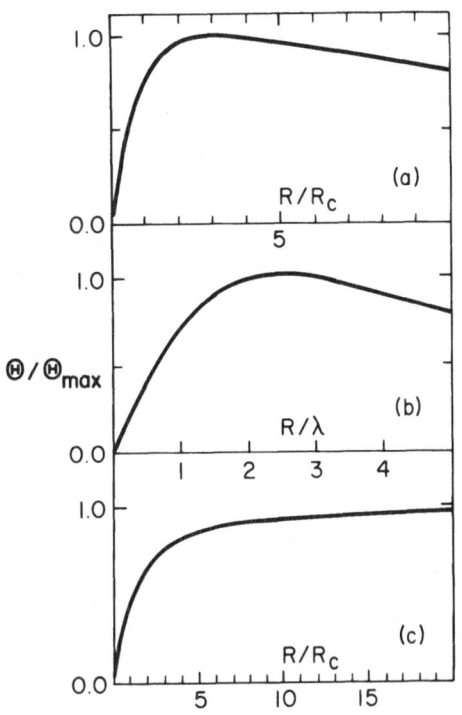

FIGURE 9. Rotation curves for (a) a spherical mass distribution with $\rho \sim R^{-3}$, (b) a thin disk with surface density $\sim \exp(-R/\lambda)$, and (c) a spherical distribution with $\rho \sim R^{-2}$.

and the density distribution of the halo

$$\rho = \rho_0 \Big/ \left(1 + (\tfrac{R}{R_c})^2 \right) \qquad\qquad (6)$$

Figure 9c shows that, in fact, the density distribution given by (6) does not give a 'flat' rotation curve - it is many core radii before Θ reaches its asymptotically constant value. For moderate values of R, say 5 to 10 core radii, the rotation velocity rises roughly linearly with R, with $\Theta \sim R^{\alpha_1}$, as is observed for many systems (Rubin, this conference) - indeed, the curve of Figure 9c bears a reasonable resemblence to those observed for late-type systems with small bulges.

The point of Figure 9 is that the approximate constancy of with R is due to the combination of all three components. That is, the value of Θ is set first of all by the spheroid, then by the disk at intermediate distances, and finally by the halo at large distances. There is, further, evidence that the rotation velocity rises to about 200 km/s within 1 kpc of the Galactic center (Lacy et al. 1980). The smooth rotation curves seen in our own and other galaxies cannot arise from a superposition of components randomly drawn from the available distribution (rotation velocities of spirals range from 50 to 350 km/s); large steps in the rotation velocity as one component yields to the next in gravitational influence do not, as far as we know, occur. The distribution of mass in galaxies, in other words, is determined during the formation process and is likely set by gravitational interaction among the components.

      Comparisons between our own and other galaxies have proven very useful in both directions. Comparison of the Galactic rotation curve with curves for other galaxies (Rubin, this conference) provides another piece of evidence in favour of the classification of the Galaxy as type SbcII (de Vaucouleurs 1979). What is at present lacking is a compilation of rotation curves of galaxies with strong bars at various orientations and inclinations - the degree to which the Galaxy is barred is proving quite difficult to pin down.

      A last uncertainty to be remarked on (there are very many more) lies in the present census of Local Group membership, which is at present incomplete because of Galactic obscuration. The presence of another massive galaxy in the Local group could destroy the simple dynamical experiment provided by the Galaxy and M31, so elegantly exploited by Lynden-Bell and others. Nothing has so far turned up in infrared or HI surveys, but we may still be in for a surprise.

To the collaborators in this work, Jim Gunn, Scott Tremaine, Tony Stark and Bob Wilson, I wish to express deepest appreciation for the learning and enjoyment I have had working with them. This research is supported by the National Science Foundation via grant AST-8009252 to Princeton University.

REFERENCES

Baldwin, J.E., Lynden-Bell, D., and Sancisi, R. 1980, MNRAS 193, 313.
Balona, L.A., and Feast, M.W. 1974, MNRAS 167, 621.
Bania, T. 1977, Ap.J. 216, 381.
Blitz, L., Fich, M., and Stark, A.A. 1982, Ap.J. (Suppl) (in press).
Dosma, A. 1901, A.J. 86, 1825.
Burton, W.B., and Gordon, M.A. 1978, Astron. Astrophys. 63, 7.
Burton, W.B., Gallagher, J.S., and McGrath, M. 1977, Astron. Astrophys.
    Suppl. 29, 123.
Caldwell, J.A.R., and Ostriker, J.P. 1981, Ap.J. 251, 61.
Crampton, D., Bernard, D., Harris, B.L., and Thackeray, A.D. 1976,
    MNRAS 176, 683.
Einasto, J., Haud, U., and Jõeveer, M. 1979, in "The Large-Scale
    Characteristics of the Galaxy", I.A.U. Symposium 84, ed. W.B. Burton,
    D. Reidel Co., p. 231.
Einasto, J., and Lynden-Bell, D. 1982, MNRAS 199, 67.
Frenk, C.S., and White, S.D.M. 1980, MNRAS 193, 295.
Frenk, C.S., and White, S.D.M. 1982, MNRAS 198, 173.
Genzel, R., Reid, M.J., Moran, J.M., and Downes, D. 1981a, Ap.J. 244, 884.
Genzel, R., Downes, D., Schneps, M.H., Reid, M.J., Moran, J.M., Kogan,
    L.R., Kostenko, V.I., Matveyenko, L.I., and Rönnäng, B. 1981b,
    Ap.J. 247, 1039.

Glass, I.W., and Feast, M.W. 1982, MNRAS 198,199.
Gunn, J.E. 1980, Phil. Trans. Roy. Soc. London, A296, 313.
Gunn, J.E., Knapp, G.R., and Tremaine, S.D. 1979, A.J. 84, 1181.
Harris, W.E. 1976, Ap.J. 81, 1095.
Hartwick, F.D.A., and Sargent, W.L.W. 1978, Ap.J. 221, 512.
Heiligmann, G.M. 1982, Ph.D. Thesis, Princeton University.
Jackson, P.D., FitzGerald, M.P., and Moffet, A.F.J. 1979, in "The Large-Scale Characteristics of the Galaxy", I.A.U. Symposium 84, ed. W.B. Burton, D. Reidel. Co., p. 221.
Kerr, F.J., Jackson, P.D., and Kerr, M. 1982, in preparation.
Knapp, G.R., Shane, W.W., van der Berg, G., Bajaja, E., Gallagher, J.S., and Faber, S.M. 1982, in preparation.
Knapp, G.R., Stark, A.A., and Wilson, R.W. 1982, in preparation.
Knapp, G.R., Tremaine, S.D., and Gunn, J.E. 1978, A.J. 83, 1585.
Lacy, J.H., Townes, C.H., Geballe, T.R., and Hollenbach, D.J. 1980, Ap.J. 241, 132.
Lynden-Bell, D., and Frenk, C.S. 1981, Observatory 101, 200.
Lynden-Bell, D., and Lin, D.N.C. 1977, MNRAS 181, 37.
Mathewson, D.S., Cleary, M.N., and Murray, J.D. 1974, Ap.J. 190, 291.
Mihalas, D., and Binney, J.J. 1981, "Galactic Astronomy: Structure and Kinematics", W.H. Freeman (San Francisco).
Oort, J.H. 1965, in Galactic Structure, ed. A. Blaauw and M. Schmidt, University of Chicago Press.
Oort, J.H. 1977, Ann. Rev. Astron. Astrophys. 15, 295.
Oort, J.H., and Plaut, L. 1975, Astron. Astrophys. 41, 71.
Quiroga, R.J. 1980, Astron. Astrophys. 92, 186.
Sanders, R.H., Wrixon, G.T., and Mebold, U. 1977, Astron. Astrophys. 61, 329.
Schmidt, M. 1965, in "Galactic Structure", ed. A. Blaauw and M. Schmidt, University of Chicago Press.
Schneps, M.H., Lane, A.P., Downes, D., Moran, J.M., Genzel, R., and Reid, M.J. 1981, Ap.J. 249, 124.
Shuter, W.L.H. 1982, MNRAS 199, 109.
Tremaine, S.D. 1982, in preparation.
de Vaucouleurs, G. 1979, in "The Large Scale Characteristics of the Galaxy", I.A.U. Symposium 84, ed. W.B. Burton, D. Reidel Co., p. 203.
Weaver, H., and Williams, D.R.W. 1974, Astron. Astrophys. Suppl. 17, 1.
Woltjer, L. 1975, Astron. Astrophys. 42, 109.
Yahil, A., Tammann, G.A., and Sandage, A.R. 1977, Ap.J. 217, 903.

A MODEL FOR THE GALAXY WITH RISING ROTATIONAL VELOCITY

Jeremiah P. Ostriker
Princeton University Observatory
Princeton, NJ 08544

John A. R. Caldwell
South African Astronomical Observatory
P. O. Box 9, Observatory 7935, South Africa

1. INTRODUCTION

Constructing mass models for our Milky Way is an old and honorable tradition in astronomy. Many of the scientists who have performed this exercise have attended this conference and would, I expect, agree with my assessment of the pitfalls. One begins with three kinds of information: a) certain fragmentary and biased bits of knowledge concerning the gravitational potential (e.g., interior rotation curve or local escape velocity); b) certain, directly determined kinematic facts (e.g., R_0, the distance to the galactic center); c) "direct" determinations of density in various components (e.g., from star counts of disc stars) or estimations of the density from measured values of light density in a given component multiplied by assumed "reasonable" values of the mass-to-light ratio. One adds to this a theoretical model for the galaxy which divides it into a (hopefully) small number of distinct components each defined by a few parameters. Then, by comparing the theoretical model with the observational data, one determines the free model parameters — i.e., the galactic mass model. The comparison requires two judgemental matters in addition to the straightforward use of Poisson's equation. First, kinematic assumptions are required such as: the galaxy (is)/(is not) axisymmetric; one should (average over)/(take envelope of) rotation curve fluctuations; the observed cutoff of high velocity stars (is)/(is not) due to escape from the galaxy. Second, one must combine various observational studies to obtain best values for input parameters and one must estimate the errors of these parameters requiring many essentially subjective judgements, and then the choice of exactly what figure of merit to use in comparing theory with observation leaves an unfortunately large amont of freedom.

It hardly needs emphasis at a meeting such as this that there are inevitably significant errors in the results and we thus present the model detailed in this paper with all due circumspection. Some final

249

W. L. H. Shuter (ed.), Kinematics, Dynamics and Structure of the Milky Way, 249–257.
Copyright © 1983 by D. Reidel Publishing Company.

results are likely to be quite good. We would be surprised if our
quoted values for R_0, or $V_{c,0}$ were in error by more than 10-15%; however, with some of the more indirectly determined quantities, such as
the local halo density, we would be surprised if the errors were less
than 30-50%. Where possible, we quote error estimates, but a more reliable method for the reader might be to compare models (cf. Table 4 of
paper II) produced by various theoretical groups or even various models
produced by our algorithm (cf. Table V, this paper). So much for the,
perhaps too obvious, discussion of pitfalls.

The method we have chosen has four principal attributes.
1)   We use all available dynamical and kinematical input data of
types a) and b) outlined in the first paragraph. Some models by
others have been based primarily on globular cluster data, some
primarily on radio rotation curves (HI/CO). We have attempted to
incorporate all relevant observations — with estimated errors —
and let the computer choose (on the basis of χ^2 minimization) which
data to weight most rather than to force the model to fit one data
set closely while ignoring discrepancies with other apparently
equally valid data bases.

2)   We have eschewed use of any information of type c), direct
density estimates, on the grounds that these are intrinsically
quite unreliable. Most of the mass is normally in stars too faint
to contribute significantly to the counts, so inevitably our "estimates" reify our personal assumptions about what is reasonable. In
any case we did not feel competent to assess the accuracy of the
published direct density determinations. For an alternative approach see Bahcall, Schmidt and Soneira (1982).

3)   The model is based on the minimal number of components (three)
believed to be astrophysically significant with respect to mass
density. Clearly a better fit to the observations could have been
obtained if a population I component or a more complex spheroidal
component had been adopted, but we felt that the obvious virtues
of simplicity in comprehension outweighed the defects in fitting.

4)   As noted above we use a standard (IMSL) routine to minimize
χ^2 with respect to variation of the theoretical parameters. In
this sense the results are "objective".

After this prologue we present in the next section a new detailed
mass model for the galaxy. For reference to details of the method the
reader is referred to paper II.

2.   MODEL DETAILS

The galactic mass distribution model developed in papers I and II
has been improved in order to take into account the strengthening evidence that the corona or "dark halo" is a significant contributor to

galactic mass interior, as well as exterior, to the sun. In particular, this conclusion follows from the observation that the rotation curve rises beyond the sun, not just in a crest due to the disc, but in a persistent slope out to at least 1.6 R_0 (Blitz 1979, Blitz, Fich and Stark 1980). A consequence is that the disc and spheroid interior to the sun will be somewhat less massive than heretofore thought. It then follows that the previously used "Oort Kz" must overestimate the disc column density or that the apparent "inner rotation curve peak" must exceed the true circular velocity at small longitudes, or both.

With these precepts in mind we revised the mass distribution model using the most up-to-date galactic structure information. The model - put, shown in Table I, is similar to that in paper II, but with the

Table I. Model Input

Quantity	Value	Error	Units
R_0	8.5	1.0	kpc
A	12.6	0.7	km s^{-1} kpc^{-1}
B	-15.7	3.0	km s^{-1} kpc^{-1}
Kz_{MAX}	7.1	1.3	10^{-9} cm s^{-2}
r_S	0.1	15%	kpc
V1	249	9	km/s (C150)
V2	193	9	km/s (C150)
V3	136	9	km/s
V4	104	9	km/s
V5	59	9	km/s
σ_{BUL}	130	20	km/s (D150)
V_{EXT}	1.08V_0	10	km/s
V_{GLOB}	226	20	km/s
$V_{MC'S}$	244	20	km/s
R_{TOT}	150		kpc (V_C flat)

following alterations (details to be given in a subsequent publication, which will also include descriptive functions for the models presented below). R_0 = 6.8 ± 1.2 kpc from globular clusters (Frenk and White 1982) and R_0 = 8.8 ± 1.0 kpc from Mira stars (Glass and Feast 1982) were folded in with the R_0 determinations from RR Lyraes and OB stars in paper II. σ_v/σ_u = .667 ± .084 (Woolley et al. 1977) replaced Erikson (1975), producing a higher $|B|$. Rohlf's (1982) rediscussion of Kz from A dwarfs (Woolley and Stewart 1967) and K giants (Oort 1960) resulted in a lower Kz estimate using the mean and dispersion of these two results. An outward correction of 7 km/s to V_{LSR} in fitting the interior recession velocities V1 to V5 was retained.

We added a new constraint fitting the central velocity dispersion of the galaxy. The Isaacman and Oort (1981) value for the bulge velocity dispersion was adopted (σ = 130 ± 20 km/s) although comparison to other galaxies indicated that it may be slightly high. Our model calculates σ_{BUL} by averaging σ in twenty shells covering R = 0 - 1 kpc, weighting the shells by ρR^2, i.e., by the number of test objects expected. The $d\sigma_{BUL}/dR$ effect has been included iteratively, but is negligible. The all-important exterior rotation data are modeled as $V_C(1.6\ R_0) = 1.08\ V_0 \pm 10$ km/s. Two further constraints on the corona are taken to be $V_C(60$ kpc$) = 226 \pm 20$ km/s from (metal poor) globular cluster kinematics (Lynden-Bell and Frenk 1981), using $V_0 = 240$) and $V_C(70$ kpc$) = 244 \pm 20$ km/s from the Magellanic Cloud orbits (Lin and Lynden-Bell 1981).

Two alternative models were run, because the inner rotation curve and the σ_{BUL} data seemed incompatible. Model C150 is simply an updated version of B(150) from paper II. It maintains as before the more massive spheroid implied by the inner rotation peak. Model D150 ignores the apparent inner rotation peak and fits σ_{BUL} instead. Tables II, III and IV give the model parameter solutions and the results for the local

Table II.   Model Solutions[*]

Parameters	C150	D150	Units
R_0	8.263	8.151	kpc
ρ_D	6.859×10^4	5.980×10^4	$M_\odot pc^{-2}$
r_D	2.70052	3.26033	kpc
r_G	2.68779	3.25115	kpc
ρ_S	1.360×10^2	1.018×10^2	$M_\odot pc^{-3}$
r_S	9.929×10^{-2}	1.004×10^{-1}	kpc
ρ_C	7.815×10^{-2}	2.003×10^{-1}	$M_\odot pc^{-3}$
r_C	3.697	2.332	kpc

[*]For reasons given in the text D150 is the preferred model. C150 is presented primarily for comparison with paper II.

Table III.   Model Output: Local[*]

Quantity	C150	D150	Units
R_0	8.26	8.15	kpc
A	12.54	12.64	km s^{-1} kpc^{-1}
B	−16.25	−16.52	km s^{-1} kpc^{-1}
V_0	238	238	km/s
V_{Eo}	675	685	km/s
Kz_{MAX}	6.79	6.19	10^{-9} cm s^{-2}
Σ_{Do}	46.3	34.5	$M_\odot pc^{-2}$
(Σ_{So})	4.0	3.2	$M_\odot pc^{-2}$
(Σ_{Co})	38.7	45.0	$M_\odot pc^{-2}$
(Σ_{effK_r})	−11.8	−12.5	$M_\odot pc^{-2}$
ρ_{So}	1.35×10^{-3}	1.09×10^{-3}	$M_\odot pc^{-3}$
ρ_{Co}	1.45×10^{-2}	1.52×10^{-2}	$M_\odot pc^{-3}$
M_{Do}	0.175	0.103	$10^{11} M_\odot$

Table III (continued)

Quantity	C150	D150	Units
M_{So}	0.310	0.238	$10^{11} M_\odot$
M_{Co}	0.538	0.703	$10^{11} M_\odot$
M_O	1.02	1.04	$10^{11} M_\odot$

*cf. note Table II.

Table IV.   Model Output: Global*

Quantity	C150	D150	Units
σ_{BUL}	156	137	km/s
V_{MAX}	245	249	km/s
V_{TOT}	241	244	km/s
$r_{D.5}$	7.20	8.71	kpc
$r_{S.5}$	7.39	7.41	kpc
$r_{C.5}$	77.6	76.6	kpc
M_{DTOT}	0.296	0.225	$10^{11} M_\odot$
M_{STOT}	0.593	0.459	$10^{11} M_\odot$
M_{CTOT}	19.4	20.0	$10^{11} M_\odot$
M_{TOT}	20.3	20.7	$10^{11} M_\odot$

*cf. note Table II.

and global galactic structure quantities of interest. Table II shows
that the corona required to fit the rising exterior rotation curve and
the large galactic mass implied by Frenk, Lin and Lynden-Bell is of
higher density and greater central condensation than indicated heretofore in any published work.

The symbols in Tables III and IV are as defined in paper II with
the exception of the parenthesized (Σ) quantities, which are the contributions to the column densities at R_0 between z = -1.5 to +1.5 kpc which
contribute to the Kz_{MAX} determined at that $|z|$. (Σ_{effKr}) is an effec-
tive mass column density contributed by the Kr term.

Table V compares the critical (solar circle) values of various
parameters in our preferred model (D150) with the values we found in
earlier papers.

Table V.   "History" of Coronal Predominance Interior to the Sun

Model	M_{Do}	M_{So}	M_{Co}	ρ_{So}	$\rho_{Co}(M_\odot/pc^3)$
Paper I	58%	37%	5%	1.3(−3)	0.2(−2)
Paper II	31%	30%	39%	1.2(−3)	1.0(−2)
B150					
D150	10%	23%	67%	1.1(−3)	1.5(−2)

Figure 1 shows the interior velocity curves, with C150 fitting all five inner recession velocity fitting points and D150 ignoring the innermost two to get a more acceptable σ_{BUL}.

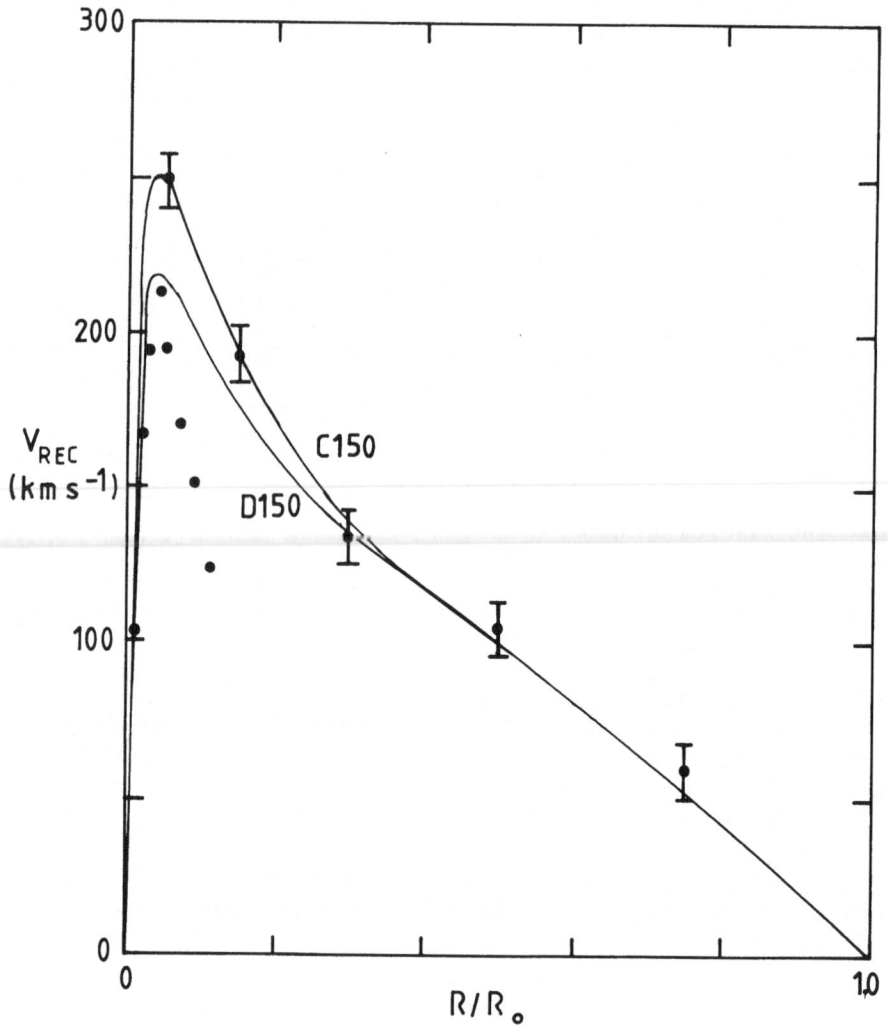

Figure 1.   Interior velocity curve

Figure 2 shows the exterior velocity curve of C150 (dashed) and D150 (solid) fitting the constraint at 1.6 R_0 from Blitz; our results have been transformed to $V_0 = 250$, $R_0 = 10$ for plotting against the data in the original Blitz diagram. Figure 3 gives the full rotation curve,

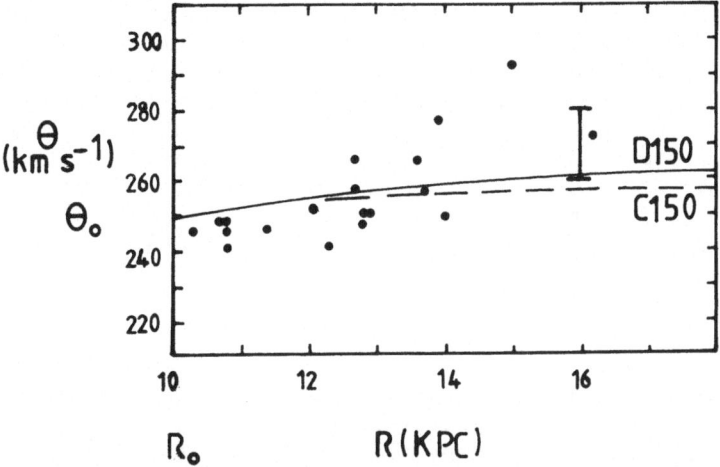

Figure 2. Exterior velocity curve.

which appears normal for an Sb spiral (cf. Bosma 1981).

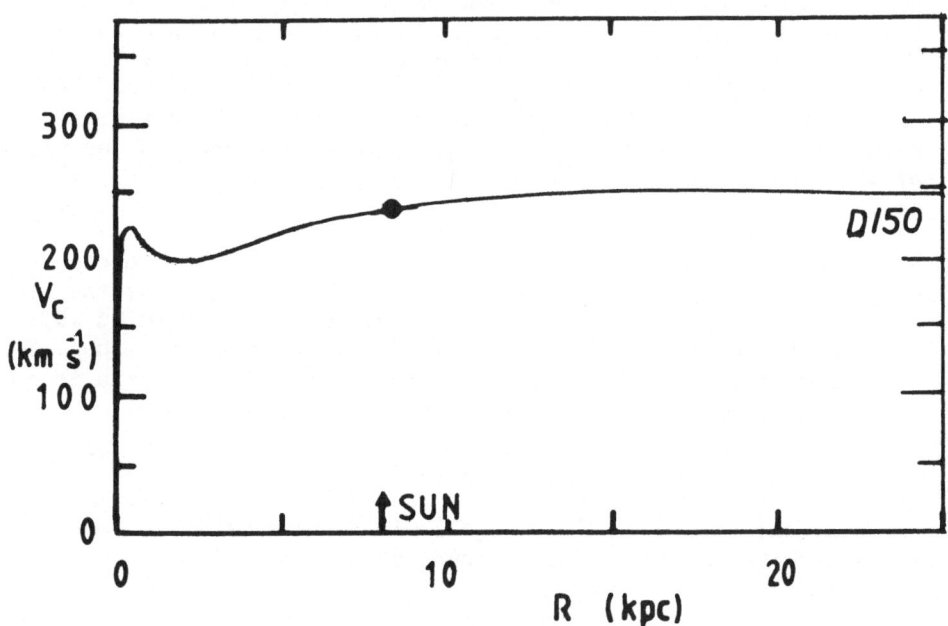

Figure 3. Galactic rotation curve.

The main conclusions of this study may now be summarized.

1)   The inner rotation peak and the measured bulge velocity dispersion are in conflict (σ_{BUL} = 156 versus 137 for models C150 and D150), with D150 clearly the preferable model. This supports the claim that the apparent recession velocities are affected by elliptical streamlines (Peters 1975) or even by bulk radial expansion (Haud 1979). The true underlying recession velocity curve might look like line D150 in Figure 1. Haud presented a model for a radial expansion field which implies a "dynamical" recession velocity pattern shown by isolated dots in Figure 1. Although no physical justification for Haud's model seems to have been published, its effect of reducing the peak V_{REC} is of the right magnitude; its average σ_{BUL} looks to be too small, however, when integrated over the entire central bulge.

2) R_0, V_0 and V_{Eo} are most consistent, with all evidence in hand, with the values 8.2 kpc, 238 km/s, and 685 km/s. This depends on $<M_V>$ (RRab $\Delta S \sim$ 5-10) = 0.6 ± 0.2 and $(m-M)_{Hyades}$ = 3.29 ± 0.05. Note that the escape velocity is roughly equal to the value 640 km/s ± 15% inferred in paper II. V_{Eo} was dropped as a constraint here because we only know a lower limit.

3)   The exterior rotation curve is all-important in deciding about the relative mass predominance of the corona interior to the sun. This is illustrated by the progression of models from paper I (exterior rotation curve falling rather fast), to paper II (falling very slowly), to the present work (rising). Although the importance of the corona to the exterior galaxy has long been appreciated, the picture of coronal dominance in the interior galaxy has become steadily more convincing, and has changed the perception of the interior galaxy rather in line with Table V. In our present work the coronal density at the sun ρ_{Co} has attained a value such as to be marginally significant in the "Oort limit" sense. The gravitational measurement of the corona by dKz/dz may thus eventually be possible. The increasing estimated coronal dominance has had a small effect of decreasing our estimated local spheroid density ρ_{So}. Nonetheless we infer ρ_{So} at least 2-3 times the recent best "observationally" estimated values (e.g., Schmidt 1975). In view of the serious difficulties of trying to estimate ρ_{So} observationally (Richstone and Graham 1981, Eggen 1979), this is not too disturbing.

4)   The total mass of the galaxy is 1.04, 1.38, 2.07 × 10^{12} $M_\odot$ out to R = 75, 100, 150 kpc. This is consistent with observed globular cluster kinematics, Magellanic Cloud kinematics, and Local Group kinematics (Einasto and Lynden-Bell 1982).

REFERENCES

Bahcall, J. N., Schmidt, M. and Soneira, R.: 1982, preprint.
Blitz, L.: 1979, Ap. J., 231, L115.
Blitz, L., Fich, M., and Stark, A. A.: 1980, IAU Symp. 87, 213.
Bosma, A.: 1981, A. J., 86, 1825.
Caldwell, J. A. R. and Ostriker, J. P.: 1981, Ap. J., 251, 61
    (Paper II).
Eggen, O. J.: 1979, Ap. J., 230, 786.
Einasto, J. and Lynden-Bell, D.: 1982, M. N. R. A. S., 199, 67.
Erikson, R. R.: 1975, Ap. J., 195, 343.
Frenk, C. S. and White, S. D. M.: 1982, M. N. R. A. S., 198, 173.
Glass, I. S. and Feast, M. W.: 1982, M. N. R. A. S., 198, 199.
Haud, U. A.: 1979, Soviet Astr. Lett., 5, 68.
Isaacman, R. and Oort, M. J. A.: 1981, Astron. Astrophys., 102, 347.
Lin, D. N. C. and Lynden-Bell, D.: 1982, M. N. R. A. S., 198, 707.
Lynden-Bell, D. and Frenk, C. S.: 1981, Observatory, 101, 200.
Oort, J. H.: 1960, B. A. N., 15, 1.
Ostriker, J. P. and Caldwell, J. A. R.: 1979, IAU Symp. 84, 441
    (Paper I).
Peters, W. L.: 1975, Ap. J., 195, 617.
Richstone, D. O. and Graham, F. G.: 1981, Ap. J., 248, 516.
Rohlfs, K.: 1982, Astron. Astrophys., 105, 296.
Sandage, A.: 1972, Ap. J., 178, 1.
Schmidt, M.: 1975, Ap. J., 202, 22.
Woolley, R. v. d. R., Martin, W. L., Penston, M. J., Sinclair, J. E.
    and Aslan, S.: 1977, M. N. R. A. S., 179, 81.
Woolley, R. v. d. R. and Stewart, J. M.: 1967, M. N. R. A. S., 136,
    329.

RADIAL DISTRIBUTION OF ATOMIC AND MOLECULAR HYDROGEN FROM PROPAGATING STAR FORMATION

Philip E. Seiden
IBM Thomas J. Watson Research Center
Yorktown Heights, New York 10598

ABSTRACT

   The stochastic self-propagating star formation model is able to
account for the division of gas between atomic and molecular hydrogen in
the Galaxy. A correlation is found between the total amount of gas
available and the size of the optical disk. The distributions of gas
and stars obtained from the model are in reasonable agreement with
observations.

INTRODUCTION

   The two-component gas model for stochastic self-propagating star
formation (SSPSF) (Seiden, Schulman and Feitzinger, 1982; Seiden and
Gerola, 1982) allows us to discuss the behavior of the gas under the
star formation process as well as the stars themselves. We assume that
the rate-limiting step in the star formation process is the creation of
giant molecular clouds; once we have a giant molecular cloud stars will
form in it quite easily. Therefore, the effective probability for cloud
formation is taken as

$$P_{eff} = P_{st} D \tag{1}$$

where P_{st} is the stimulated probability of the SSPSF model and D is the
local density of atomic hydrogen. If a cloud is formed in a region the
gas in that region is changed from atomic to molecular. The molecular
hydrogen is then allowed to return to HI with a time constant τ.

   The most important result of the two-component gas model is the
coupling of gas and stars such that a negative feedback is introduced
that controls the star formation rate and, as will be seen below, the
density of HI. This feedback allows the galaxy to evolve in an
approximate steady state. For a more detailed description and
justification of the model see Seiden (1983).

   The consequences of the feedback for HI can be seen in Figure 1

W. L. H. Shuter (ed.), Kinematics, Dynamics and Structure of the Milky Way, 259–263.
Copyright © 1983 by D. Reidel Publishing Company.

which was obtained from an SSPSF simulation that

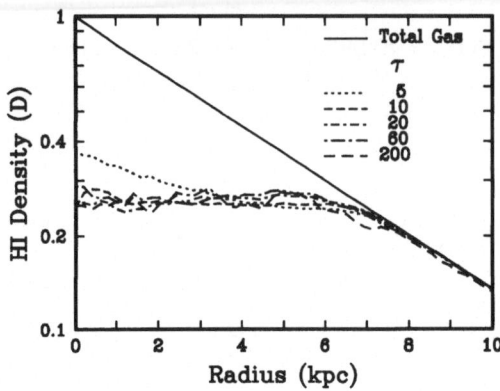

Figure 1. HI gas density for models with an exponential total gas disk. The scale of the ordinate is normalized to unity at R=0.

includes a gas density that declines exponentially as a function of radius. For large enough τ the density of HI is flat, independent of τ and radius, until the total gas density drops below the value at which HI is pinned. This is just what is observed (Rogstad and Shostak, 1972; Burton and Gordon, 1978; Newton, 1980) HI radial distributions are generally flat in the optical region of a galaxy (although there is sometimes a hole in the center as in our Galaxy).

APPLICATION TO THE GALAXY

We now apply this model to the galaxy. Figure 2 shows a plot of the radial distribution of gas in our Galaxy.

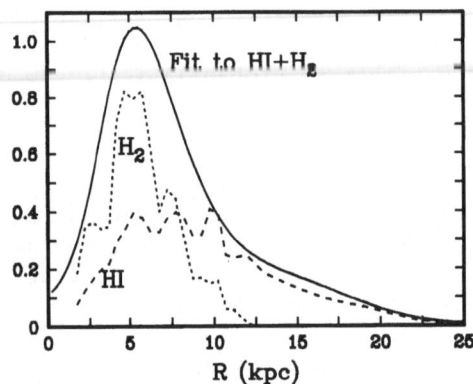

Figure 2. The radial dependence of atomic and molecular hydrogen for the galaxy. The solid line is a fit to the sum of the two dashed curves. The scale of the ordinate is in arbitrary units.

The molecular hydrogen and HI inside the solar circle are from Burton and Gordon (1978), and HI outside the solar circle from Kulkarni, Blitz and Heiles (1982). Although the distribution of each component is probably good there is still some question as to the absolute scales due to uncertainties in the conversion of CO to hydrogen (Liszt, Delin and Burton, 1981) and the optical depth of HI (Dickey and Benson, 1982).

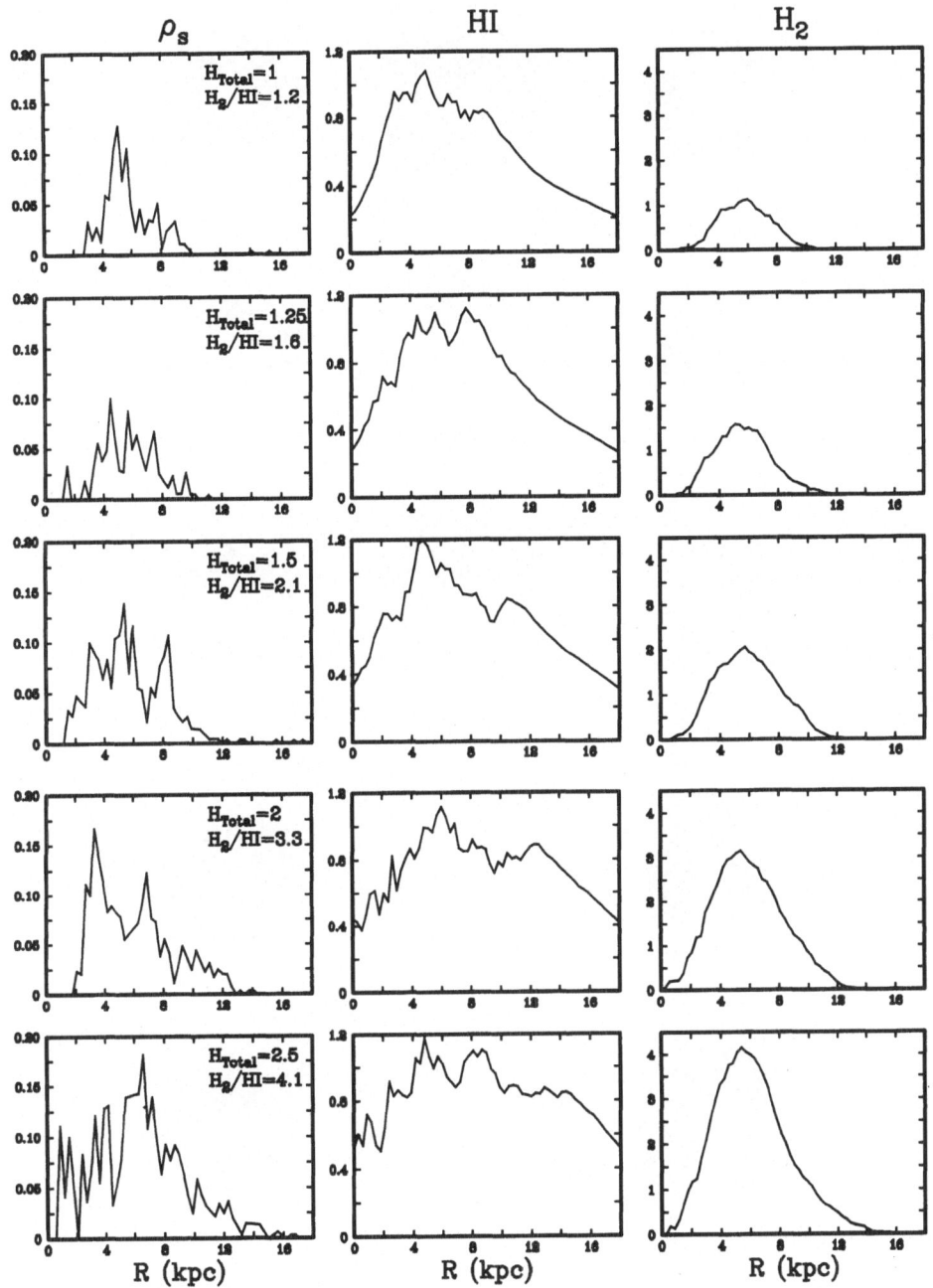

Figure 3. Radial distribution of the star formation rate and atomic and molecular hydrogen for five values of the total gas density (arbitrary units). The resulting H_2/HI peak values are shown in each left-hand frame.

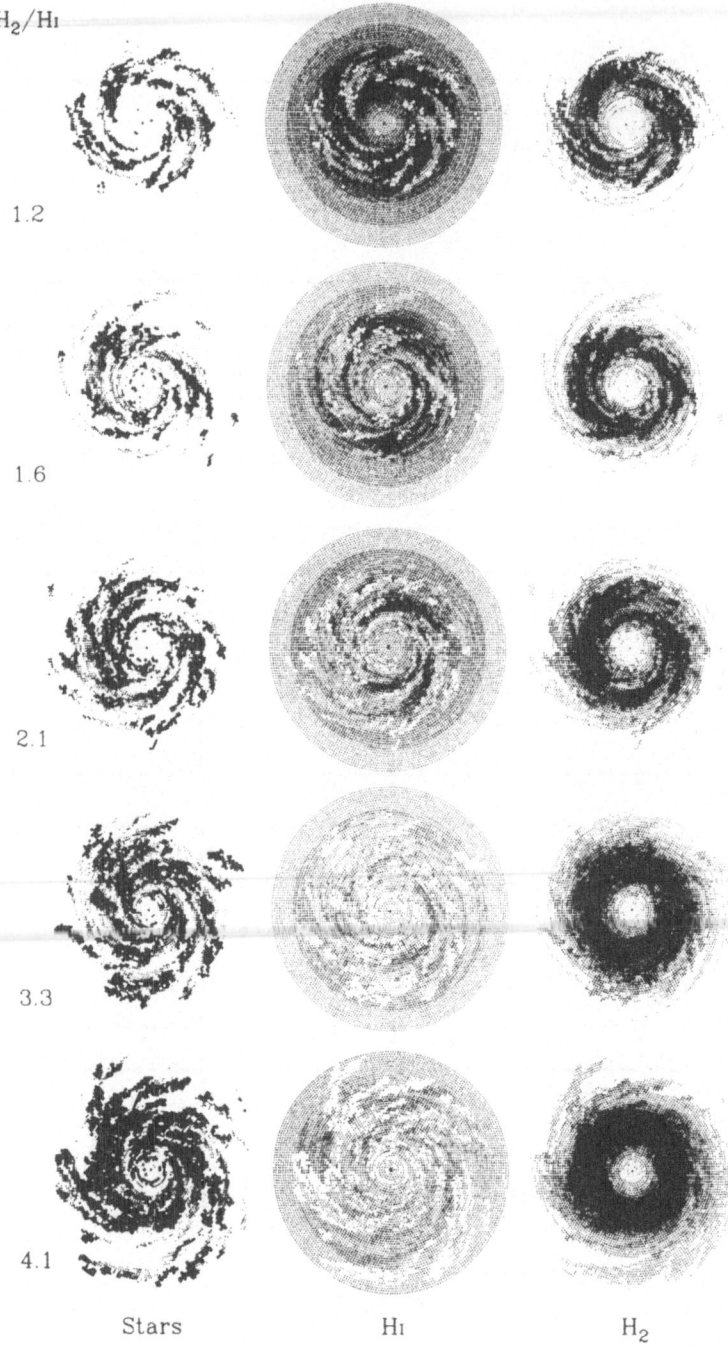

Figure 4. Examples of the stars and atomic and molecular hydrogen for each run of Figure 3. The gray scale for the gas is arbitrary in each case.

Therefore, we have arbitrarily chosen a relative H_2/HI ratio equal to the mean of the values suggested by various investigators. The solid line in Figure 2 is a fit to the sum of molecular plus atomic hydrogen. This will be used as the radial distribution of the total gas for the simulations of the galaxy. Since SSPSF itself will divide the gas between molecular and atomic hydrogen we will cover the range of the uncertainties by scaling this distribution over a factor of 2.5, i.e. we let the total gas content of the galaxy vary by a factor of 2.5 but keep the same radial distribution.

The results for the star formation rate and atomic and molecular hydrogen are shown in Figure 3. For all values of the total gas the value of HI is pinned to about 0.8-0.9 as long as the total gas density is at least this value. The five runs shown cover the range of H_2/HI at the peak of 1.2-4.1.

From the star formation rate column we see that the optical radius of the Galaxy is directly connected with the total gas density. For the top row the optical radius is just over 10 kpc, a value that is probably too small. For the bottom three rows the optical edge is in the range between 13.5 to 17 kpc, very reasonable values.

Figure 4 shows representative time steps for each of the models. Note that the spiral arms are visible in both atomic and molecular hydrogen as well as in the stars, although the contrast in the stars is much greater than in the gas.

Lastly it must be stressed that a requirement of the model is that the lifetime of the molecular gas be long, of at least the order of a hundred million years. This long time is necessary to insure strong feedback (Seiden, 1983).

REFERENCES

Kulkarni, S. R., Blitz, L., Heiles, C. 1982, Astrophys. J., (Part 2, in print).
Burton, W. B., Gordon, M. A. 1978, Astron. and Astrophys. 63, p.7.
Dickey, J. M., Benson, J. M. 1982, Astrnom. J. 87, p.278.
Liszt, H. S., Delin, X., Burton, W. B. 1981, Astrophys. J. 249, p.532.
Newton, K. 1980, Mon. Not. R. Astr. Soc. 191, p.615.
Rogstad, D. H., Shostak, G. S. 1972, Astrophys. J. 176, p.315.
Seiden, P. E. 1983, Astrophys. J. (to be published).
Seiden, P. E., Gerola, H. 1982, Fund. Cosmic Physics 7, p.241.
Seiden, P. E., Schulman, L. S., Feitzinger, J. V. 1982, Astrophys. J. 253, p.91.

CLOUD-PARTICLE GALACTIC GAS DYNAMICS AND STAR FORMATION

William W. Roberts, Jr.
University of Virginia

Galactic gas dynamics, spiral structure, and star formation are
discussed in the context of N-body computational studies based on a
cloud-particle model of the interstellar medium. On the small scale,
the interstellar medium appears to be cloud-dominated and supernova-
perturbed. The cloud-particle model simulates cloud-cloud collisions,
the formation of stellar associations, and supernova explosions as
dominant local processes. On the large scale in response to a spiral
galactic gravitational field, global density waves and galactic shocks
develop with large-scale characteristics similar to those found in
continuum gas dynamical studies. Both the system of gas clouds and the
system of young stellar associations forming from the clouds share in
the global spiral structure. However, with the attributes of neither
assuming a continuum of gas (as in continuum gas dynamical studies) nor
requiring a prescribed equation of state such as the isothermal condition
so often employed, the cloud-particle picture retains much of the detail
lost in earlier work: namely, the small-scale features and structures
so important in understanding the local, turbulent state of the interstellar medium as well as the degree of raggedness often observed superposed on global spiral structure. The results have important implications toward the goal of bridging the gap between global and local
mechanisms of star formation and understanding how a global density wave
and galactic shock may interact with local stochastic processes to provide for a more unified picture of star formation.

INTRODUCTION

Extragalactic spirals often show a degree of order on the large
scale through their luminous spiral arms. If our own Milky Way System
should exhibit a similar degree of order on the large scale, it might be
much more easily perceived by an observer with a bird's eye view located
external to our Galaxy. Enhanced star formation occurs along spiral arms
and dark dust lanes can often be traced along their lengths. Gas dynamical studies of the past decade and a half, based on the hypothesis that
the gaseous interstellar medium can be regarded as a continuum on the

265

W. L. H. Shuter (ed.), Kinematics, Dynamics and Structure of the Milky Way, 265–276.
Copyright © 1983 by D. Reidel Publishing Company.

large-scale, have indicated that the dark dust lanes may be related to
global galactic shock waves formed in the gas along spiral arms
(Fujimoto, 1968; Roberts, 1969). These continuum gas dynamical studies
have shown that galactic gravitational fields, such as may be due to
spiral density waves (Lin and Shu, 1964, 1966; Lin and Lau, 1979) or to
central bars, can rather easily drive such shocks (Huntley, Sanders, and
Roberts, 1978; Roberts, Huntley, and van Albada, 1979; Sanders and
Tubbs, 1980; and van Albada and Roberts, 1981) and that the shock itself
might provide a possible triggering mechanism for the formation of young
stars along spiral arms.

On the other hand, as has been clear for many years and particularly
recent years, our Galaxy's interstellar medium is not at all like a
continuum on the small scale (see e.g., Levinson, 1980; Bania and
Lockman, 1982). Rather the local interstellar medium is clumpy with
clouds, turbulent, and perturbed by star formation activity. Of course,
the same may be very true on the small scale within extragalactic systems,
even those possessing the most coherent global spiral structures. It is
now generally believed that a great deal of the mass of interstellar gas
is tied up in dense clouds of rather small volume filling factor, ranging
in density from perhaps 20 particles per cm^3 in the diffuse HI clouds,
long known from absorption against background stars, to 1000 or more
molecules per cm^3 in the dense molecular clouds, believed to constitute
much of the mass in the inner regions of the Galaxy. On the other hand,
it is clear that most of the volume of interstellar space may contain
much more tenuous, high temperature gas than do the clouds. The possibility is widely discussed that expanding supernova remnants may fill
the major part of the interstellar volume with an ionized gas so hot
(perhaps $10^5 - 10^6$ K) and rarefied (Cox and Smith, 1974; Chevalier,
1977; Jones et al, 1979; Heiles, 1979a, b; Hu, 1980) that it would
not be subject to a large-scale galactic shock, for example, when passing
through a spiral arm (Scott, Jensen, and Roberts, 1977; Shu, 1978).

Because of this large range of densities and temperatures, it is
important to incorporate the small-scale inhomogeneities of the interstellar medium into the large-scale picture of spiral structure (also
see McKee and Ostriker, 1977; McCray and Snow, 1979). For a deeper
understanding, it is necessary to reconcile these two apparently-divergent
pictures - the global picture on the one hand, the local clumpy picture
on the other - and to try to understand the relationship between global
triggering mechanisms of star formation, such as large-scale density waves
and galactic shocks on the one hand, and local mechanisms and stochastic
processes of star formation on the other (also see Miller, Prendergast,
and Quirk, 1970; Mueller and Arnett, 1976; Elmegreen and Lada, 1977;
Gerola and Seiden, 1978; Seiden and Gerola, 1979; and Comins, 1981).
How might such global and local mechanisms of star formation act in
concert with each other? Here we discuss some recent work which under-
takes to address some of these motivating questions (Levinson and Roberts,
1981; Roberts, Hausman, and Levinson, 1982).

CLOUD-PARTICLE MODEL OF THE INTERSTELLAR MEDIUM

With the goal of better understanding both the global and local pictures and their relationship with one another, while retaining much of the local detail lost in earlier continuum gas dynamical studies, we consider a cloud-particle model for the interstellar medium (also see Schwarz, 1979). In contrast to the assumption of a continuum of gas, as in continuum gas dynamical studies, or the requirement of a prescribed equation of state such as the isothermal condition so often employed, the cloud-particle model has, as its fundamental building blocks, a discrete system of gas clouds and a corresponding system of young stellar associations. For the cases to be discussed in this paper, the initial state for the model is adopted as follows. Initially, 20,000 clouds are distributed randomly in space within the model galactic disk of radius 12 kpc; 10,000 in each half plane with symmetry about the origin. The mean free path for the cloud system is one of the most important physical parameters to be considered. Estimates on the basis of observations of diffuse HI clouds (Spitzer, 1978) and HI self absorption clouds (Levinson, 1980) yield an average value for the mean free path in the range $\lambda \sim 100 - 200$ pc. We adopt a value for the cloud cross section in the computations in order to provide for an equivalent mean free path in this range consistent with observations. Initially, each cloud is assigned an initial velocity taken as a composite of (a) the mean rotation velocity (u_{x0}, u_{y0}) at that radius and (b) a random gaussianly-distributed part $(\Delta u_x, \Delta u_y)$ which provides the cloud system with a mean initial velocity dispersion adopted at a level consistent with observations (a one-component mean of about 6 km/s). In a similar manner, the sample of young stellar associations is initially distributed in the model disk.

The dynamics of the cloud-dominated, "stellar association"-perturbed interstellar medium are followed in this cloud-particle picture from its prescribed initial state through two-dimensional, N-body computational simulations. In the presence of a spiral galactic gravitational field, we compute the dynamical time evolution of the system of clouds and the corresponding system of young stellar associations, both those present initially and those continually forming from the clouds, by a 3-step cyclic procedure:

Step 1. Dynamical propagation of the clouds and young stellar associations over a computational time step Δt.

Step 2. Simulation of the cloud-cloud collisions that occur over that time period Δt.

Step 3. Formation of new associations of protostars that are triggered over that time period Δt via the local mechanisms of (a) cloud-cloud collisions and (b) cloud interactions with existing young stellar associations in their active stages, e.g. impulsive supernova explosions.

Step 1 involves the propagation of each cloud of the N-cloud system according to the following dynamical equations, written with respect to the frame of reference of the spiral pattern rotating at an angular pattern speed Ω_p:

$$\frac{d\vec{x_i}}{dt} = \vec{v_i} \; , \qquad\qquad\qquad i = 1, N$$

$$\frac{d\vec{v_i}}{dt} = \Omega_p^2 \vec{x_i} - 2 \vec{\Omega}_p \times \vec{v_i} - \vec{\nabla} U \; , \qquad i = 1, N$$

where $\vec{x_i}$ and $\vec{v_i}$ denote the position and velocity of cloud i, and U denotes the gravitational potential field driving the cloud system. Similarly, each young stellar association is propagated in time, from the point where it is triggered, to the location where it reaches its most active stages with supernova explosions. The underlying gravita-tional potential field U consists of two components: (a) an axially-symmetric, thin-disk component (Toomre, 1963) in differential rotation with a maximum rotation velocity of 250 km/s occurring at a radius of 8 kpc, and (b) a two-armed spiral perturbation, sinusoidal in polar angle about the disk, to represent the spiral structure. With an ampli-tude adopted to vary with radius as prescribed by Roberts et al. (1979), the perturbing force represents a 5% - 10% perturbation superimposed on the axially-symmetric force field of the disk over the radii of 5 - 10 kpc. The spiral pattern is adopted to have spiral arms with 10^0 pitch angle and a pattern speed of rotation of Ω_p = 13.5 km/s/kpc.

Step 2, the simulation of cloud-cloud collisions, is based on the criteria that two clouds collide if: (a) they are approaching, and (b) at the end of a particular time step they are found closer than the collision radius R_{coll} (also see Brahic, 1977). These clouds are regarded as "soft" clouds to the degree that the collision radius R_{coll} is adopted to be the same as the cloud radius R_c. Thus clouds can interpenetrate upon collision. Futhermore, a collision is taken to be dissipative in the sense that two clouds with incoming "pre-collision" relative velocity c_r will have an outgoing "post collision" relative velocity $f_r c_r$, where $0 < f_r < 1$.

Step 3 simulates the formation of stellar associations and their activity. A new stellar association is prescribed to be triggered at the collision center of a cloud-cloud collision with probability P_{cl-cl}. A delay time of $r\tau_D$ is allowed before this association of protostars can become an active stellar association with giant HII regions and supernovae. τ_D represents the maximum delay time prescribed, and r is a random number selected between zero and unity for each stellar associa-tion ($0 < r < 1$). When a supernova explosion occurs, neighboring clouds within the radius of the supernova shell are given impulsive velocity boosts. Such activity by the population of stellar associations acts as an energy source for the cloud system. Furthermore, a cloud impulsed by one supernova explosion is allowed to form a new association of proto-stars after a time delay of $r\tau_D$ with probability P_{st-cl}. Thus in this cloud-particle model, stellar associations can be triggered via two local mechanisms of star formation (a) cloud-cloud collisions and (b) cloud interactions with existing active stellar associations. Colliding clouds and "impulsed" clouds that serve as "stellar association"-forming partici-pants in the model are required to wait a composite delay time $r\tau_D + r'\tau_{D'}$ from the time of the collision and "impulse" respectively before participating in further "stellar association"-forming events. Maximum delay times τ_D and $\tau_{D'}$ up to 20 Myr have been used in the computational simulations; for each "stellar association"-forming event, a random number generator selects the random numbers r and r', $0 < r < 1$, $0 < r' < 1$.

In this threefold cyclic procedure, both the propagation of clouds and stellar associations (step 1) and the collisions of clouds (step 2) are carried out as fully deterministic steps. On the other hand, step 3 is a fully probabilistic step by the fact that new stellar associations are triggered probabilistically with probabilities P_{cl-cl} and P_{st-cl} respectively in cloud-cloud collisions and cloud interactions with active stellar associations. Moreover in step 3, a stellar association becomes active with HII regions and supernova explosions only after the probabilistically-selected time delay $r\tau_D$, and a "stellar association"-forming cloud can participate again in further "stellar association"-forming events only after the probabilistically-selected composite time delay $r\tau_D + r'\tau_{D'}$. Thus, we denote this model as a "quasi-deterministic-cloud-particle" (QDCP) model.

This QDCP model can be contrasted with the "fully-deterministic-cloud-particle" (FDCP) model of Levinson and Roberts (1981), representing the early stages of development of the present work. At that time, only the "cloud-cloud collision"-induced mechanism of star formation was present (P_{st-cl} = 0.0); physically-realistic time delays were not yet incorporated (τ_D = 0.0, $\tau_{D'}$ = 0.0); and each cloud-cloud collision was assumed to trigger exactly one "supernova-producing" event with unit probability (P_{cl-cl} = 1.0). The present QDCP model has also been refined in several other ways; one of which involves the implementation of a feedback mechanism to insure the proper energy balance for the dissipative cloud system and to maintain a physically-realistic rate of energy transfer to the cloud system from the energetic system of stellar associations. The greater the energy of dissipation in cloud-cloud collisions, the lower the mean velocity dispersion to be expected for the cloud system. On the other hand, the greater the impulses on clouds by active stellar associations, the higher the cloud system's velocity dispersion. In order that the proper global energy balance be achieved to maintain the cloud system's velocity dispersion Δu at a level consistent with observations (a one component mean of about 6 km/s), we allow the computer program to activate a feedback mechanism, when needed, through which the magnitude of the impulses on clouds by active stellar associations can be adjusted until a steady-state is reached.

RESULTS: TIME EVOLUTION OF THE GAS CLOUD SYSTEM, GLOBAL SPIRAL
STRUCTURE, AND LOCAL CLUMPINESS

Illustrative sample results from the computational simulations will
now be discussed. We first focus on one case for which cloud-cloud
collisions are considered as the dominant local mechanism for star
formation (P_{cl-cl} = 0.5, P_{st-cl} = 0.0) and time delays of 20 Myr for
τ_D and $\tau_{D'}$ are adopted. This QDCP case permits stellar associations
to become active with HII regions and supernova explosions after time
delays $r\tau_D$, randomly distributed between 0 and the maximum delay time
of τ_D = 20 Myr, with a statistical mean of about 10 Myr.

From its initially-uniform distribution, the cloud system evolves
in time, driven by the spiral galactic gravitational field. Figure 1
provides a view of the distribution of gas clouds in the model galactic
disk at a sample time epoch of 250 Myr during the evolution. The clouds
in the outermost regions of the disk lie at galactic radii of 12 - 14 kpc.
Only one half of the total number of computed clouds are plotted. Furthermore, those clouds located in the lower half plane are the axially-
symmetric images of those computed in the upper half plane.

Figure 1. Time snapshot of
the distribution of gas clouds
in the disk of the cloud-
particle model at the sample
time epoch of 250 Myr during
the time evolution. Global
density waves and galactic
shocks form in the gas cloud
system on the large scale in
response to the underlying
spiral galactic gravitational
field despite the turbulence
and chaotic activity on the
small scale.

One of the fundamental issues under study in this work is that of
gravitationally-driven global galactic shocks and density waves and the
degree to which they can form and be maintained globally despite the
chaos and turbulence of the local processes of cloud-cloud collisions and
cloud interactions with active stellar associations. At this time epoch

of 250 Myr, spiral compression waves are clearly evident in the distribution of gas clouds. In fact, already by the earlier time epoch of 100 Myr, a spiral pattern became apparent in the response. The ridge of the compression waves lies along the potential minimum of the background spiral potential field. The cloud system is undergoing high differential rotation in the clockwise sense with respect to the slower rotating spiral wave pattern. The peaks of these spiral-arm compression waves are much narrower than the interarm troughs; a result in agreement with the nonlinear wave responses associated with the galactic shocks computed in continuum gas dynamical studies. Such shock-like compression waves therefore seem to be capable of forming on the large scale despite the local chaos and turbulence inherent in cloud-cloud collisions and cloud interactions with active stellar associations.

Figure 2 illustrates in more detail the nature of the computed distribution of gas clouds and galactic flow characteristics, measured with respect to the clockwise-rotating frame of the spiral wave pattern. Plotted versus phase of the spiral gravitational potential field are the

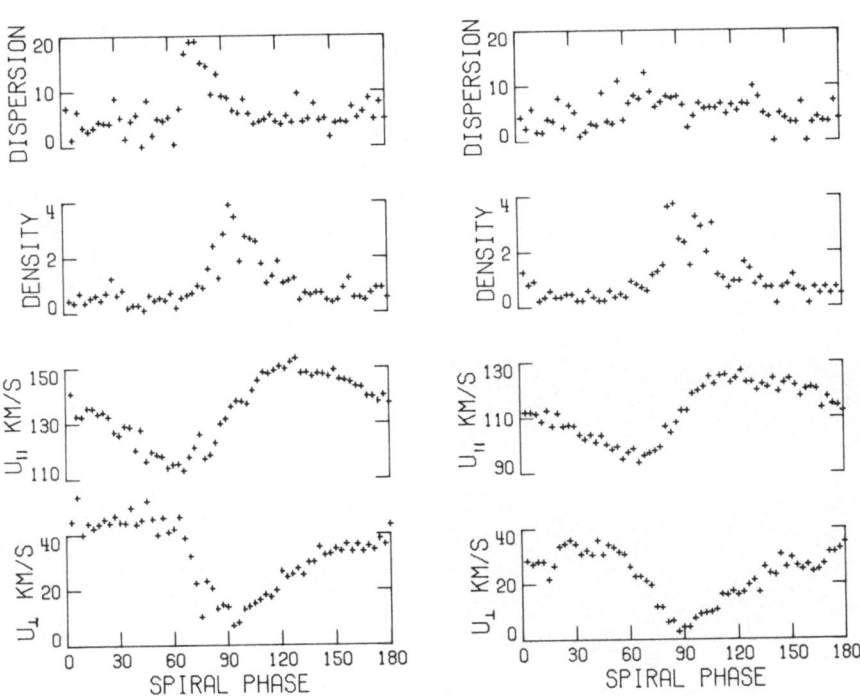

Figure 2. Components of velocity (perpendicular and parallel to spiral equipotential contours), density, and velocity dispersion of the gas cloud system plotted versus spiral phase in two representative annuli, at galactic radii of 8 kpc (left panel) and 10 kpc (right panel) at the sample time epoch of 250 Myr.

components of flow velocity perpendicular and parallel to spiral equipotential contours, the normalized number density distribution of clouds
denoted as density, and the velocity dispersion for those clouds in two
representative annuli, 500 pc wide and extending one half revolution about
the disk at radii of 8 and 10 kpc (left and right panels respectively).
Pluses (+) indicate mean quantities of velocity, density, and velocity
dispersion of the clouds contained in each bin along the annuli. Each
plus (+) represents about 10 clouds on the average. With respect to the
direction of spiral phase perpendicular to the spiral equipotential contours, the pluses (+) are separated in distance by about 70 pc on average
in these annuli. This distance of separation is about 50% of the average
mean free path $\lambda \sim 100 - 200$ pc for the system of clouds over the computed disk. Consequently, the rapid decline in the normal component of
flow velocity (bottom frames) occurs over a width of the order of 5 mean
free paths. This width of several mean free paths is typical of measured
shock structures in terrestrial gas flows. The density enhancement measured from maximum to mean is of the order of 3:1 to 4:1; whereas the
contrast in density between arm and interarm is as large as 5:1. It is
interesting to note that the peak of cloud number density occurs down-
stream toward the subsonic edge of the shock velocity jump. The largest
enhancement in velocity dispersion occurs at about the point of steepest
decline in the normal component of shock velocity from supersonic to
subsonic.

One important issue, that has played a large role in motivating these
investigations, centers on the local clumpiness of the interstellar medium.
In order to study aspects of the local clumpiness and characteristics of
the turbulence present in the cloud-particle model, we have developed
auxiliary computer programs with specific computer graphics capabilities
for the display of the results calculated with the main computational
program. Figure 3 provides a photographic color-intensity display of the

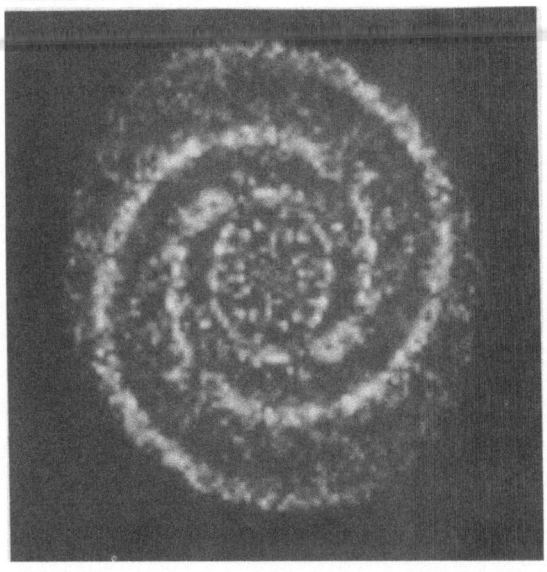

Figure 3. Photographic color-
intensity map of the density
distribution of the cloud system
over the disk at the sample time
epoch of 250 Myr. The regions
of highest cloud concentrations
are light-blue in color. The
cloud conglomerations provide
a raggedness and degree of disorder on the small scale that
permeates the global spiral
structure.

(reproduced from a color print)

distribution of clouds in the model disk. The light-blue color depicts the regions of high concentrations of clouds. Enhanced portions of the spiral arms at concentrations above the mean have widths on the order of 1 kpc. Cloud concentrations appear here with a local clumpiness covering a range of scales. There are cloud concentrations with clumps up to several hundred pc in linear dimension for example in the spiral arms. We are currently carrying out investigations to determine what relationship these large "cloud-concentration" clumps may have with the giant molecular cloud conglomerations observed in the Galaxy.

We can understand the local clumpiness of the modeled interstellar medium as a manifestation of turbulent processes. Indeed, in contrast to continuum gas dynamical studies, the cloud-particle picture attempts to incorporate ingredients of the dominant local thermophysical processes of cloud-cloud collisions and stellar association activity. These processes are characteristically turbulent. The photographic display in Figure 3 illustrates to a degree the turbulent state of the medium simulated. In viewing a time-sequence of such photographic displays, certain aspects of the local chaos and turbulence, as well as the evolution of the local clumpiness, can be followed through the small-scale fluctuations that appear and evolve, superposed on the global spiral structure. The detailed character of this turbulent state, simulated with the cloud-particle model, is still under investigation. We hope to determine the extent to which it might be associated with the corresponding raggedness and degree of disorder on the small scale that are often observed as characteristics permeating the global spiral structures of real galaxies.

RESULTS: COMPUTED DISTRIBUTION OF ACTIVE STELLAR ASSOCIATIONS AND GLOBAL SPIRAL STRUCTURE

Figure 4 shows the distribution of stellar associations which become

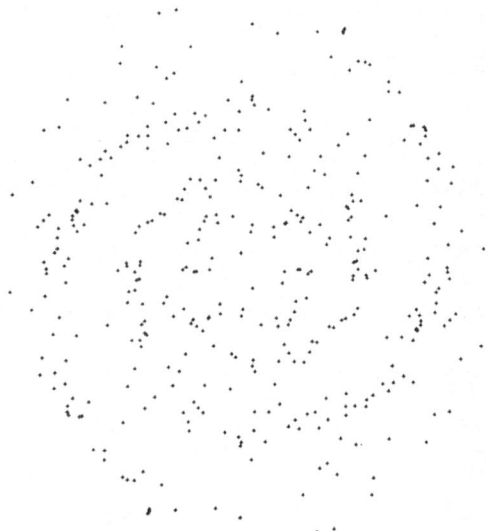

Figure 4. Time snapshot of the distribution of those young stellar associations which reach their active stages (supernova explosions) during the 2 Myr time interval centered on the sample time epoch of 250 Myr.

active with HII regions and supernovae during a 2 Myr time interval centered about the sample time epoch of 250 Myr. The location of each stellar association is calculated deterministically, based on the original position of the collision center of the cloud-cloud collision that triggered it, plus the distance propagated during the time delay before it evolved to its active stages. A spiral pattern is clearly evident in this distribution of active stellar associations. It is interesting to note the degree of raggedness exhibited. To a large extent, this raggedness is a direct consequence of the probabilistic time delay $r\tau_D$ allowed to help account for the time of evolution necessary for associations of protostars, from the time of their initiation, to reach their active stages. Also playing an important role in the degree of raggedness is the second probabilistic time delay $r'\tau_{D'}$ over which "stellar association"-forming clouds must wait before they can participate again in further "stellar association"-forming events.

Other important questions might also be raised. For example, we might ask: at what location would the largest proportion of the energy from stellar associations be dumped back into the cloud system of the interstellar medium? In order to address this question and at the same time exhibit more clearly the physical effects induced by the time delays τ_D and $\tau_{D'}$, we will compare the present case ($P_{cl-cl} = 0.5$, $P_{st-cl} = 0.0$, $\tau_D = \tau_{D'} = 20$ Myr) with a second case ($P_{cl-cl} = 1.0$, $P_{st-cl} = 0.0$, $\tau_D = \tau_{D'} = 0$ Myr). Of those cases considered by Roberts et al (1982), this second case is the QDCP case closest to the fully deterministic limit. With zero time delays prescribed, the associations of protostars forming at a particular time become active stellar associations with HII regions and supernovae in the next computational time step (= 2 Myr, in this case).

Figure 5 shows photographic color-intensity maps of the cloud distributions in these two cases together with the distributions of those stellar associations (white dots) which reach their active stages in the 2 Myr time interval centered on the sample time epoch of 250 Myr. The left panel for the case ($P_{cl-cl} = 0.5$, $P_{st-cl} = 0.0$, $\tau_D = \tau_{D'} = 20.0$ Myr) is therefore just a superposition of the distributions displayed in figures 3 and 4. In the right panel for the case corresponding to little or no time delay ($P_{cl-cl} = 1.0$, $P_{st-cl} = 0.0$, $\tau_D = \tau_{D'} = 0.0$ Myr), we see a much more well-delineated spiral pattern in the computed distribution of active stellar associations. These stellar associations would perhaps be representative of the most massive protostar complexes whose self gravitation is so strong that the star formation process proceeds rapidly without much time delay from the stages of initiation in cloud-cloud collisions. These are just the stellar associations that would be expected to evolve very rapidly through their active stages of producing giant HII regions and supernovae. This distribution (right panel), perhaps representative of the most massive stars and their associated giant HII regions, coincides more closely with the ridge of the gas cloud distribution (light-blue color). Investigations of these and other cases are currently underway toward the goal of determining how such concentrations might be associated with the giant cloud complexes, dust regions, luminous stellar associations, and giant HII regions often observed as prominent manifestations in the spiral arms of many spiral galaxies.

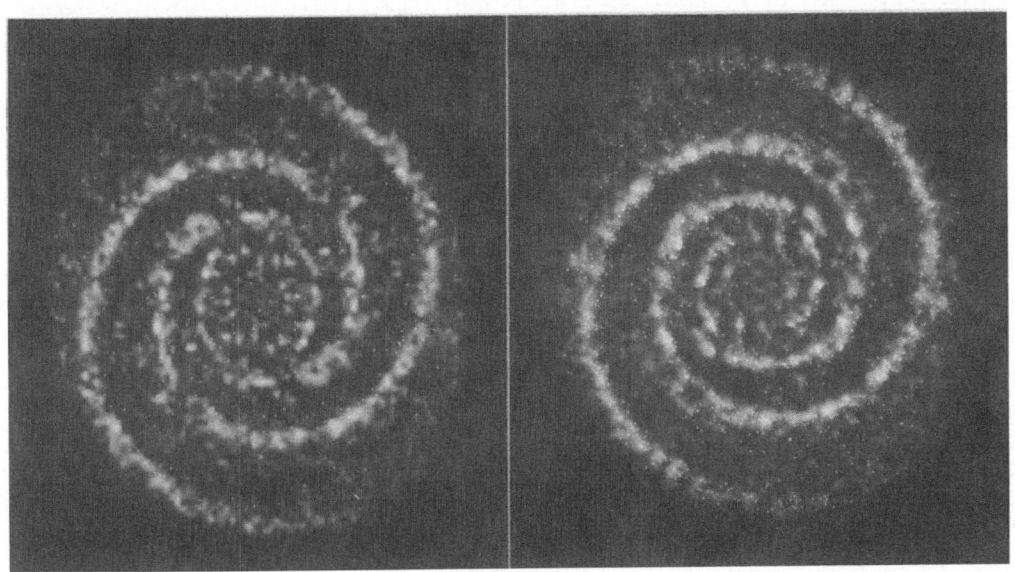

Figure 5. Distribution of young stellar associations (white dots) which become active with supernovae during the 2 Myr time interval, centered on the sample time epoch of 250 Myr, superposed on the density distribution of the corresponding cloud system (photographic color-intensity map) from which the stellar associations are triggered. Two cases: left panel, with probabilistic time delays; right panel, without probabilistic time delays. The sample of active stellar associations in the right panel would perhaps be representative of the most massive protostar complexes whose self-gravitation is sufficiently strong that the active stages of HII regions and supernovae proceed very rapidly after formation.
(reproduced from color prints)

ACKNOWLEDGEMENTS

    This work was supported in part by the National Science Foundation under grant AST-7909935; the author also received support under NASA contract NAS1-15810 while in residence at the Institute for Computer Applications in Science and Engineering (ICASE) at the NASA Langley Research Center.

REFERENCES

Bania, T.M., and Lockman, F.J. 1982 (submitted for publication).
Brahic, A. 1977, Astr. Ap., 54, 895.
Chevalier, R.A. 1977, Ann. Rev. Astr. Ap., 15, 175.
Comins, N. 1981, M.N.R.A.S., 194, 169.
Cox, D.P., and Smith, B.W. 1974, Ap. J. (Letters), 189, L105.
Elmegreen, B.G., and Lada, C.J. 1977, Ap. J., 214, 725.
Fujimoto, M. 1968, in Proc. IAU Symp. No. 29, ed. M. Arkeljan
       (Yerevan: Armenian Academy of Science), p. 453.
Gerola, H., and Seiden, P.E. 1978, Ap. J., 223, 129.
Heiles, C. 1979a, Ap. J., 229, 533.
Heiles, C. 1979b, Ap. J., 235, 833.
Hu, E.M. 1980, Bull. A.A.S., 12, 468.
Huntley, J.M., Sanders, R.H., and Roberts, W.W. 1978, Ap. J., 221, 521.
Jones, E.M., Smith, B.W., Straka, W.C., Kodis, J.W., and Guitar, H. 1979,
       Ap. J., 232, 129.
Levinson, F.H. 1980, Ph.D. Dissertation, Univ. of Virginia.
Levinson, F.H., and Roberts, W.W. 1981, Ap. J., 245, 465.
Lin, C.C., and Lau, Y.Y. 1979, Stud. in Appl. Math., 60, 97.
Lin, C.C., and Shu, F.H. 1964, Ap. J., 140, 646.
Lin, C.C., and Shu, F.H. 1966, Proc. Nat. Acad. Sci., 55, 229.
McCray, R., and Snow, T.P. 1979, Ann. Rev. Astr. Ap., 17, 213.
McKee, C., and Ostriker, J.P. 1977, Ap. J., 218, 148.
Miller, R.H., Prendergast, K.H., and Quirk, W.J. 1970, Ap. J., 161, 903.
Mueller, M.W., and Arnett, W.D. 1976, Ap. J., 210, 670.
Roberts, W.W. 1969, Ap. J., 158, 123.
Roberts, W.W., Hausman, M.A., and Levinson, F.H. 1982 (in preparation).
Roberts, W.W., Huntley, J.M., and van Albada, G.D. 1979, Ap. J., 233, 67.
Sanders, R.H., and Tubbs, A.D. 1980, Ap. J., 235, 803.
Schwarz, P. 1979, Ph.D. Dissertation, Univ. of Canberra.
Scott, J.S., Jensen, E.B., and Roberts, W.W. 1977, Nature, 265, 123.
Seiden, P.E., and Gerola, H. 1979, Ap. J., 233, 56.
Shu, F.H. 1978, in Proc. IAU Symp. No. 77, ed. E.M. Berkhuijsen and
       R. Wielebinski (Dordrecht: Reidel), p. 139.
Spitzer, L. 1978, Physical Processes in the Interstellar Medium (New
       York: Wiley), p. 227.
Toomre, A. 1963, Ap. J., 138, 385.
van Albada, G.D., and Roberts, W.W. 1981, Ap. J., 246, 740.

SPIRAL MODES AND THE MILKY WAY

C.C. Lin
Department of Mathematics, Massachusetts Institute of
Technology, Cambridge, Massachusetts 02139   USA

In this short communication, I intend to comment on the studies of the
structure of the Milky Way in view of current progress made in the
density wave theory of spiral modes. I shall first make a brief report
of these developments.

SPIRAL MODES

The density wave theory began with the desire to explain the grand
design of the spiral structure observed in many galaxies. To this end,
the hypothesis of quasi-stationary spiral structure was made. In
essence, this QSSS hypothesis states that such grand designs can be
described by a superposition of spiral modes (see Lin, 1965, p. 68).
Substantial progress has recently been made along these lines (see
Lin & Bertin, 1981; Haass, Bertin, and Lin, 1982; Haass, 1982). Methods
for calculating such spiral modes have now been well developed. Indeed,
they can be used to calculate models for spiral galaxies of various
morphological types. By changing one parameter in the model, one can
vary the spiral modes to correspond to normal spiral galaxies of types
Sa, Sb, and Sc. By changing another parameter, one can demonstrate the
transition from normal spirals to barred spirals.

A movie has been prepared by Jon Haass (shown at the Vancouver
meeting) to demonstrate the vascillation of the spiral structure when
three spiral modes (two with two arms and one with three arms) are
superposed. When this movie was shown earlier at a meeting of the
American Astronomical Society, Vera Rubin identified one range of
frames as similar to the photograph of a galaxy (D100) in her collec-
tion. The galaxy appears to have two spiral arms with a predominant
spur superposed.

For further details, including a picture of the galaxy D100, the
reader should consult the references cited above.

The theory has been developed not only to provide methods for the

W. L. H. Shuter (ed.), Kinematics, Dynamics and Structure of the Milky Way, 277–281.
Copyright © 1983 by D. Reidel Publishing Company.

calculation of the spiral modes, but also to understand the mechanisms for the maintenance and the excitation of the spiral modes. The detailed analyses of the two simplest cases – the pure trailing spirals, and the rudimentary barred spirals – contribute to the essentials of our understanding.

The use of the theory of pure trailing spirals for the description of normal spiral modes has been summarized in earlier reviews (Lin and Lau, 1979; Bertin, 1980). The maintenance of the spiral mode is ascribed to long and short trailing waves, propagating in opposite directions. The excitation mechanism is attributed to the WASER process studied in detail by James Mark (1976). In this process, there is a transfer of angular momentum associated with a trailing spiral wave propagating outwards from the corotation circle (where the spiral wave pattern co-rotates with the material). Goldreich and Tremaine (1978) showed that the WASER mechanism can also be described in the co-moving representation (with local approximation) used by Goldreich and Lyndon-Bell in the fluid description. It is expected that the same can be achieved in the stellar dynamical theory of Julian and Toomre (1965). A welcome unification of the two approaches is thereby accomplished in this case.

Similar theoretical developments have been achieved by G. Bertin (to be published) for the case of barred spirals. The simplest of a barred spiral consists of a leading wave and a trailing wave, instead of the two trailing waves in the previous case. There is a similar WASER mechanism. A leading wave approaching the corotation circle is refracted as a trailing wave; but it excites, at the same time, a pair of trailing waves propagating in opposite directions from the corotation circle. Thus, there is re-enforcement of the inward propagating trailing wave, yielding an amplification of the spiral mode. There is transfer of angular momentum associated with the outwards propagating trailing wave. To distinguish the present case from the previous one, we shall designate the mechanism studied by Mark as WASER of Type I and that studied by Bertin as WASER of Type II. The latter can be shown to be closely related to the process of swing amplification investigated by Toomre (1981). This unified picture removes the lingering concern whether the impressive transient growth of the swing amplifier might imply that there are excitation mechanisms which are missed out in the study of normal modes in the QSSS perception (cf. discussions recorded in the Proceedings of IAU Symposium 84, p. 152). In the end, there does not seem to be any additional mechanism of great significance. The role of particle resonance has been studied in a separate context (Bertin and Haass, 1982).

Modal calculations have not only included those with two arms ($m = 2$) but also modes with $m = 1, 3, 4$ (see Bertin et al, 1977). Such modes cannot be so easily excited except through internal mechanisms, as visualized in the QSSS hypothesis. Besides the galaxy D100 mentioned above, other recent observational studies also give strong support to this point of view. Iye et al (1982) found that, in the spiral galaxy

NGC4254, the prominent Fourier components are <u>odd</u> harmonics (m = 1, 5, 3) rather than even harmonics (m = 2, 4). In a statistical study, Elmegreen and Elmegreen (1982) found that a substantial fraction of isolated galaxies have grand design.

THE MILKY WAY

Let us now turn to the discussions of the Milky Way. Jon Haass has now improved on the calculation of modes in the Milky Way presented at IAU 84 (see poster paper he presented at this meeting). Prominent among the conclusions is the existence of a one-armed spiral. This would cause significant <u>asymmetry</u> in the determination of kinematical distances. For example, in the interpretation of the data on the warps of the Milky Way and on the location of distant HII regions one need not be unduly concerned about asymmetrical distortions, if kinematical distances are used.

Prominent in the evidence for the existence of density waves in the Milky Way is the observation of streaming motions by both radio and optical methods. A model for the spiral gravitational field has been constructed for the solar vicinity. This model is perhaps reasonably accurate within 3 kpc. of the Sun. Confidence in the model is built up because of the success of the use of the model for the analyses of a variety of observations. In particular, the study of the migration of stars provides a quite stringent test of the overall picture of the local structure of the spiral gravitational field. The stars sample the gravitational field over different trajectories and over different periods of time. Thus, such analyses prove the spiral structure of the Milky Way over an area within several ($\sim$ three) kiloparsecs of the Sun, during the past one or two galactic years. Four sets of investigations have been made:
- A. The original program of Strömgren, including the following studies:
  - (i) the sample of 25 stars (Yuan, 1969) and
  - (ii) the enlarged sample of over 300 stars (Grosbol, 1977);
- B. The Gepheids (Wielen);
- C. Moving groups (Yuan, 1977; Yuan and Waxman, 1977);
- D. Vertex deviation in the distribution of stellar motions (Yuan, 1971).

For further details of a survey of the various analyses, the reader is referred to a paper by Lin and Yuan (1978) presented at a symposium in honor of Bengt Strömgren. We remark here that the model will not be significantly modified by the changes in the equilibrium model of the Milky Way such as those presented at this meeting. A brief explanation will be given below.

The flow field in this adopted model was published as Fig. 5 in Lin, Yuan, and Roberts, 1974, and as Fig. 3 in Lin, Yuan, and Roberts, 1978. This flow field includes an <u>expansive</u> motion in the solar vicinity, with the K parameter approximately equal to 5 km/sec-kpc.

Indeed, the density wave theory, which implies <u>compressions</u> at galactic
shocks, demands the existence of expansions in between shocks just to
maintain the nearly periodic behavior of the interstellar medium as it
circulates around the galaxy. This general expansive motion in the
solar vicinity could help to reduce the energy estimate of the local
"explosions" needed to account for the motion of local young stars, if
not to replace it entirely.

Especially noteworthy among the applications is the use of the
associated streaming motions for the description of the stellar motions
observed from the solar vicinity. The inward motion in the Perseus
direction is well-known and can be accounted for by the density wave
theory (see Fig. 1 and Fig. 4 of Lin, Yuan, and Roberts, 1978 and refer
to the earlier work of Roberts cited). In connection with the discussions of the galactic model at this conference, one should perhaps pay
special attention to the longitude directions on the two sides of the
anti-center. For example, in Fig. 4a of the 1978 LYR paper just cited,
one can easily see that, at a galacto-centric distance $\tilde{\omega}$ of 12.5 kpc.,
the apparent angular velocity of rotation of the galaxy is about
20 km/sec.-kpc. because of the streaming motions produced by the spiral
gravitational field. This value would correspond to a <u>constant</u> circular
velocity of 250 km/sec., even though the analysis was mode on the basis
of the 1965 Schmidt model, in which the angular velocity is 18.8 km/sec.-
kpc. at $\tilde{\omega}$ = 12.5 kpc. The difference of 1.2 kpc./sec.-kpc. is indeed
quite small! Thus, from the point of view of the density wave theory,
it is important <u>to construct a plausible model of the spiral gravita-</u>
<u>tional field</u>, if one wishes to have any confidence in the study of the
outer regions of the Milky Way within a few kiloparsecs of the Sun.
The use of circular models can be quite misleading. Incidentally, the
corotation circle of the two-armed mode calculated by Jon Haass lies at
a galacto-centric distance of about 16 kpc. Optical spiral arms may
therefore be expected to terminate at or close to such distances,
presumably at a longitude lying in the third quadrant.

Let us now make a brief comment on the general question whether
the theoretical spiral structure discussed here would be significantly
modified by the changes in the model of the Milky Way presented at this
conference. The answer is clearly "no". For the changes largely
concern the addition of a large halo or corona in the mass distribution,
and involve very small changes in the distribution of the angular velocity in the inner parts of the disk. The extra mass clearly produces
little or no influence on the gravitational field inside. The only
change in perception is that the Milky Way now has a very large ellip-
soidal component, with a rather small region of intense star formation
(< 13 kpc.) in a relatively small galactic disk. To be sure, the
change of the distance scale from 10 kpc. to 7 kpc. would also require
corresponding changes of all scales, but such changes would not imply
qualitative changes of any kind. We may thus continue to have confidence
in the local spiral gravitational field present, on the basis of the
evidence summarized in the 1978 paper of Lin and Yuan, and further
evidence presented by Wielen (1978).

REFERENCES

Only selected references are listed below. Review articles are marked with asterisks. Other references may be located through these articles.

Bertin, G.: 1980[*], Phys. Reports 61 (1), pp. 1-69 (217 references).
Bertin, G. and Haass, J.: 1982, Astron. and Astrophys. 108, pp. 265-273.
Elmegreen, Debra M. and Elmegreen, B.: 1982, submitted to the Astrophysical Journal.
Goldreich, P. and Tremaine, S.: 1978, Astrophys. J. 222, pp. 850-858.
Haass, Jon: 1982, Ph.D. Dissertation, Department of Mathematics, Massachusetts Institute of Technology.
Haass, J., Bertin, G., and Lin, C.C.: 1982, Proc. Nat. Acad. Sci. USA, June Issue.
Iye, Masanori et al: 1982, Astrophys. J. 256, pp. 103-111.
Lin, C.C.: 1965, in Proceedings of the Fourth Summer Seminar in Applied Mathematics, American Mathematical Society; published in 1967 as Volume 9 in Lectures in Applied Mathematics: Relativity Theory and Astrophysics (2), Galactic Structure, pp. 66-97.
Lin, C.C. and Bertin, G.: 1981[*], in Plasma Astrophysics, T.D. Guyenne and G. Levy eds. (European Space Agency) SP-161, pp. 191-205.
Lin, C.C. and Lau, Y.Y.: 1979[*], Studies in Appl. Math. 60, pp. 97-163 (110 references).
Lin, C.C. and Yuan, C.: 1978, Astronomical Papers dedicated to Bengt Strömgren, A. Reiz and T. Andersen eds., published by Copenhagen University Observatory, pp. 369-386.
Lin, C.C., Yuan, C., and Roberts, W.W., Jr.: 1978, Astron. and Astrophys. 69, pp. 181-198.
Toomre, A.: 1977[*], Ann. Rev. Astron. Astrophys. 16, pp. 437-478 (145 references).
Toomre, A.: 1981, The Structure and Evolution of Normal Galaxies, S.M. Fall and D. Lyndon-Bell eds., Cambridge University Press, pp. 111-136.
Wielen, R.: 1978, IAU Symposium No. 84: The Large-Scale Characteristics of the Galaxy, W.B. Burton, ed., Reidel Publishing Company, pp. 133-144.

A MODAL APPROACH TO SPIRAL STRUCTURE: TWO EXAMPLES

Jon Haass
Department of Mathematics
Massachusetts Institute of Technology
Cambridge, Massachusetts 02139 USA

Two galaxy models are examined as illustrations of the modal approach to the problem of spiral structure. First a simple model is considered to demonstrate the effect of superposition of modes. Next a possible model for the Milky Way is shown for the purpose of exploring the effect on the unstable modes of three different rotation curves; falling, flat or rising.

SUPERPOSITION OR EVOLUTION?

By analogy to other, simpler dynamical systems, like vibrating strings or membranes, it is natural to consider spiral structure in terms of superposed disturbances. In the linear theory any spiral form can be decomposed into a sum of individual spiral modes with different radial and angular dependence. However, for useful application, it is best if a few modes already give a good representation of the spatial structure. Of course, the ability to model a "snapshot" of the system doesn't guarantee that the time dependence will be as well reproduced. An example of this would be a tidally forced bisymmetric spiral (e.g. the response calculations of Toomre [1981]). The current shape may be very close to a single mode of the unperturbed system yet the actual pattern might be transient, damping in a revolution or two. This is mentioned as a warning to modellers (myself included) who might like to understand the grand design spirals, such as M81 or M51, as nearly permanent modes. Until more information is available on the velocity dispersion in the disk population as well as the fraction of the mass that can be ascribed to the disk component within the optical image, "The jury is still out!".

Nonetheless, it is still instructive to be able to visualize the effect of superposing linearly three spiral modes and following the resultant "evolution". A simple two-component model consisting of a Kuzmin disk and Plummer sphere was analyzed in this fashion. A family of linearly unstable spiral modes was calculated for the model equilibrium. From these were chosen the first three armed mode and the first

W. L. H. Shuter (ed.), Kinematics, Dynamics and Structure of the Milky Way, 283–287.

two bisymmetric modes, which were initially imposed at small amplitude
(the relative amplitudes chosen in the ratio of their growth rates).
The system was then photographed as the modes rotated and grew according
to their different pattern speeds and growth rates. A sampling of the
frames from the movie so produced is shown in the figure below.

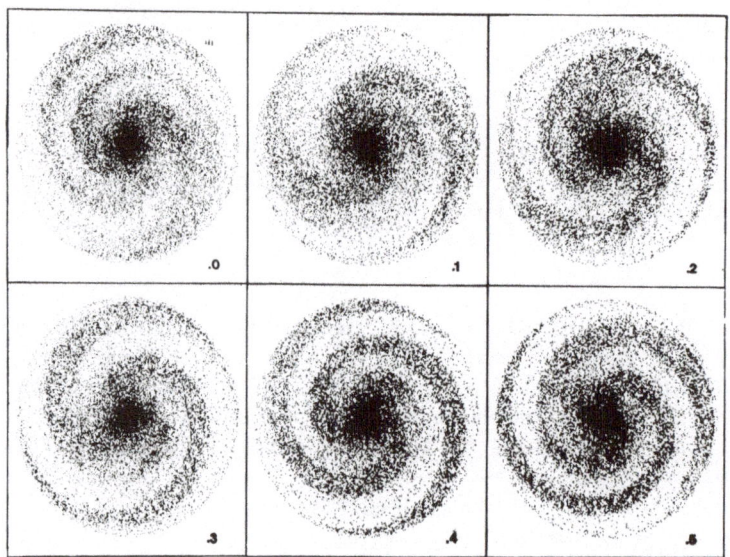

Figure 1.   Scaling the radius of the disk to 20 kpc. yields the time
scale shown in billions of years. Note that the "stars" were placed
statistically according to the density distribution calculated via a
linear modal analysis, so the amplitudes are only relative.

     For this type of model, characterized by its small central nucleus,
the m=3 modes grow about twice as rapidly as the m=2 modes. Higher m
modes, in this fluid model, grow even faster but are more limited in
their spatial extent hence would tend to further confuse only the central region of the model. Furthermore, in a stellar model, the inner
resonance would act to damp many of these bothersome instabilities.

     Already, with only three modes superposed, the forms are interest-
ing while retaining a good deal of globally coherent structure. The
interaction of all the modes admitted by this model (there are about 20
modes with various arm number that have significant growth rate and
escape having an inner resonance) would probably result in a rather
chaotic looking spiral. Perhaps it is no coincidence that many small
galaxies (like M33) are not so regular in their structure.

     Of course, no real galaxy will ever be so perfectly axisymmetric as
the theoretical models usually considered, consequently some modes may
be favored resulting in a simpler pattern. The question now seems to be:
How are these simple patterns produced from such a zoo of unstable
modes? Further work in this direction should investigate the effects of
initial perturbations and the possible nonlinear coupling between modes.

MILKY WAY ROTATION: FALLING, FLAT OR RISING?

A good deal of work on the local streaming motion has been carried
out using the Schmidt [1965] rotation curve. For example Lin, Shu and
Yuan [1969] showed how a spiral pattern could modify the local gas flow.
Their adopted spiral disturbance, based on the local dispersion relation
for density waves, still finds use today (see Bash and Yuan this meeting). Because of the interest in this model and to give an updated
analysis of the spiral pattern in terms of unstable modes, a poster was
prepared displaying the existence of both one and two armed modes. But
it was evident at the meeting that unlike external galaxies the rotation
velocity is not well known for our own galaxy. The analogy to other
galaxies and recent observations presented by Blitz suggest instead a
flat or even rising rotation curve beyond the solar circle. So I have
taken the same disk model presented at Vancouver and added extra
spherical mass to alter the rotation curve. The different models are
graphically presented in the following figure.

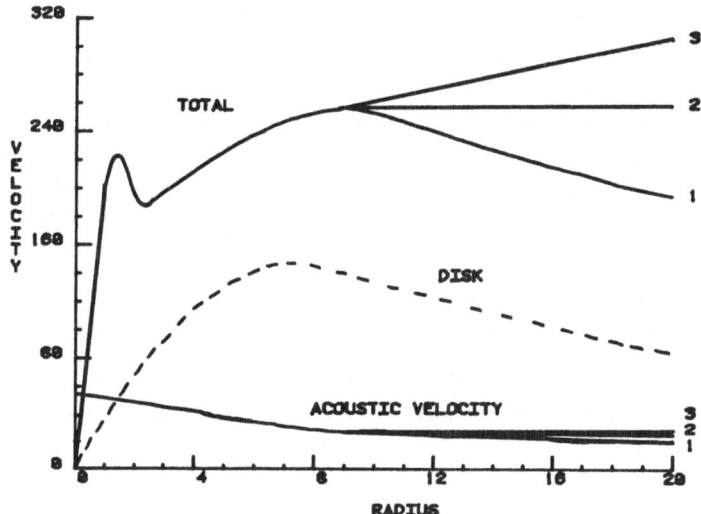

Figure 2. The rotation curve and effective acoustic velocity of the
disk stars are shown for three different models. The disk components'
contribution to the rotation curve is fixed and shown as the dashed
curve.

Since less than 30% of the total mass is considered to reside in a
thin disk and the assumed acoustic velocity was chosen to be just marginally larger than that required for axisymmetric stability, a simple
fluid asymptotic analysis can be justifiably used for the modal calculations (see Lin and Lau [1979] for the details of this approach).

The Milky Way seems to have a small but dense nucleus creating a
rapidly rising rotation curve. An important effect of this is the
existence of an inner Lindblad resonance for all but a few of the one

armed modes. By a careful choice of the disk density and velocity dispersion, two of the two armed modes are also spared the damping effect expected of the resonance (but not reproduced in the fluid model). Higher m modes can formally be calculated but are deemed less trustworthy since the Q-barrier fails to shield their resonance. An impression of the spiral structure expected for this model is shown in the next figure.

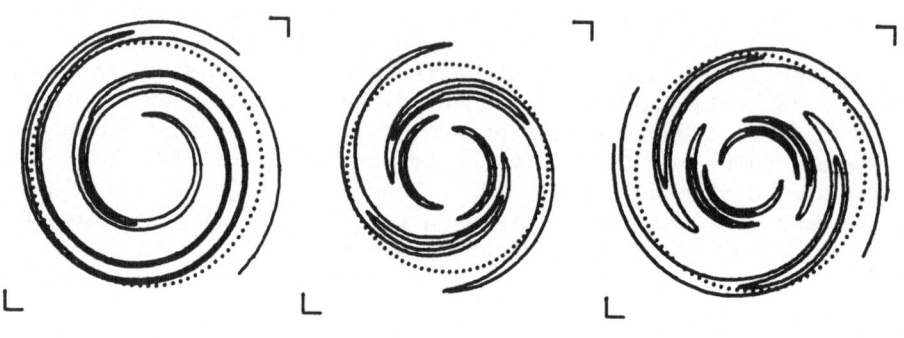

Figure 3. Contours of the perturbation density (reported at 1/3 and 2/3 of the maximum) are given for the modes in model 1. The circle marks the corotation radius. Note that it is the second two arm mode that corresponds most closely in pattern speed to that assumed in the star migration studies.

Modal information for the three models is compared in the table below. Of more interest is the calculated shape of the spiral modes. Although the corotation radius changes from model to model, the pitch angle remains nearly the same inside of 10 kpc., as it should. The first two armed mode for each case is shown in the next figure.

	Pattern Speed (km/sec-kpc)	Growth Rate (km/sec-kpc)	Corotation (kpc)	Model
M=1	12.7	1.1	16.5	1
	13.9	1.2	17.8	2
	15.1	1.0	20.0	3
M=2	15.9	.7	14.1	1
	16.2	.8	15.0	2
	17.4	1.0	16.1	3
	13.2	.6	16.1	1
	13.8	.5	17.9	2
	15.5	.3	19.2	3

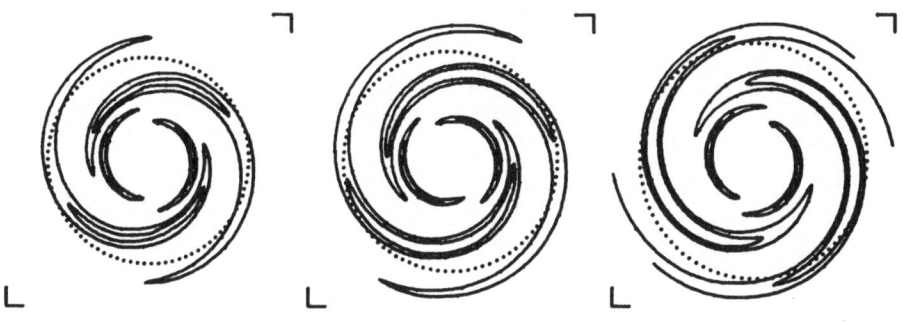

Figure 4. The same spiral mode is shown for the three models (in ascending order). The first frame is a repeat from figure 3.

This exercise suggests that the proposed spiral structure can be produced for a range of rotation curves, provided an appropriate choice of the disk mass and velocity dispersion is made. It would be interesting to explore these variations as well. Such a program is underway for a model galaxy similar to that used in producing the movie. When the modal behavior as a function of the equilibrium values of rotation, density and pressure is well explored, it may be possible to use this dynamical information to place limits on some of the currently unobservable properties of galaxies.

REFERENCES

Lin, C.C. and Lau, Y.Y.: 1979, Studies in Appl. Math. 60, pp. 97-163.

Lin, C.C., Shu, F. and Yuan, C.: 1979, Astrophysical Journal 155, pp. 721-746.

Schmidt, M.: 1965, Stars and Stellar Systems V: Galactic Structure, A. Blaauw and M. Schmidt eds., Chicago University Press, pp. 513-530.

Toomre, A.: 1981, The Structure and Evolution of Normal Galaxies, S.M. Fall and Lynden-Bell eds., Cambridge University Press, pp. 111-136.

A THEORY OF BENDING WAVES WITH APPLICATIONS TO DISK GALAXIES[1]

James W-K. Mark
Lawrence Livermore National Laboratory
Livermore, CA, 94550, U.S.A.

Abstract: A theory of bending waves is surveyed which provides an explanation for the required amplification of the warp in the Milky Way. It also provides for self-generated warps in isolated external galaxies. The shape of observed warps and partly their existence in isolated galaxies are indicative of substantial spheroidal components. The theory also provides a plausible explanation for the bending of the inner disk (< 2 kpc) of the Milky Way.

I. INTRODUCTION

The bending of the disk of the Milky Way has been observed earlier by Burke (1957), Kerr (1957) and Westerhout (1957), and has generated considerable interest for theoretical explanation (e.g. Kahn and Woltjer 1959, Elwert and Hablick 1965, Lynden-Bell 1965, Avner and King 1967, Hunter and Toomre 1969). The work of Hunter and Toomre and that of Fujimoto and Sofue (1976, 1977), Spight and Grayzeck (1977) all suggested that the most attractive reason for the bending was due to tidal distortion by the LMC. But quantitative comparison was satisfactory only if the tidally generated warps could somehow be amplified "by a factor approaching 3" (cf. p. 773 of Hunter and Toomre 1969).

Toomre (1966), briefly with moment equations, and Kulsrud et al. (1971) and Mark (1971) with detailed stellar dynamics, reported on the possibility of a "fire-hose" type instability in the stellar disk which could provide for self-generation and/or amplification of bending waves. A small warp in the more massive stellar disk could drive larger amplitude warps in the outer gaseous disk. This mechanism appeared useful for providing wave amplification in the Milky Way. It became more interesting in the light of subsequent observations of warps in rather isolated galaxies (Sancisi 1976, Bosma 1978; Rogstad et al. 1974 found it difficult to conclude that the warp of M83 was tidal). Also some stellar disks showed warps (Tsikoudi 1977, 1980; van der Kruit 1979). However this mechanism depended upon galaxy disks possessing sufficiently large stellar velocity dispersions in the plane relative to

W. L. H. Shuter (ed.), Kinematics, Dynamics and Structure of the Milky Way, 289–302.
Copyright © 1983 by D. Reidel Publishing Company.

that out of the plane. The only obvious observational test based on
stellar dispersive speeds in the solar neighborhood is suggestive of
stability or near marginal stability, making the question of the
allowed amount of residual wave amplication more difficult to ascertain
(we recall that only a factor of 3 amplification was needed in the
Milky Way). Although difficult to test, the possibility exists that
such a mechanism plays a role in some presently observed galaxy warp
or in the warps at earlier states of galaxy evolution. For example,
it might represent one of the possible constraints which caused disk
galaxies to maintain a certain ratio of stellar velocity dispersion in
their disk. Perhaps dispersions in the plane were needed to stabilize
Jeans instability while the vertical dispersion had to maintain a
proper ratio to these latter in order to stabilize the "fire-hose"
instability.

More recent theoretical studies of bending waves in disk-like
systems have been motivated firstly by the observational discovery of
warps in numerous external galaxies (Roberts 1966, Rogstad et al.
1974, 1976, Roberts and Whitehurst 1975, Sancisi 1976, Bosma 1978 and
Weliachew et al. 1978) showing that the phenomena is rather widespread
as well as occurring also in isolated galaxies. Interest has also been
generated by the conjecture that the phenomena of warps might be more
easily understood because of the occurrance of extended spheroidal
"halo" matter (Polyachenko 1977, Binney 1978, Saar 1978, Sanders 1978,
Polyachenko and Shukhman 1979, Tubbs and Sanders 1979). Also, other
possibly bent disks in or around galaxies require dynamical explanation, for example that in the inner few kiloparsecs of the Milky Way
(Burton and Liszt 1978; Liszt and Burton 1978, 1980; Sinha 1979).
Warps in isolated galaxies (cf. Sancisi 1976, Bosma 1978) are indicative of the need for internal excitation mechanisms. These recent
theoretical descriptions of warped structures either lack detailed
dynamical justification (particularly discussions of properties of
such bending waves), or, they assume a priori a very dominant "halo"
(sometimes even triaxial in shape, Binney 1978), while the very question of the mass and distribution of spheroidal matter is a subject
for detailed observational determination.

Based on the earlier studies of bending waves with self-gravity
by Hunter and Toomre (1969), Kulsrud et. al. (1971), Mark (1971), and
based on a new approach to the gravitational potential theory for waves

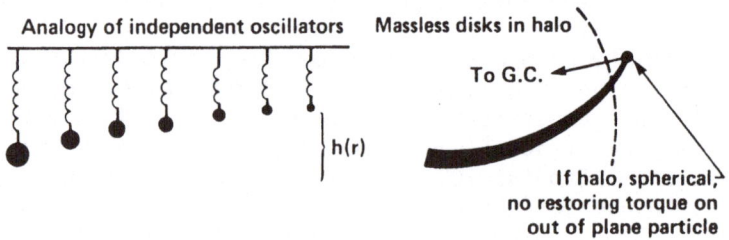

Fig. 1. Model of kinematical bending waves.

in disks (cf. Mark 1976c), the recent study of Bertin and Mark (1980) further examines the effects of unstable, two-stream interactions with a spheroidal subsystem (the disk is like "a flag waving in the wind"). This subsystem could be a massive "halo" but the theory allows it also to be just the extensions of a "bulge" (less massive). We find that sensible explanations of self-excited warps or driven waves are possible without necessarily postulating "haloes" which are either excessively large or triaxial (or even flattened). A crude heuristic description of our detailed dynamical theory is included in Section II. The very fact that warps could also be driven by small spheroidal subsystems or other mechanisms (e.g. "fire-hose") makes it at first more difficult to connect the properties of halo matter with that of warps in isolated galaxies.

On the other hand, by a novel use of a wave energy or angular momentum relation (cf. Eq. 1 below), Bertin and Mark (1980) showed that just the typically observed <u>qualitative shapes of galaxy warps</u> are <u>indicative of extensive spheroidal matter</u> within the radii of the observed warps (cf. Section III). Preliminary confirmation with a few details has been carried out by Lake and Mark (1980) for the Milky Way and some external galaxies; additional suggested examples are given in the paper by Bertin and Casertano (1982). Since our theory does not assume substantial haloes to begin with, it could provide independent evidence for halo sizes. As we will emphasize in Sections II and III, our arguments depend only on very few basic concepts of the theory and are thus largely independent of the detailed mathematical analyses. In fact they only depend on bending wave propagation properties in the disk and do not entail the details of whether self-generated warps interact with the halo or with fire-hose instabilities in the stellar disks.

Our theory of bending waves also seems to provide a plausible dynamical rationale (cf. Section IV and Blitz et. al. 1981) regarding the interpretation that the inner gaseous disk (< 2 kpc) of the Milky Way is bent out of the plane of the main galaxy disk (Burton and Liszt 1978, Liszt and Burton 1978, 1980, Sinha 1979). In this case it is due to an analogous two-stream type instability between the inner disk and the bulge (or spheroidal matter with this smaller length-scale). Further theoretical work and comparisons with observations would be useful.

II. A HEURISTIC DISCUSSION OF THE BENDING WAVE THEORY

For details on the analytical derivations of the bending wave theory as used here, we refer the reader to Mark (1971), Bertin and Mark (1980). We will not attempt to summarize these extensive calculations here. Rather, we provide discussions of much simplified models as a guide to the basic physics involved.

If we artificially "turn-off" the self-gravity in the disk, then we are left with so-called "kinematic waves". Their nature is best illustrated in the case of a massless flat disk imbedded in a spherical halo (cf. Figure 1). Particles displaced from the disk will remain in orbits inclined from the original flat disk. There are no actual torques on the particle orbits so that the appearance of vertical oscillations is purely kinematical. If the individual particle displacements are systematic over the disk (such as due to initial external driving), they could form a warped disk. In such a kinematic wave, each particle's bending amplitude above the initial plane bears no particular universal relation to that of particles at other radii except where it relates to the initial external forcing. Also lack of coherence of frequencies at different radii results in eventual "washing-out" of the warp due to phase-mixing.

Restoring of self-gravity to the disk changes the vertical oscillation frequency as well as allows horizontal propagation of wave information. This self-gravitating bending wave can be crudely illustrated by the model of Figure 2. Let us denote the height of bending $\Delta z = h(r,\theta,t) = \text{Re}\{\exp[i(\omega t - m\theta + \int kdr)]\}$ in the usual cylindrical polar coordinate system (r,θ,z) where t is the time [$k(r)$ can be a complex-valued wavenumber containing both phase and amplitude information]. Clearly these waves would exhibit a dispersion relation $D(\omega,m,k) = 0$ and group velocity $c_g = -d\omega/dk$. Even without entering into details, we can surmise that conservation of wave energy implies that the flow of wave energy through a fixed radius $2\pi r\, c_g\, \mathscr{E} \sim$ constant, where $\mathscr{E}$ is wave energy density. From general principles we expect $\mathscr{E} \propto \Sigma(r)\, h^2$, where $\Sigma(r)$ is the local surface mass density of the disk. Also $c_g \propto \Sigma$ and thus we expect

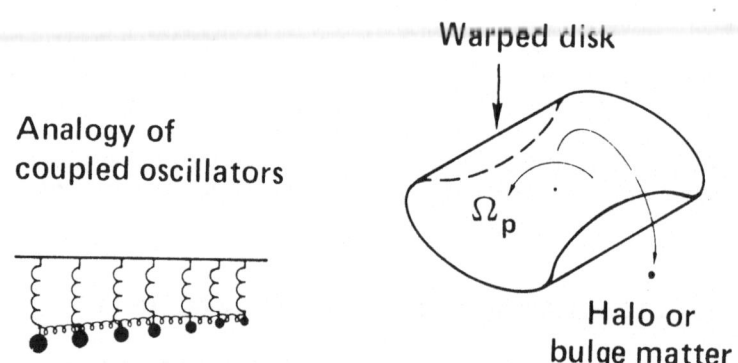

Fig. 2.   Model of bending waves with self-gravity.

$$r \ \Sigma^2 \ h^2 \ \underset{\sim}{} \ \text{constant}, \quad \text{or}, \quad h \propto [r^{1/2} \ \Sigma(r)]^{-1}. \tag{1}$$

This approximate relation depends only on self-gravity and very little on the details of the calculations (cf. Bertin and Mark 1980). The fact that this relation seems to be obeyed in a number of applications (cf. Sections III-IV) to warped and bent disks is an indication of the importance of disk self-gravity. If we recall from electrostatics that self-forces of disk charges are important relative to distributed volume charges, we are then not too surprised that self-gravity should be important unless this disk mass is rather miniscule. In fact even in the extreme case of Saturn's rings relative to the mass of Saturn, preliminary evidence suggests bending waves where self-gravity affects the wave amplitude relation (Shu et al. 1982).

A somewhat more detailed (but still heuristic) calculation could equate the vertical accelerations

$$\left[\frac{\partial}{\partial t} + \Omega(r) \frac{\partial}{\partial \theta} \right]^2 h = \nu_z^{\ 2}(r) \ h + g_z \tag{2}$$

where $\Omega(r)$ is the angular frequency of rotation of the disk, and $\nu_z^{\ 2}(r)$ represents the kinematic oscillation frequency of Figure 1. The restoring self-gravity can be estimated by using the physical picture at the left frames of Figure 3. First we might analyze the wavy bent disk into two planes with densities $\Sigma/2$ and attraction $g_z \sim G\Sigma$ (G = gravitational constant). Clearly the real stiuation has an additional angular projection factor (h/wavelength) or kh. Thus $g_z \sim - G\Sigma|k|h$ for the warped disk. Equating to equation (2) gives a dispersion relation

$$[\omega - m\Omega(r)]^2 = \nu_z^{\ 2}(r) + 2\pi \ G\Sigma(r)|k(r)| \tag{3}$$

where the extra 2π factor can be known only by detailed calculations (cf. Mark 1971, Bertin and Mark 1980). The right hand frame of Figure 3 illustrates this dispersion relation for $\nu_z = 0$. Additional effects such as disk thickness, response of stellar disk including vertical mode

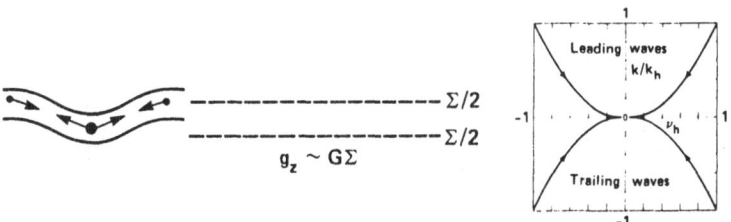

Fig. 3.   a (Left frames): Illustrates restoring gravitational force.
b (Right frame): Dispersion relation in one special case.

structure and resonances were discussed by Mark (1971). This paper also
gives some details of the "fire-hose" instability mentioned already in
the introduction. Bertin and Mark (1980, particularly App. D.) summarizes recent applications where response of gaseous disk is emphasized. From the above, it is clear that the difference in amplitude
of response of gas and star disks is due mainly to their relative in-
ertias; but in addition, smaller effects due to dispersive velocities
enter. The theory does allow occasional observable warping of stellar
disks also, and these appear to have been detected in some galaxies
(Tsikoudi 1977, 1980, van der Kruit 1979).

In the above discussion the halo, if present, provides an inactive
basic gravitational field. On the other hand, the warped disk has an
azimuthal component of the gravitational force which exerts a torque
on the particles of the halo that pass close enough to the disk (cf.
Figure 4a). This interaction allows a two stream instability (cf.
Bertin and Mark 1980) which amplifies wave disturbances in the disk.
For example, a driven wave excited by passage of a companion could be
so amplified as it propagates inwards and as it is reflected and
becomes outward propagating (Figure 4b). According to Bertin and Mark
(1980), this could be sufficient to explain the amplitude discrepancy
(Hunter and Toomre 1969) in driving of our galaxy's major warp by the
Large Magellanic cloud. The amount of wave amplification per wave
crossing time is sufficiently small that equation (1) remains roughly
valid. Bertin and Casertano (1982) performed further theoretical con-
firmations of the disk-halo interactions by means of a gas-dynamical
model (cf. also Yoshii and Fujimoto 1981 who specified an actual gaseous halo rather than using gas-dynamics as a model; and Nelson 1976
whose discussion is relevant to the gas disk at intermediate radii).

On the other hand, if the boundary conditions are suitable, bending modes (Hunter and Toomre 1969, Kulsrud et al. 1971, Mark 1971,
Bertin and Mark 1980) could be set up which are now self-generated
because of this destabilizing interaction with the halo. Some such
modes are conceptually illustrated in Figure 5. We emphasize that
these bending modes are standing waves in the radial direction but
propagate in the azimuthal direction with rotation frequency $\Omega_p = R_e(\omega)/m$.

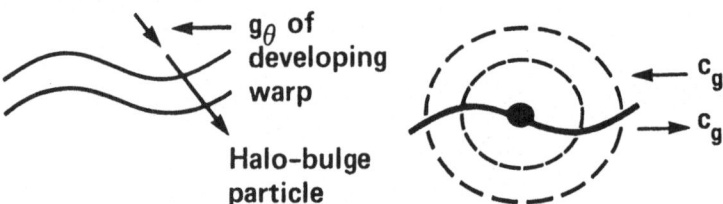

Fig. 4.   a (Left frame): The disk-halo interaction. b (Right frame):
Driven responses amplify as they propagate.

Higher bending modes could also exist

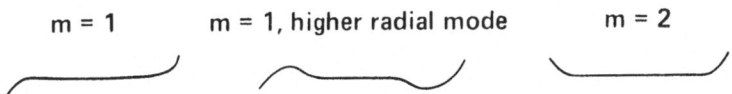

m = 1   m = 1, higher radial mode   m = 2

Fig. 5. Higher order bending modes could also exist.

For standing waves, the range of allowable outer conditions have not all been delineated. But it is sufficient (Hunter and Toomre 1969) to have a rather sharp cutoff in the outer disk density; and some galaxies indeed have unusually sharp outer boundaries (van der Kruit and Searle 1981a,b) which are more than the minimum required for standing waves. It would be helpful to pursue further theoretical developments similar to that carried out for spiral structure (e.g., cf. Lau et al. 1976, Mark 1976 a-c, 1977, Bertin et al. 1977, Bertin and Mark 1978).

III. WARPS AND EXISTENCE OF SUBSTANTIAL HALOES IN GALAXIES

Even very simple considerations are suggestive that the theoretical picture of self-gravitating warps is consistent with substantial haloes inside the region where the disk warps exist as a diagnostic. This relationship follows (cf. Bertin and Mark 1980) from the simple scaling law (1) relating height of bending $h(r)$ to the disk surface mass density $\Sigma(r)$. For the sake of argument, let us first presume that the observed flat rotation curves (cf. Rubin 1978) in galaxies represent somehow a mass distribution all restricted to the disk. Then $\Sigma(r) \sim 1/r$ in the outer parts where the rotation curves are near flat and where warps are observed. From equation (1), this implies a contradiction in that warps have a height relation $h(r) \propto r^{-3/2}$ which should look conceptually like Figure 6a. Of course observed warps have more typically the shape like Figure 6b which is indicative of a much more rapid fall off of disk mass according to equation (1). Thus, we find that the shape of warps and our theory together give an independent argument that flat rotation curves indeed are suggestive of substantial haloes (nb. our argument here is independent of detailed disk-

Fig. 6.   a (Left frame): Warp in disk with slowly decaying surface density.   b (Right frame) Rapidly decaying disk density.

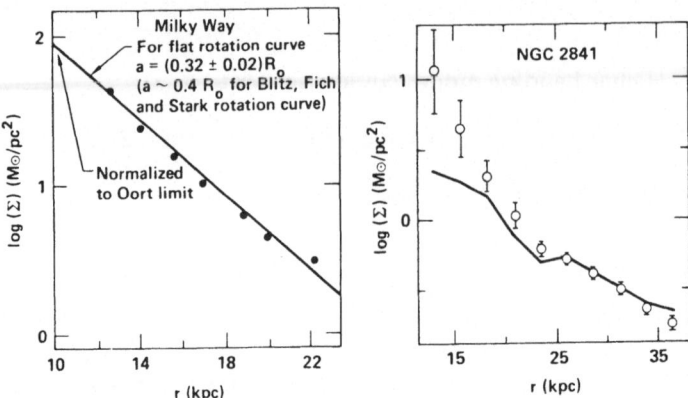

Fig. 7.   a (Left frame) The surface density $\Sigma(r)$ against radius in our galaxy as derived from equation (1) using Henderson's (1978) data for the warp. The straight line is an exponential with a scale length of 3.4 kpc, normalized to the Oort limit at the solar radius (10 kpc). b (Right frame): The surface density against radius for galaxy NGC 2841 as derived from equation (1) using the data of Bosma (1978). The solid line is the observed neutral hydrogen surface density. The deviations at distances less than 20 kpc may be due to the contribution of the optical disk.

halo interactions). The remainder of this section discusses this point further in regard to the Milky Way and one external galaxy, NGC 2841.
        Figure 7a plots surface mass density against radius in the Milky Way galaxy as derived from equation (1) using the data of Henderson (1978). If we temporarily adopt a flat rotation curve with the circular velocity beyond the solar radius (R_0) equal to 250 km/s, we find the resulting points are well-fit by an exponential with a scale length of 3.4 ± 0.2 (R_0/10 kpc) kpc. This procedure is superior to determinations of the disk profile from fitting HI velocity profiles, because nothing is initially assumed about the mass distribution as a function of radius. (Of course, in both cases a rotation curve must be assumed or measured so as to derive distances from the observed line-of-sight velocities.) Assuming the same flat rotation curve, Knapp et al. (1978) derived a scale-length of 3.2(R_0/10 kpc) kpc from such fitting of HI profiles with an assumed exponential mass distribution. Using the Oort (1960) limit to normalize the density at R_0, we obtain a mass model close to that of Caldwell and Ostriker (cf. Ostriker 1982). If we use the most recent rotation curve of Blitz et al. (1979) and Blitz (1979), the scale length is 4.0 ± 0.2 (R_0/10 kpc) kpc, the disk mass integrated to 20 kpc is approximately 7.6×10^{10} $M_\odot$ (changing by $\sim$ 10% depending on whether a central gap is included), and the combined mass of the halo and bulge out to 20 kpc is 3.1×10^{11} $M_\odot$. If this new rotation curve is used to recompute the halo component in the Caldwell-Ostriker model (leaving their bulge and

disk parameters fixed), we find that the fraction of the gravitational force due to the halo at R_0, 1.5 R_0 and 2 R_0 is 17%, 42% and 64%, respectively. These studies regarding the warp in the Milky Way should be refined in particular to take into account the new data of Henderson et al. (1982). The "scalloped" outer edges of the warp might well be due to an additional superposition of higher modes with m > 1 (cf. Fig. 5).

Next, we examine the galaxy NGC 2841, which has been observed by Bosma (1978). Figure 7b compares the surface density of neutral hydrogen with the surface density derived from the observations of the amplitude of the warp, normalized to the HI surface density in the outer parts. (All distances in NGC 2841 assume H_0 = 75 km/s/Mpc.) If there is no systematic variation of the mass-to-light ratio with radius in the observed optical disk and the mass-to-light ratio is less than 6 (twice the values obtained from stellar population models, cf. Larson and Tinsley 1978), the surface photometry (van Houten et al. 1954) suggests that the density of the optical disk is negligible in the region from $\sim$ 20 to 30 kpc; this allows the normalization of the derived density from the warp to the HI surface density in the outer parts. However, the optical disk may be the source of the extra mass at radii less than 20 kpc. In any case, the agreement in the range $\sim$ 20-35 kpc indicates that the HI surface density dominates the total surface density in this region. This is only $\sim$ 1% of the local surface density needed to maintain the observed flat rotation curve (Bosma 1978), which implies that the mass density of the halo is inversely proportional to the square of the distance as may be derived from the rotation curve in this region.

Although these bending wave results were derived with the assumption of small inclinations in the warp, it is encouraging that the theory gives a fairly reasonable description of the large inclinations in galaxy M33 (cf. Lake and Mark 1980 for further discussions). Preliminary applications to several other galaxies can be found in Bertin and Casertano (1982).

For galaxies such as NGC 2841, there is the tantalizing possibility that with better data it may be possible to measure a mass-to-light ratio in the disks of external galaxies using the bending wave theory. This measurement makes possible a computation of the core radius of the halo component, an important parameter for constraining theories of its origin and composition (Peebles 1979). This is possible if the warp is measured over a range where it samples the mass distribution of the optical disk, and extends out to the region where the HI surface density dominates and provides the normalization. This project is ideally and uniquely suited for study with the VLA telescope, when accurate 21-cm line observations become possible.

IV.   THE INNER DISK WITHIN 2 KPC OF GALACTIC CENTER

Independent analyses of the HI and CO within 2 kpc of the Galactic center have shown that the gas distribution is best described as a disk-like structure with a rotation axis inclined 20-25° to that of the Galaxy (Burton and Liszt 1978, Liszt and Burton 1978, 1980, Sinha 1979, cf. our Fig. 8a). Detailed models of the gas as an expanding circular disk (Burton and Liszt 1978, Liszt and Burton 1978) or a thin ellipti-cal bar (Liszt and Burton 1980) describe the observations equally well, but they are without any dynamical basis. We propose that the bending wave theory surveyed here can provide a plausible dynamical explanation for the tilt, and that the predictions of the theory are independent of the kinematic model of the gas distribution. We show that the predicted relation between the disk surface density distribution and the height of the bending are satisfied for both the disk-like and bar-like models of the gas.

The original motivation for the theory was to provide an explana-tion for the warps in the outer parts of spiral galaxies (cf. Section I). The model shows that the warp arises as a self-excited bending wave which is driven unstable because of its motion through the slowly rotating halo. In this section we present some evidence that a similar model can provide a dynamical explanation for the tilt in the gas disk near the center of our Galaxy. In this case the gas disk moves through the more slowly rotating bulge and an analogous instability results in a "flapping" of this disk. Direct observations of the bulge component

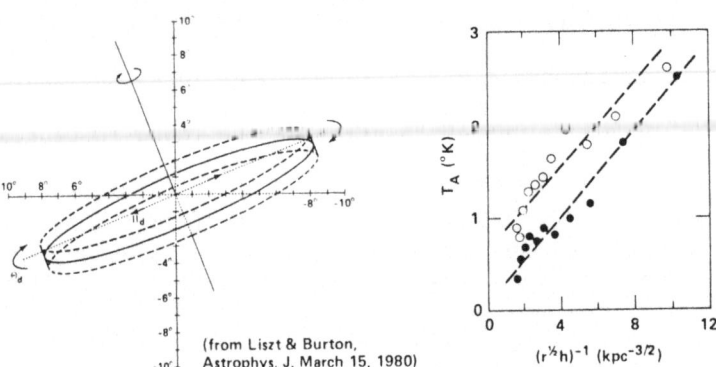

Fig. 8.   a (Left frame): The inner tilted gas disk (< 2 kpc) as depicted by Liszt and Burton. b (Right frame): Plot of HI antenna temperature (a quantity proportional to gas density) as a function of the quantity $(r^{1/2}h)^{-1}$ for the circular model (filled circles) and bar-like model (open circles) described in the text. The dashed lines are least squares fits to the data.

of the galactic nucleus using OH masers related to Mira variables (Baud 1978), and infrared radiation from population II stars (Maihara et al. 1978), indicates the presence of a slowly rotating spheroidal distribution of stars with a large scale height, comparable to what is seen in other galaxies. Liszt and Burton model their CO and HI observations of the central regions of the Galaxy by a massive disk or bar ($> 10^9 M_\odot$) with a gaussian scale height of 100 pc and typical rotational velocities of 200-300 km s^{-1}. Thus, the conditions of a thin rapidly rotating disk embedded within a larger, spheroidal mass distribution considered by the theory are applicable within the central two kiloparsecs of the Galaxy.

The bending wave theory (cf. Section II) predicts the scaling law given in equation (1). This scaling law arises because the bending wave propagates with a nearly constant rate of outflow of wave energy through circles of radius r. Although interactions with the bulge slowly amplify this rate of outflow, the amount of amplification per cycle of wave propagation is small and it is the short cycle time which determines the wave growth within the age of the Galaxy.

To examine the constancy of the product $\Sigma\, r^{1/2}\, h$, we rely on the detailed models of the nuclear disk for the geometric and kinematic properties of the disk. Burton and Liszt assume a constant surface density in their models and although there is excellent morphological agreement between the predicted and observed moment maps, the agreement between the contour gradients of their model and the observations is not particularly good. We assume that the observed contour gradients provide information on the actual run of density of the disk. The observed antenna temperature (T_A) along the line b=-ℓ tan 22° should provide a good indication of the run of surface density because: i) This locus of points is the axis of kinematic symmetry (corresponding very nearly to the projected major axis of the circular disk model) and observations along it provide the greatest contrast between the gas in the nuclear disk and the Galaxy at large. ii) The gas is surely optically thin and thus the antenna temperature of the gas should be directly proportional to the volume density. iii) The Burton and Liszt models show good agreement with the observations under the assumption of a disk of constant thickness. If this assumption is approximately correct, the observed HI antenna temperature should give the run of surface density.

To determine the surface density, we used the improved HI observations (Liszt and Burton 1980) along the line b= - ℓ tan 22°. This line intersects the midplane of the disk-like and bar-like models along a unique locus of points, and we determined the apparent LSR velocity along the locus for both models at 0.5° intervals. From the moment map of the HI observations, we found the HI antenna temperature on both sides of b=0 and averaged the results. Because the data are satisfactorily modeled by a planar distribution with a constant inclination angle, i, the product $\Sigma\, r^{1/2} h \propto T_A\, r^{3/2} \sin i$ for the disk-like model. For the bar-like model, $r^{1/2} h = r^{3/2} \sin i \cos \Theta_0$ where Θ_0 is the angle between the major axis of the elliptical streamlines and the projection of the line $b=-\ell \tan 22°$ on the disk.

In Figure 8b we plot T_A as a function of $(r^{1/2}h)^{-1}$ for both models. The points do not differ markedly from a straight line in either case, as expected from the theory. The dashed lines are the least squares fit to the data for each model. The data plotted in Figure 8b include the HI emission for $\ell \geq 2°$. Observable CO emission extends to nearly $\ell=2°$, and at lower longitudes the H_2 mass is likely to dominate the surface density of the nuclear disk. It is not possible to determine the H_2 mass from CO observations in the nuclear disk with any confidence, and we have therefore excluded the region of $\ell < 2°$ from our analysis.

It seems reasonable to interpret Figure 8b as implying that the run of density in the nuclear disk agrees with the predictions of the bending wave theory independent of the kinematics of the model used to describe the data. The largest uncertainty probably comes from the assumption of a constant thickness of the gas layer. The disk-like model could have an increasing scale height due to decreased self-gravity of the disk while the bar-like model may have the opposite tendency if it is really a triaxial ellipsoid rather than a thin disk. These possibilities can be tested only with better observations and more refined model making. Nevertheless, for the models we consider, which satisfactorily describe the present observations, the bending-wave theory provides a plausible dynamical basis for the tilt of the nuclear disk.

As can be seen from the references cited, the work described here is the result of a collaborative effort with G. Bertin, L. Blitz, G. Lake and R. P. Sinha. We also appreciate discussions with A. Bosma, W. B. Burton, A. P. Henderson, F. Kerr, P. C. van der Kruit, C. C. Lin, H. S. Liszt, J. P. Ostriker, V. C. Rubin, R. Sancisi, M. Schwarzschild, F. H. Shu, G. de Vaucouleurs, G. Westerhout.

REFERENCES

Avner, E. S., and King, I. R.: 1967, Astron. J. 72, p. 650.
Baud, B.: 1978, Ph.D. Thesis, University of Leiden, The Netherlands.
Bertin, G. and Casertano, S.: 1982, Astron. Astrophys. 106, p. 274.
Bertin, G. and Mark, J. W-K.: 1978, Astron. Astrophys. 64, p. 389
Bertin, G. and Mark, J. W-K.: 1980, Astron. Astrophys. 88, p. 289.
Bertin, G., Lau, Y. Y., Lin, C. C., Mark, J. W-K., and Sugiyama, L.:
    1977, Proc. Natl. Acad. Sci. U.S.A. 74, p. 4726.
Binney, J.: 1978, Monthly Notices Roy. Astron. Soc. 183, p. 779.
Blitz, L.: 1979, Astrophys. J. Letters 231, p. L115.
Blitz, L., Fich, M., and Stark, A. A.: 1979, in IAU Symp 87.
Blitz, L., Mark, J. W-K., and Sinha, R. P.: 1981, Nature 290, p. 120.
Bosma, A.: 1978, Ph.D. Thesis, University of Groningen, The
    Netherlands.

Burke, B. F.: 1957, Astron. J. 62, p. 90.
Burton, W. B. and Liszt, H. S.: 1978, Astrophys. J. 225, p. 815.
Elwert, G. and Hablick, D.: 1965, Z. Astrophys. 61, p. 273.
Fujimoto, M. and Sofue, Y.: 1976, Astron. Astrophys. 47, p. 263.
Fujimoto, M. and Sofue, Y.: 1977, Astron. Astrophys. 61, p. 199.
Henderson, A. P.: 1978, in IAU Symp. 84, p. 493.
Henderson, A. P., Jackson, P. D. and Kerr, F.: 1982, Astrophys. J.,
     in press.
Houten, C. J. van, Oort, J. H., and Hiltner, W. A.: 1954, Astrophys.
     J. 120, p. 439.
Hunter, C. and Toomre, A.: 1969, Astrophys. J. 155, p. 747.
Kahn, F. D. and Woltjer, L.: 1959, Astrophys. J. 130, p. 705.
Kerr, F. J.: 1957, Astron. J. 62, p. 93.
Knapp, G. R., Tremaine, S. D., and Gunn, J. E.: 1978, Astrophys. J.
     83, p. 1585.
Kruit, P. C. van der: 1979, Astron. Astrophys. Suppl. 38, p. 15.
Kruit, P. C. van der, Searle, L.: 1981a, Astron. Astrophys. 95,
     p. 105.
Kruit, P. C. van der, Searle, L.: 1981b, Astron. Astrophys. 95,
     p. 116.
Kulsrud, R. M., Mark, J. W-K., and Caruso, A.: 1971, Astrophys.
     Space Sci. 14, p. 52.
Lake, G. and Mark, J. W-K.: 1980, Nature 287, p. 705.
Larson, R. B., and Tinsley, B. M.: 1978, Astrophys. J. 219, p. 46.
Lau, Y. Y., Lin, C. C., and Mark, J. W-K.: 1976, Proc. Natl. Acad.
     Sci. U.S.A. 73, p. 1379.
Liszt, H. S. and Burton, W. B.: 1978, Astrophys. J. 226, p. 790.
Liszt, H. S. and Burton, W. B.: 1980, Astrophys. J. 236, p. 779.
Lynden-Bell, D.: 1965, Monthly Notices Roy. Astron. Soc. 129, p. 299.
Maihara, T., Oda, N., Sugiyama, T., and Okuda, H.: 1978, Pub. Astron.
     Soc. Japan 30, p. 1.
Mark, J. W-K.: 1971, Astrophys. J. 169, p. 455.
Mark, J. W-K.: 1976a, Astrophys. J. 203, p. 81.
Mark, J. W-K.: 1976b, Astrophys. J. 205, p. 363.
Mark, J. W-K.: 1976c, Astrophys. J. 206, p. 418.
Mark, J. W-K.: 1977, Astrophys. J. 212, p. 645.
Nelson, A. H.: 1976, Monthly Notices Roy. Astron. Soc. 174, p. 661.
Oort, J. H.: 1960, Bull. Astron. Inst. Netherlands 15, p. 46.
Ostriker, J. P.: 1982, this conference.
Peebles, P. J. E.: 1979, Lecture notes for L'Ecole d'Eté de Physique
     Théoretique, Les Houches.
Polyachenko, V. L.: 1977, Sov. Astron. Lett. 3, p. 51.
Polyachenko, V. L., and Shukhman, I. G.: 1979, Sov. Astron. 23,
     p. 407.
Roberts, M. S.: 1966, Astrophys. J. 144, p. 639.
Roberts, M. S. and Whitehurst, R. N.: 1975, Astrophys. J. 201,
     p. 327.
Rogstad, D. H., Lockhart, I. A., and Wright, M. C. H.: 1974,
     Astrophys. J. 193, p. 309.
Rogstad, D. H., Wright, M. C. H., and Lockhart, I. A.: 1976,
     Astrophys. J. 204, p. 703.

Rubin, V. C.: 1978, in IAU Symp. 84, p. 211.
Saar, E. M.: 1978, in IAU Symp. 84, p. 513.
Sancisi, R.: 1976, Astron. Astrophys. 53, p. 159.
Sanders, R. H.: 1978, in IAU Symp. 84.
Sinha, R. P.: 1979, Ph.D. Thesis, University of Maryland.
Shu, F. H., Cuzzi, J. N., and Lissauer, J. J.: 1982, work in
      progress.
Spight, L. and Grayzeck, E.: 1977, Astrophys. J. 213, p. 374.
Toomre, A. 1966: Lectures in Geophysical Fluid Dynamics at the Woods
      Hole Oceanographic Institution.
Tsikoudi, V.: 1977, Ph.D. Thesis, Univ. of Texas, Austin.
Tsikoudi, V.: 1980, Astrophys. J. Suppl. 43, p. 365.
Tubbs, A. D. and Sanders, R. H.: 1979, Astrophys. J. 230, p. 736.
Weliachew, L., Sancisi, R., and Guelin, M.: 1978 Astron. Astrophys.
      65, p. 37.
Westerhout, G.: 1957, Bull. Astron. Inst. Netherlands 13, p. 201.
Yoshii, Y., and Fujimoto, M.: 1981, Astron. Astrophys. 104, p. 142.

I _____ Work performed under the auspices of the U.S. Department of Energy
      by Lawrence Livermore National Laboratory under Contract Number
      W-7405-ENG-48

ON COROTATING HIGH-Z HI

Felix J. Lockman
National Radio Astronomy Observatory[1]
Charlottesville, Virginia

ABSTRACT

There is evidence in surveys of HI in the inner Galaxy for gas more than 500 pc from the plane that shares the rotation of material in the plane. The percentage of HI with $|z| > 500$ pc at the subcentral points rises approximately monotonically from less than 2% at $R < 3$ kpc to $\gtrsim 15$% near the solar neighborhood. Some cloudy structure is observed. Some of the high-z gas has a larger velocity dispersion than the HI confined more closely to the plane.

INTRODUCTION

It is a commonplace that galactic HI, interior to the solar circle, is confined to a narrow layer with a width between half-brightness points, $z_{1/2}$, of about 250 pc (Schmidt, 1957; Kerr, 1969; Jackson and Kellman, 1974). The layer has been modeled by a Gaussian distribution in z, by a combination of two Gaussians for a cloud plus intercloud medium (Baker and Burton, 1975; they find $z_{1/2} = 190$ pc for the clouds and 280 pc for the intercloud gas), and by a Gaussian with exponential wings (Celnik, Rohlfs and Braunsfurth, 1979). These studies have given a reasonably accurate description of the average properties of the HI layer with one exception: the observed HI intensity at $|z| > 400$ pc lies above every model prediction, even when the model includes exponential wings (e.g., Figures 5 and 7 of Celnik, Rohlfs and Braunsfurth, 1979). There is HI fairly far from the plane whose properties have not yet been described.

This paper presents some initial results of a study of corotating HI away from the plane that might properly be called "halo" gas. Its distance is known exactly because it lies along the locus of subcentral points in the first quadrant of galactic longitude.

[1]The National Radio Astronomy is operated by Associated Universities, Inc., under contract with the National Science Foundation.

W. L. H. Shuter (ed.), Kinematics, Dynamics and Structure of the Milky Way, 303–313.
Copyright © 1983 by D. Reidel Publishing Company.

THE VERTICAL DISTRIBUTION OF THE HIGHEST VELOCITY GAS

Figure 1 gives a synoptic view of the vertical distribution of galactic HI. Contours of brightness temperature are drawn in velocity-latitude coordinates for a constant longitude. Emission at positive velocities comes from gas interior to the solar circle; that at negative velocities from gas outside the solar circle on the far side of the Galaxy from the sun. The innermost gas visible at longitude $45°44$ lies at the subcentral point, $R = R_0 \sin \ell = 7.125$ kpc from the galactic center (throughout this paper $R_0 = 10$ kpc, $\theta_o = 250$ km s^{-1}). The velocity at the subcentral point (ssp) is 70 km s^{-1} (Burton and Gordon, 1978), and is the maximum or terminal velocity given by galactic rotation. It is marked with a dashed line that curves in at large $|b|$ due to projection effects. Streaming motions may give gas somewhat displaced from the subcentral point the highest velocity. However, streaming motions in the Galaxy are rather small ($\lesssim 10$ km s^{-1}; Burton, 1971), and their influence on this work is negligible. Also, the terminal velocities used here are always derived from observation and do not depend on an assumed rotation curve.

Negative velocity gas shows the well-known warp (see Kulkarni et al., this volume). It is difficult to determine properties of gas in this area because outside the solar circle distances depend entirely on an adopted rotation curve; small deviations from the assumed rotation make large distance errors. This is especially true far from the galactic center where kinematic distance discrimination is quite poor. For example, at $\ell = 45°44$ with a flat rotation curve, gas 20 kpc from the galactic center differs by only ~4 km s^{-1} (LSR) from gas at $R = 21$ kpc. Moreover, any deviation from the assumption of corotation produces large errors in the derived thickness of the warped layer. Nonetheless, there is evidence that beyond the solar circle the HI layer thickens and may have localized extensions reaching $|z|$ ~2 kpc (e.g., Kepner, 1970).

HI is more easily quantified in the inner Galaxy. Gas near the terminal velocity comes from a small region around the subcentral point, and the assumption of corotation can be checked by simply observing at a constant $V_{LSR} \cos(b)$. Observations indicate that there are not significant amounts of high latitude HI rotating faster than material in the plane (this would show up, for example, as a bowing of the high velocity contours in Figure 1 to higher velocity with increasing $|b|$). Therefore, any HI observed near the terminal velocity above the plane is corotating. A halo rotating more slowly than the disk (as suggested by deBoer and Savage, 1982) would not be observed at the terminal velocity.

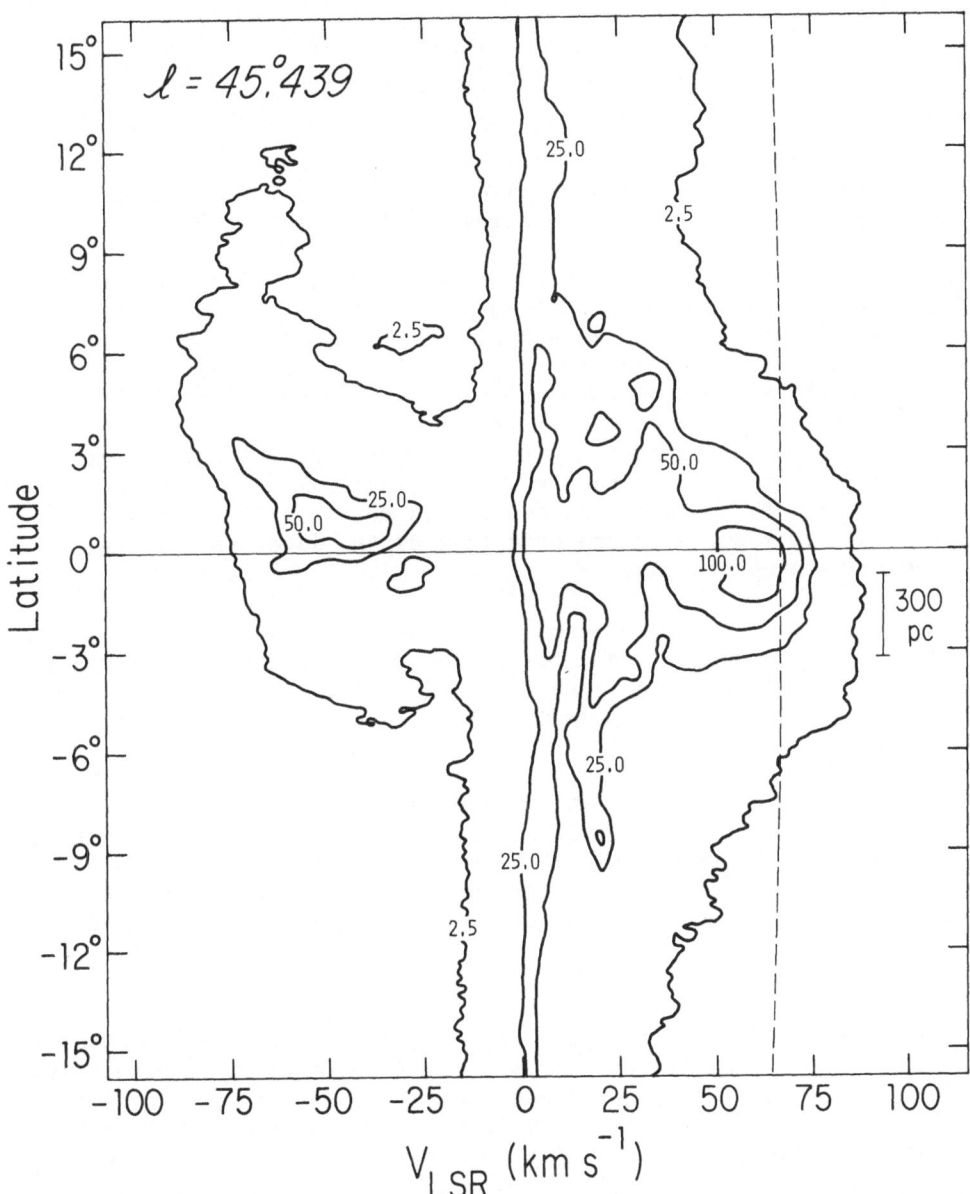

Figure 1: Velocity-latitude diagram for HI at longitude 45°439 where
the subcentral point is 7.125 kpc from the galactic center. Dashed
line marks the terminal velocity; the linear scale refers only to
material at the terminal velocity or greater. The HI layer is much
thinner in the bright emission near V_t than in the faint emission near
85 km s^{-1}.

The bright gas near the terminal velocity lies in a narrow layer; at 70 km s^{-1} $z_{1/2}$ = 300 pc. But it is clear that this is not true at higher velocities. Faint HI near 85 km s^{-1} changes brightness very little over ~6° in latitude, i.e., over 700 pc in z. The highest velocity HI on the wing of the profile has a broader vertical distribution than lower velocity HI does.

This is shown in more detail in Figure 2, which is made from the data of Figure 1. Normalized profiles of T_b vs. z are drawn at four velocities, from the terminal velocity, to 85 km s^{-1} where the brightness temperature at any latitude is always <5 K. The half-thickness of the layer changes from 300 pc to 850 pc between V_t and the profile wing. In a multicomponent ISM, the gas in the wings of a profile will contain the largest proportion of high velocity-dispersion gas, and it is precisely this gas that is expected to have the thickest vertical distribution. Figures 1 and 2 therefore show that there is a component of galactic HI with a high velocity dispersion and a correspondingly thick vertical distribution that is corotating with material directly in the plane. It may extend to $|z| \sim 1000$ pc. Although the high-z gas is more visible at high velocities, it is actually the brightest at the lower velocities, consistent with corotation.

This material, which appears on v-b maps as a straightness of the faint contours compared to the bright ones, can be seen in virtually all previously published surveys (e.g., Burton and Verschuur, 1973; Weaver and Williams, 1974), provided that they cover sufficient latitude.

Not all longitudes show the high-velocity, high-z emission. Figure 3 shows a v-b map for 15°96, corresponding to R_{ssp} = 2.75 kpc. Gas near the ssp is once again confined in b, but the faint gas at higher velocities also forms a thin layer. At this location there is no corotating high-z material bright enough to have been detected.

THE AMOUNT OF HALO HI

From now on I will use "halo" for all HI more than 500 pc from the galactic plane. This distance is ≳5σ for a Gaussian layer with a $z_{1/2}$ of 250 pc, and thus marks an appropriate division between a thin layer and a more extensively distributed medium. The amount of halo gas is determined by integrating the HI profiles from V_t to the highest velocity containing detectable emission. For these calculations I used the 300-foot data from the Arecibo-Green Bank survey of the vertical distribution of HI (Bania and Lockman, 1983) and other observations made with the 140-foot telescope at longitudes 15°96, 20°49 and 45°44. The Arecibo data from Bania and Lockman cover only $|b| < 3°$ and provide little extra information on the halo.

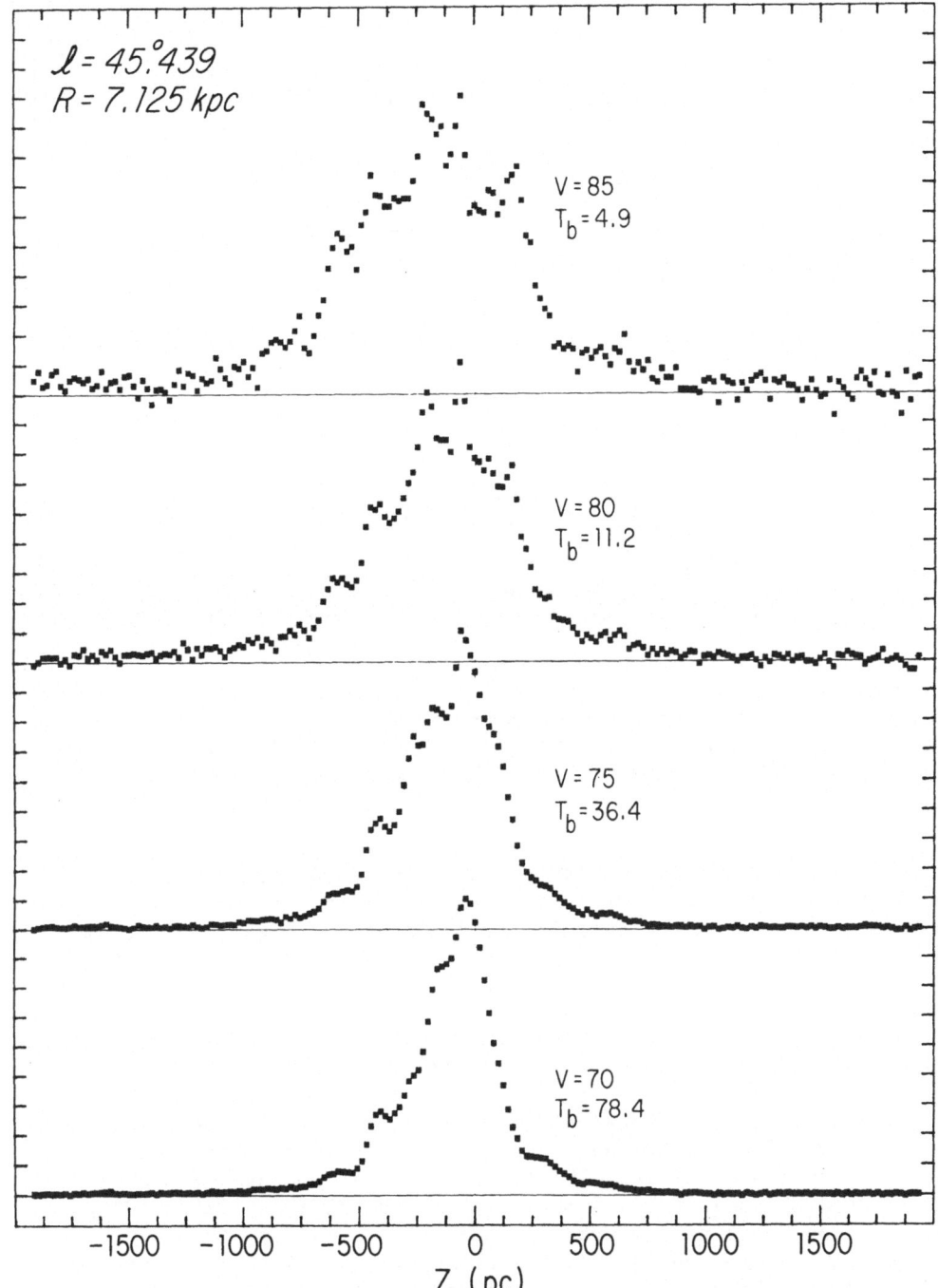

Figure 2: The brightness temperature of HI vs. distance from the galactic plane at several velocities for l = 45°.44. Gas near the terminal velocity (70 km s⁻¹) has z₁/₂ = 300 pc; at higher velocities the layer thickens, reaching a z₁/₂ of 850 pc at 85 km s⁻¹.

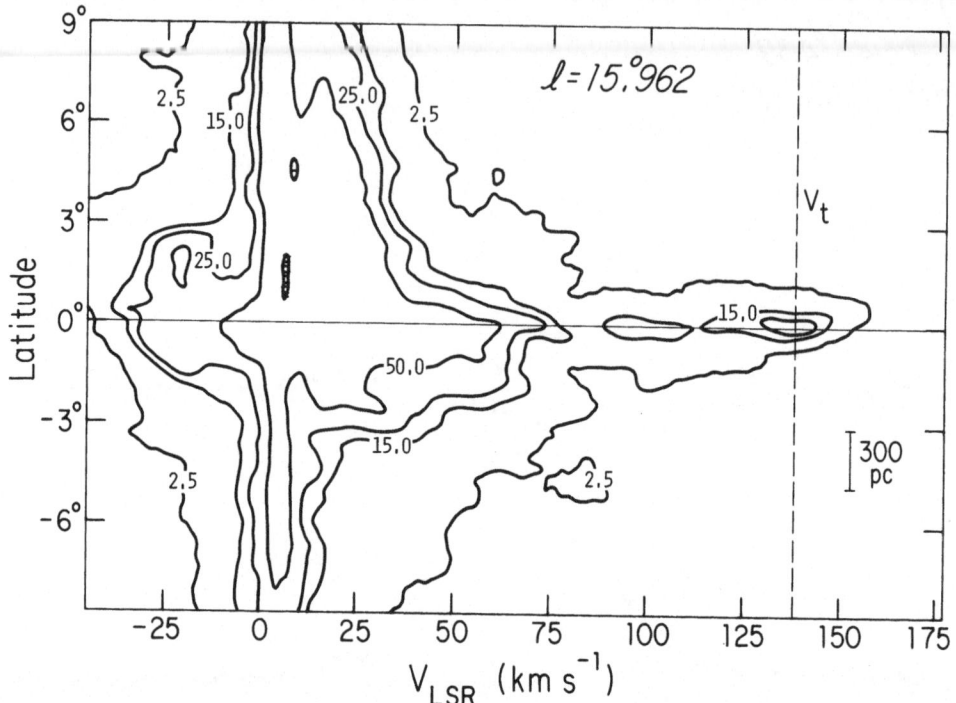

Figure 3: Velocity-latitude diagram for HI at longitude 15.°962 where the subcentral point is 2.75 kpc from the galactic center. Both the bright and faint HI near V_t are confined to a narrow layer.

The derived column density of HI in the halo depends sensitively on the precise choice of V_t: if V_t is chosen too low, gas is included that does not lie at the ssp; if it is chosen too high then the amount of halo gas is underestimated. Thus the halo content is given as a percentage of the total observed HI emission in the velocity interval. Results are shown in Figure 4. Lower limits obtain where the vertical coverage of the surveys was insufficient to cover all high latitude gas (even though in most cases the data cover $|b| < 10°$).

There is little if any halo gas interior to R = 3 kpc. Here the main HI layer is also thinner than in the rest of the inner Galaxy (e.g., Jackson and Kellman, 1974). Substantial amounts of high-z gas appear suddenly near R = 3.5 kpc: 17% of the HI is at $|z| > 500$ pc. The fraction drops near 4 kpc and then rises steadily, though irregularly, moving outward to the solar circle. Figure 4 indicates that at least 15% of the HI near the solar neighborhood lies more than 500 pc from the plane. This limit is consistent with the values inferred from other studies (e.g., Bohlin, Savage and Drake, 1978; Hobbs et al. 1982) which, incidentally, suggest that the true fraction at R $\gtrsim$ 8 kpc may be twice the limits given here.

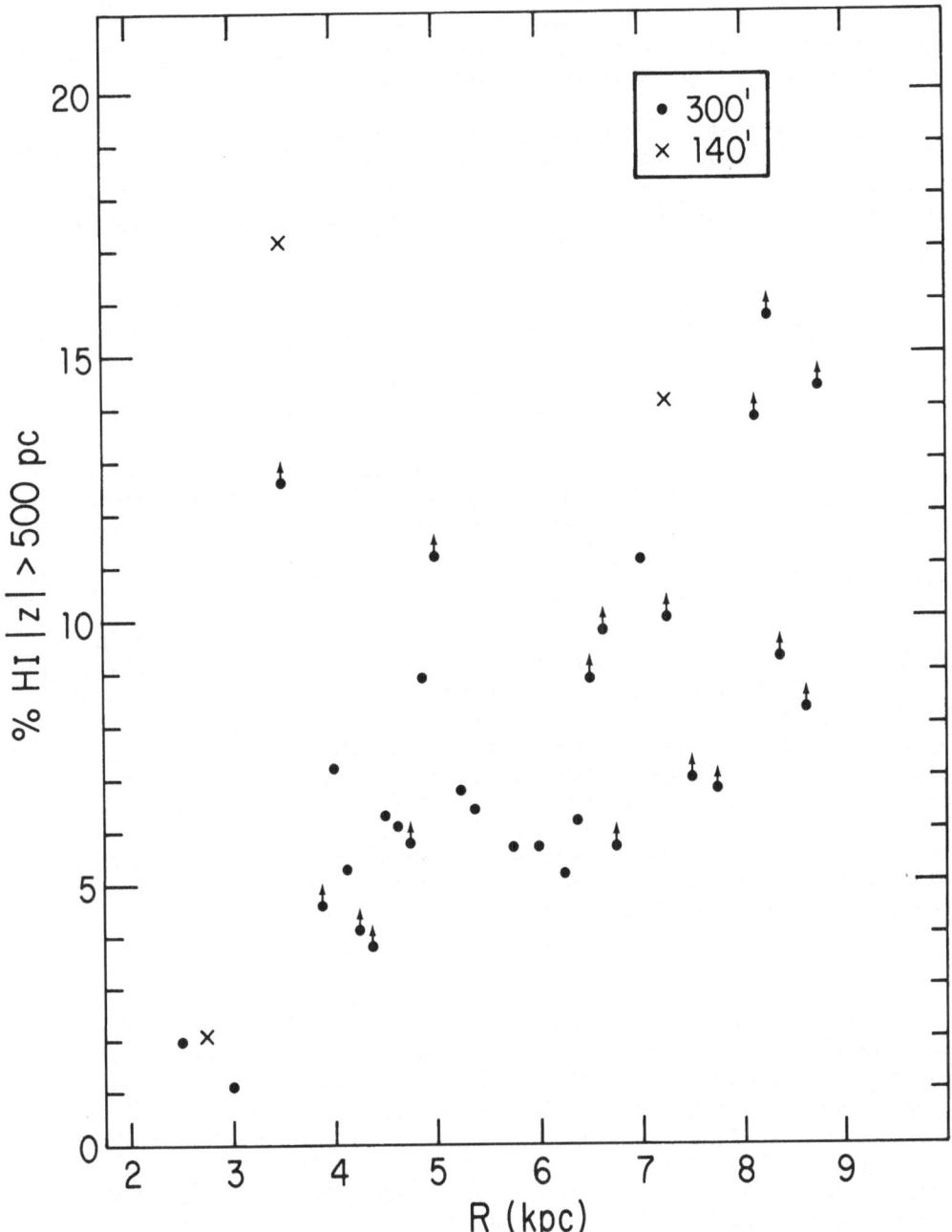

Figure 4. The percentage of HI with |z| > 500 pc vs. distance from the galactic center. Lower limits indicate longitudes where the emission extends beyond the latitude coverage of the observations.

Point to point variations in the percentage of halo gas can be large. In addition, T_b does not always smoothly decrease with increasing $|z|$. Both these facts make it likely that cloudy structure is being observed. This is most apparent at $R \geqslant 5$ kpc.

The density of HI near the plane is fairly flat over $4 < R < 14$ kpc (e.g., Burton and Gordon, 1978) but decreases interior to 4 kpc. Thus the total amount of halo HI may be even smaller at $R \leqslant 3$ kpc than Figure 4 suggests. It is possible that there is no HI with $|z| > 500$ pc and $R \leqslant 3$ kpc.

The maximum extent of detectable HI varies from position to position. Some locations, like those around $R = 8$ kpc, show definite emission where the data end ~1 kpc from the plane (but see the caution in the following section). Other positions, such as those near $R = 5$-6 kpc show no evidence that HI has been detected much beyond 1 kpc. It is possible that the maximum extent, like the fraction in the halo, varies with distance from the galactic center. Additional data, covering a larger latitude range with more sensitivity than available in any survey yet made, are needed to clarify these points.

BUT IS IT REAL?

The problem of stray 21-cm radiation, entering a telescope's sidelobes some angle away from the main beam, has plagued high-latitude HI studies. (Those who wish to develop the appropriate paranoia about this effect should examine Figure 5 in Kalberla, Mebold and Reich, 1980). The main contribution to stray radiation comes from bright, low velocity HI near the plane that is present at all longitudes and extends tens of degrees in latitude. Owing to differing projection of the earth's motion, the emission will appear Doppler-shifted with respect to the LSR in the direction of the main beam. This emission makes high-latitude observations of faint HI subject to considerable uncertainty (see also Heiles, Stark and Kulkarni, 1981).

The problem is not so severe, however, for the material studied here. First, the halo emission is often comparatively bright (Figure 2). Second, the terminal velocity is usually so large that other emission at its velocity is found only in a relatively small fraction of the sky (contrast the angular extent of high and low velocity gas in Figures 1 and 3). For example, bright ($T_b \geqslant 10$ K) HI at 100 km s^{-1} is found only in two bands visible from Green Bank, each covering about 35° in longitude and about 7° in latitude. Only one of these bands is above the horizon at any time. In contrast, bright hydrogen near zero velocity is found at all longitudes in a band >50° wide. Thus, for a main beam efficiency of 70% and an average HI brightness temperature of 25 K, we expect ~2 K in stray radiation at low velocities, but only $\lesssim$0.1 K from gas near 100 km s^{-1}. The latter temperature is below the noise level of the data used here, while the former value is approximate agreement with that derived from more elaborate considerations (Kalberla, Mebold and Reich, 1980). Even

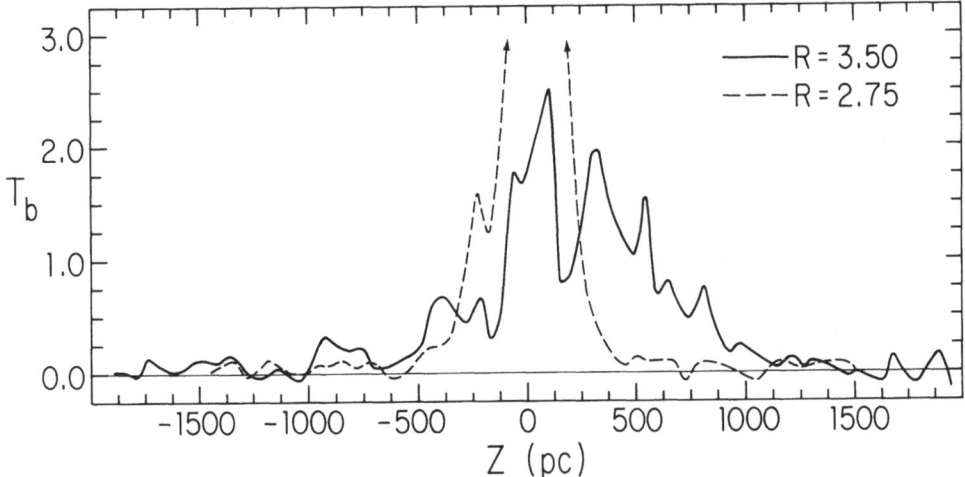

Figure 5: T(z) averaged over the velocity range 135–145 km s⁻¹ at longitudes 20°.49 (solid line) and 15°.96 (dashed line). Around z = 500 pc one radius shows halo HI, the other does not.

under the worst possible configuration of near sidelobe and HI, stray radiation seems incapable of providing the amount of halo HI that is observed at the high velocities. The terminal velocity of 100 km s⁻¹ used in this example is reached at $\ell \sim 35°$; lower longitudes have potentially less of a stray-radiation problem because V_t is larger and the proportion of the sky filled with competing emission even lower.

Further tests support this conclusion. The location of many of the near sidelobes, which result from diffraction, scales with the inverse of the telescope diameter. The profile of halo HI, however, appears almost identical at several longitudes where data from 300-foot, 140-foot, and 85-foot (Weaver and Williams, 1974) telescopes are available. Moreover, if the apparent halo gas were coming from near sidelobes, then the pattern of faint HI would be a map of the sidelobes. In fact, the observed halo seems to follow a different distribution, one that varies from longitude to longitude.

Finally, the absence of a significant halo at R < 3 kpc can be used to test for stray radiation in the most direct way. Figure 5 shows T(z) for R_{ssp} = 3.5 kpc (solid line; ℓ = 20°.49) where there seems to be a large percentage of halo gas, and R_{ssp} = 2.75 kpc (dashed line; ℓ = 15°.96) with little or no halo gas. Both curves include gas in the velocity range 135 to 145 km s⁻¹; this covers the bright gas near V_t at the lower radius and the faint profile wings at the higher radius. The data were taken sequentially on the 140-foot telescope; points at identical z are separated on the sky by about 4°5. At R = 3.5 kpc there is significant HI at z > 500 pc, while at R = 2.75 kpc there is

only the nearly Gaussian wings of the thin central layer. Because the
two longitudes are so close on the sky, and because the velocity
interval is identical, contribution from stray radiation should be
similar. But one radius shows halo gas and the other does not. This
may be the most convincing evidence that the phenomenon discussed here
is not entirely instrumental.

It is more difficult to be certain about the higher latitude data
($\ell \gtrsim 55°$; $R \gtrsim 8$ kpc) where the terminal velocity is relatively low
($\lesssim 50$ km s^{-1}) and stray radiation is of increasing concern. Results
for this longitude interval should be treated with caution at present.
Nonetheless, there is often emission >5 K at $|z| \sim 500$ pc, and $T(z)$ is
again not monotonic. In addition, the amount of halo gas inferred from
the 21-cm emission is completely reasonable and in agreement with that
deduced from measurements at other wavelengths (e.g., Bohlin, Savage
and Drake, 1978).

In summary, despite justifiable skepticism, there is a strong case
for the existence of this gas. Moreover, its absence, in light of
existing evidence for a neutral halo, might pose more of a problem than
the claims made here.

There are several other experimental methods that can separate
stray radiation from that detected via the main beam. These include
using the horizon to block stray radiation by observing at extremely
low elevations; "cleaning" of the spectra by forcing them to agree with
the low-resolution but very low sidelobe Crawford Hill HI survey,
(Heiles, 1983), and by detailed study and modelling of the telescope
beam pattern as done by Kalberla, Mebold and Reich (1980) for the 100 m
telescope. These tests, and a new high sensitivity HI survey covering
a large latitude range, are in progress.

SUMMARY

An interesting fraction of the HI in the inner Galaxy lies more
than 500 pc from the plane, in approximate corotation with material
below it. The fraction of halo HI is very small at $R \lesssim 3$ kpc, but
increases almost monotonically to values $>15\%$ near the sun. The
existence of this gas is entirely consistent with evidence for a halo
from a variety of sources including optical and UV absorption studies
(e.g., Munch and Zirin, 1961; Bohlin, Savage and Drake, 1978; Hobbs
et al., 1982), analyses of galactic synchrotron emission (e.g., Brown,
this volume), and studies of the evolution of high latitude supernova
remnants (Chevalier, 1982). Even the cold HI clouds studied in
absorption in the solar neighborhood have a component ($\sim 12\%$ of the
total) that lies at $|z| > 500$ pc (Dickey et al., this volume), and this
population should be well-confined to the plane. Furthermore, it is
possible that the halo HI is distributed approximately like the
electrons that produce pulsar dispersion measure; a recent analysis of
the dispersive medium suggests that its scale height increases
monotonically from the interior of the Galaxy to the solar neighborhood

(Harding and Harding, 1982), just as the halo HI does. Corotation of the halo, to a distance of at least 1 kpc, is also supported from other considerations (Weisheit, 1978; Savage and deBoer, 1981; York $et\ al.$, 1982).

Changes in $z_{1/2}$ with velocity (Figure 2) indicate the complexity of interstellar HI. The low velocity dispersion, bright gas is confined to a relative small z, while the predominent gas at higher velocities extends to larger z. It is likely that a full multi-component model will have to be constructed, more elaborate than the best previous work (Baker and Burton, 1975), in order to quantify the properties of each component.

REFERENCES

Baker, P. L. and Burton, W. B. 1975, Ap. J., 198, 281.
Bania, T. M. and Lockman, F. J. 1983 (in preparation).
Bohlin, R. C., Savage, B. D. and Drake, J. F. 1978, Ap. J., 224, 132.
Burton, W. B. 1971, Astron. Ap., 10, 76.
Burton, W. B. and Verschuur, G. L. 1973, Astron. Ap. Suppl., 12, 145.
Burton, W. B. and Gordon, M. A. 1978, Astron. Ap., 63, 7.
Celnik, W., Rohlfs, K. and Braunsfurth, E. 1979, Astron. Ap., 76, 24.
Chevalier, R. A. 1982, Ap. J. (in press).
deBoer, K. S. and Savage, B. D. 1982, Ap. J. (in press).
Harding, D. S. and Harding, A. K. 1982, Ap. J., 257, 603.
Heiles, C., Stark, A. A. and Kulkarni, S. 1981, Ap. J. (Letters),
    247, L73.
Heiles, C. 1983 (in preparation).
Hobbs, L. M., Morgan, W. W., Albert, C. E. and Lockman, F. J. 1982,
    Ap. J. (in press).
Jackson, P. D. and Kellman, S. A. 1974, Ap. J., 190, 53.
Kalberla, P.M.W., Mebold, U. and Reich, W. 1980, Astron. Ap., 82, 275.
Kepner, M. 1970, Astron. Ap., 5, 444.
Kerr, F. J. 1969, Ann. Rev. Astr. Ap., 7, 39.
Munch, G. and Zirin, H. 1961, Ap. J., 133, 11.
Savage, B. D. and deBoer, K. S. 1981, Ap. J., 243, 460.
Schmidt, M. 1957, B.A.N., 13, 247.
Weaver, H. and Williams, D.R.W. 1974, Astron. Ap. Suppl., 17, 1.
Weisheit, J. 1978, Ap. J., 219, 829.
York, D. G., Blades, J. C., Cowie, L. L., Morton, D. C., Songaila, A.
    and Wu, C. 1982, Ap. J., 255, 467.

THE MOUNT WILSON HALO MAPPING PROJECT

Allan Sandage
Mount Wilson and Las Campanas Observatories of the
Carnegie Institution of Washington

ABSTRACT

Four parts of the long-range Mount Wilson halo mapping project are designed to find the density gradient, the flattening, and the metallicity distribution of the halo at various heights above the plane. The plan is described, and first results on the photometry to $B \sim 20$ mag in SA28, 55, 82, and 107 are illustrated. Two methods are described to study the chemical difference between the old disk and the high velocity halo, one using local stars with a range in W velocities, and the other by observing faint stars directly _in situ_.

I. INTRODUCTION

A long-range study of the Galactic halo has been in progress at Mount Wilson for the past 8 years. It has four goals. (1) We hope to provide new evidence on the nature of the halo stars, such as their age and metallicity distributions at various heights above the plane. In every respect, are all halo stars merely the field representatives of metal-poor globular cluster stars, or are there other stellar components of different age in the halo field as well? (2) What are the B-V and V-R color distributions, and the corrections to the Seares blue magnitudes to $B \sim 19$ in the Mount Wilson Catalog of 139 Selected Areas (Seares, Kapteyn, and van Rhijn 1930) for the 10 Selected Areas in the halo mapping program? (3) What is the density gradient and the flattening of the halo in the $\ell = 0°$ and $\ell = 180°$ meridional plane, and (4) Is there a chemical gradient in the halo, or merely a gross metallicity difference between old disk and extreme halo stars, with each component having a constant (but different) metallicity distribution over their respective scale-heights (Searle and Zinn 1978; Sandage 1981)? This program is an extension of the earlier Basel halo mapping carried forward by W. Becker and his colleagues since the middle 1960's. Their work is the foundation upon which the present program is based.

W. L. H. Shuter (ed.), Kinematics, Dynamics and Structure of the Milky Way, 315–324.

II.   NATURE OF THE HALO STARS

Since 1950 when an evolutionary understanding of the globular cluster stars was beginning, the general assumption has been that the stellar content of the Galactic halo is like that of classical globular clusters.   This assumption, first based primarily on the presence of field RR Lyrae stars at all halo heights, has recently gained new support from studies of (a) the color-magnitude diagrams of faint high latitude field stars, and (b) the metallicity distributions of nearby high velocity stars that are just now crossing the plane and which have a large range in the W velocities.   Part of the modern evidence for this view is summarized in Figures 1-3.

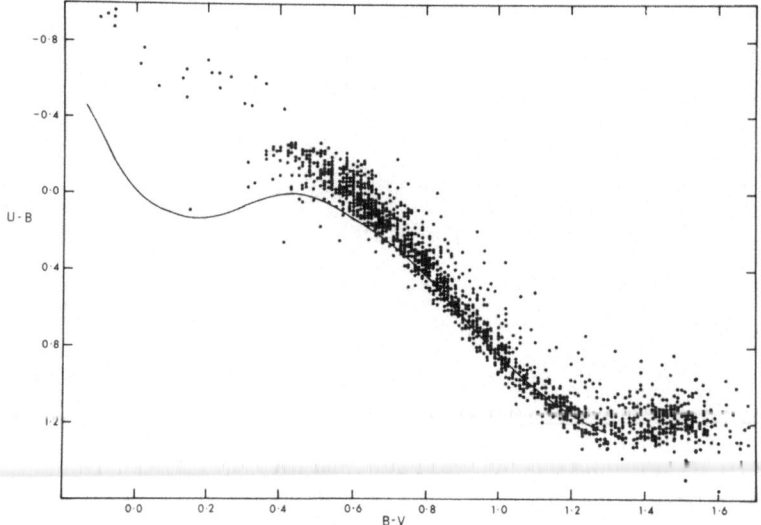

Figure 1.   Distribution of ∿1700 stars from an unpublished photometric catalog of the first 100 Lowell Proper Motion fields.

Figure 1 is from an unpublished photoelectric survey made in the 1960's (Sandage and Kowal) of the first 100 Lowell Proper Motion fields of all stars brighter than listed magnitude 13, bluer than color class 2.   Recall that the Lowell survey contains proper motions larger than $\mu = 0.27$ sec yr^{-1}, hence all stars more distant than 40pc (∿70% of the 1700 plotted points) have transverse velocities of at least $S_T = 50$ km s^{-1}, and most are higher because the proper motions are larger than the catalog lower limit.   The Hyades fiducial line is shown as solid.   Bluer than B-V = 0.8, it forms the <u>lower</u> envelope to the distribution.

That many very high velocity stars ($S_T > 200$ km s^{-1}) exist in this sample is known from the total space motions of those stars whose radial velocities are known (cf., Sandage 1969, 1981). The ultraviolet excess values for these stars range up to 0.3 mag, which corresponds to [Fe/H] $\simeq$ -2.5. The upper envelope in Figure 1 (not drawn) gives the maximum underabundance limit for the U-B/B-V blanketing vector (Sandage and Eggen 1959; Wildey, Burbidge, Sandage, and Burbidge 1962). This upper envelope agrees with the U-B/B-V ultraviolet excess data for the most metal-poor clusters M15 and M92 (Sandage 1970, Fig. 15), a result which is an extension of Roman's (1950) central discovery of the UV excess of high velocity field subdwarfs, and the subsequent identification of the same UV excess values for stars in the globular clusters themselves (Sandage and Walker 1955).

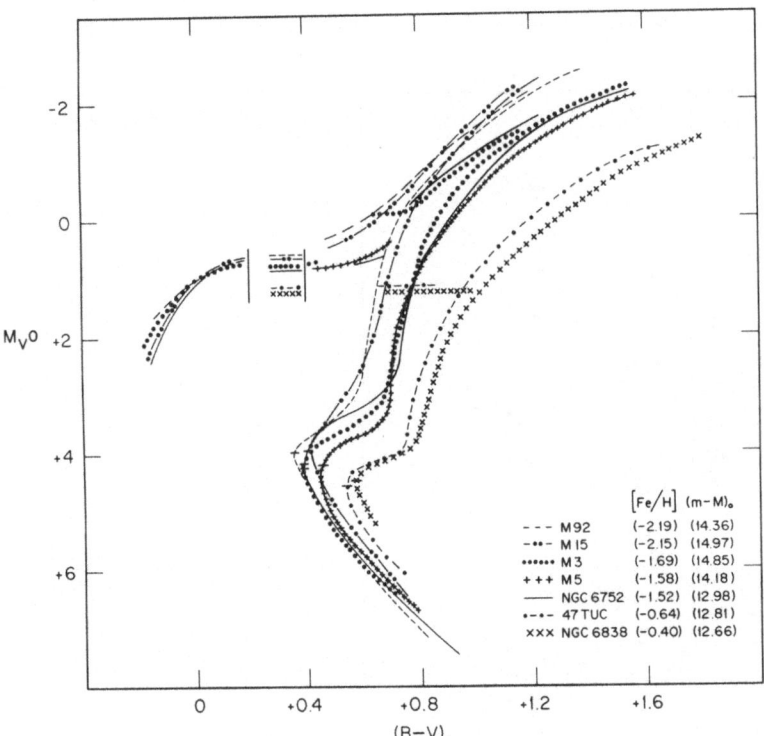

	[Fe/H]	(m−M)$_0$
--- M92	(−2.19)	(14.36)
−••− M15	(−2.15)	(14.97)
••••• M3	(−1.69)	(14.85)
+++ M5	(−1.58)	(14.18)
—— NGC 6752	(−1.52)	(12.98)
•−•− 47 TUC	(−0.64)	(12.81)
xxx NGC 6838	(−0.40)	(12.66)

Figure 2. Composite CM diagram (Sandage 1982) of seven globular clusters of different [Fe/H] values as listed in the code. Note the progressively redder main-sequence termination point, between 0.36 ≤ B-V ≤ 0.60, as the metallicity increases.

    The two important aspects of the distribution in Figure 1 are
(1) the blue color limit at B-V = 0.36 for the F and G subdwarfs, and
(2) the lower-envelope limit line to these stars that has the shape of
the blanketing vector, starting at B-V = 0.6 and reaching the tip of
the distribution at B-V = 0.36.

    This blue color limit extends an earlier result (Sandage 1964),
and is the most positive identification that these field high velocity
stars are like the globular cluster main-sequence stars at the turn off
point, and hence have the same age as globular clusters. This conclusion follows from Figure 2 by examining the color distribution of the
main-sequence termination of the seven plotted clusters whose metal
abundances range from [Fe/H] = -2.2 to -0.4. These termination colors
range from 0.36 to 0.60, consistent with the item (2) of Figure 1 just
discussed.

    An example of the second recent proof that the halo field is like
the globular clusters is shown in Figure 3, which is the color-magnitude
diagram for SA45 from unpublished photometry. The characteristic

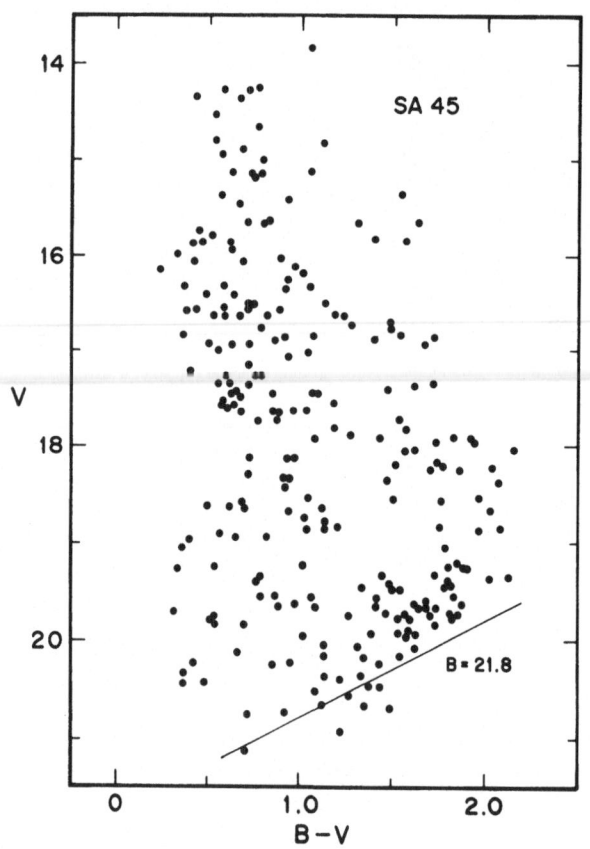

Figure 3. CM diagram
of SA45 (not part of
the halo mapping) from
unpublished photometry
by Sandage and Katem.

feature of this and of every other CM diagram for all high latitude
fields measured to date is the sharp edge to the color distribution at
B-V ~ 0.36, again similar to Figures 1 and 2. The same result is present
in each of the Basel fields (Becker and Fenkart 1976; Becker et al. 1976)
when their RGU catalogs for SA51, 54, 57, 71, 82, 94, and 107 are transformed to V, B-V magnitudes and plotted (David Sandage, private communication).

To extend the argument still further requires measurement of the
distribution of metallicities at each apparent magnitude (via the
distribution of UV excess values) so as to compare with the known distribution of globular cluster metallicities at each height (cf.,
Kukarkin 1974; Zinn 1980; Smith 1982). Such observations are in progress as part of the Mount Wilson mapping program.

III. THE HALO MAPPING PLAN

Figure 4 shows the Selected Areas in the present program. They
are close to the ℓ = 0°, ℓ = 180° meridional plane and are at galactic
latitudes 90°, ±70°, ±50°, ±35°. Many of these areas are in the original Basel survey.

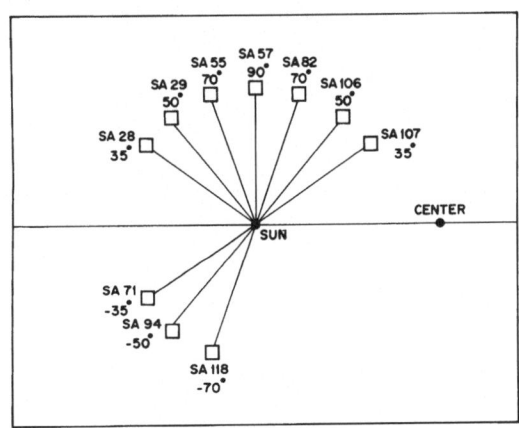

Figure 4. The halo mapping
Selected Area fields in the
ℓ = 0°, ℓ = 180° meridional
galactic plane.

The program consists of (1) measuring photoelectric standards to
B ~ 18 in each of the fields in BV and R wavelengths, (2) measuring BV
and R magnitudes for all stars to B ~ 22 from 5m Palomar photographic
plates, taken with a 5 mag Pickering/Racine wedge over a 15 x 15 arc-
minute field, (3) forming photographic catalogs and CM diagrams from
these complete data from which the density gradients and the halo
falttening for the F and G subdwarfs should be found by restricting the
data to a color range of 0.36 ≤ B-V ≤ 0.7 and using Figure 2 as an
absolute magnitude calibration, (4) measuring photoelectric UV excess

values to V = 19 for all stars in the stated color range, from which the distribution of metallicity at each height to $Z \approx 6$ kpc can be found. The way these data can then be used to find the nature of the chemical gradient between disk and halo follows from a model previously discussed (Sandage (1981).

   Progress to date is as follows.   (1)   The BVR photoelectric calibration in each of the 10 fields is complete from measurements made
since 1972 with the Mount Wilson 2.5m Hooker reflector.   (2)   BVR photographic iris photometry is nearly complete to B = 22 over the 15-arc-
minute field of the Palomar 5m reflector, and the candidate catalogs
exist.   (3)   Stars in these catalogs and in the wider Basel fields in
the F and G color range have begun to be measured in UBV for the UV
excess distributions using a new automated photometer on Mount Wilson.

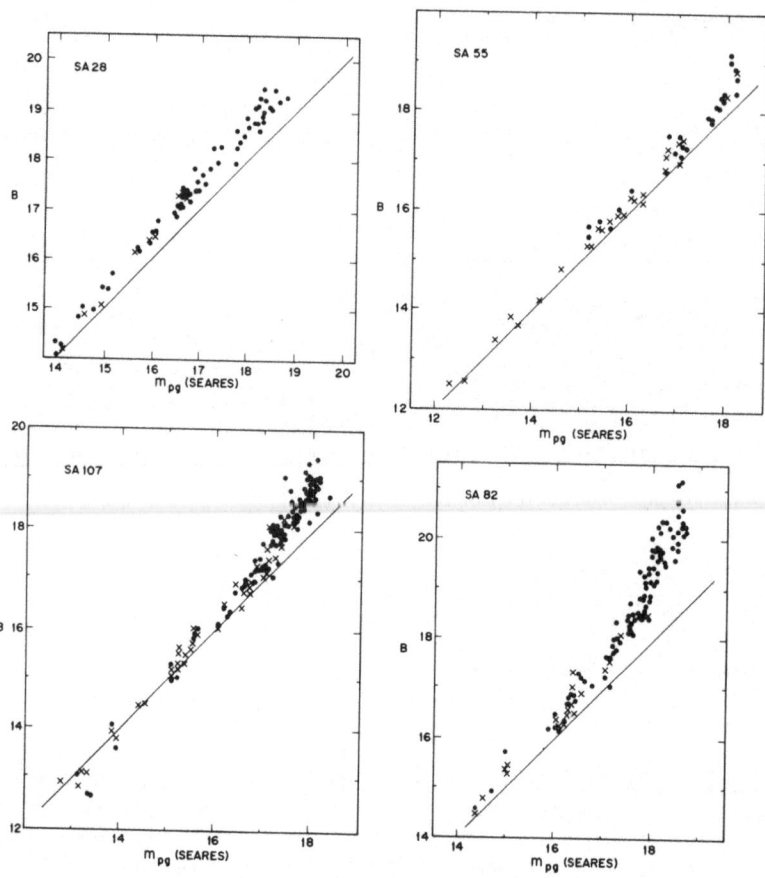

Figure 5.   Comparison of the new B magnitudes with the m_{pg} values of Seares from the <u>Mount Wilson Catalog of 139 Selected Areas</u>.

The results of corrections to the Seares magnitude scales for
SA28, 55, 82, and 107 are shown in Figure 5. Crosses are the direct
photoelectric measurements, dots are the interpolated photographic
values. The well-known correction to the Seares scale fainter than
B ≃ 15 is evident in these areas, as discovered in SA57, 61, and 68 in
1950 by Stebbins, Whitford, and Johnson (1950). The correction varies
from area to area. The largest we have found so far is in SA82 where it
reaches 2.5 mag at m_{pg} (Seares) ≃ 18.5. As the faint magnitudes in the
Mount Wilson Catalog have provided the standards for a large number of
statistical investigations of galactic structure in the past, as well as
the early standards for photometry of individual stars in galaxies by
Hubble and by Baade, it is clear that, due to these corrections, some of
the early results will need modification. A well-known example is that
Hubble's(1936) estimates of the apparent magnitudes of the knots and of
the resolved stars in nearby galaxies are too bright, generally by more
than 2.5 magnitudes.

IV. THE DENSITY GRADIENT IN THE HALO

If the halo is composed only of globular cluster-like stars (i.e.,
the oldest first generation stars formed that still exist) then the
density gradient of halo field stars toward the galactic poles should be
identical with that of the distant RR Lyrae variables. Many studies of
the RR Lyrae gradient have been made; the most complete are those by
Perek (1951), Plaut and Soudan (1963), Plaut (1966), and by Kinman and
colleagues summarized by Kinman (1972). All these studies show the
RR Lyrae gradient is very shallow, of order d log Z/dZ ≃ -0.1 dex kpc^{-1}.
What, in fact, is expected if the halo follows a Hubble-type intensity
law for spheriods of form $\rho(R) \sim R^{-n}$, where 3 < n < 4?

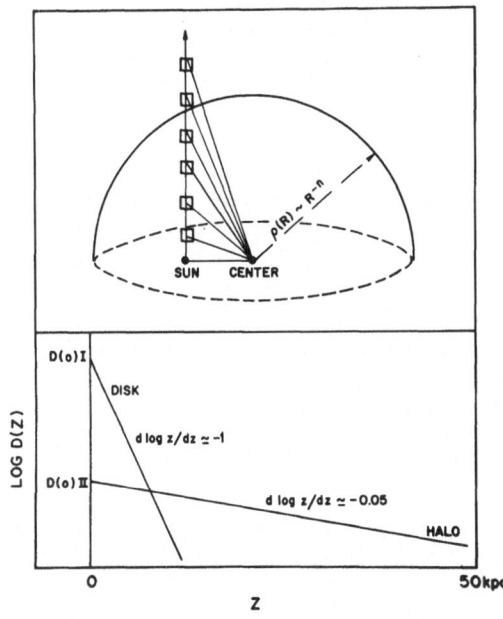

Figure 6. Illustration
of the calculation of
an approximation to a
logarithmic density
gradient using a Hubble
law for the halo den-
sity, for any given value
of the Hubble exponent
n.

Figure 6 shows the simple conceptual model of how the expected gradient in the pole can be approximated. Numerical calculation of the expected density at any given height above the sun, given a distance to the center of R_0 = 10 kpc, shows that a logarithmic gradient is a very good approximation, having values of d log Z/dZ = -0.05, -0.065, and -0.10 dex kpc^{-1} for $\underline{n}$ = 3, 4, and 6, giving scale heights of 9.4, 6.7, and 4.3 kpc respectively. To within the accuracy of the present RR Lyrae star data, a Hubble Law with $\underline{n}$ between 3 and 5 is possible, but the present data are very poor. (Also shown in Figure 5b is the gradient of ∿-1 known to apply to the intermediate population-I old disk.)

It is the purpose of the present halo mapping project to (1) determine both the halo gradient for the F and G subdwarfs to compare with the RR Lyrae data, and (2) to find the density ratio $D(0)_I/D(0)_{II}$, left arbitrary in Figure 6 . One of the mysteries of the current data in the literature is that the Basel "halo" gradient is nowhere near as shallow as -0.05 or even -0.10. Fenkart (1967, 1968) gives far-field values of -0.22 dex kpc^{-1} for SA57 and -0.35 dex kpc^{-1} for SA54, which are far too steep if the halo is really globular cluster-like (§II). Possibilities are that the Basel separation of disk and halo should be modified, or that small color errors have large effects on the Basel densities. We hope that restriction of the density calculation to the F and G stars in CM diagrams similar to Figure 3 for the 10 program areas, and hence restriction to a narrow range of absolute magnitude, will solve this problem, but if we cannot find a shallow enough gradient, then we have not sampled a globular cluster-like halo.

V. THE CHEMICAL GRADIENT

a) Inferred from local stars in the plane: There is current debate whether a true chemical gradient exists in the halo or whether it is merely a transition between the disk and the halo component of different metallicity, which imitates a gradient in the height range of 0 < Z ≤ 3 kpc perpendicular to the plane. The problem is, of course, important in studies of the early galactic collapse and the subsequent chemical enrichment (Eggen et al. 1962).

Figure 7 summarizes the available data on nearby subdwarfs whose space motions are known (via photometric parallaxes, known proper motions, and radial velocities). Plotted is the W velocity (perpendicular to the plane) against the UV excess (reduced to B-V = 0.6, cf., Sandage 1969). An explanation of the data sample is given elsewhere (Sandage 1981). There is a clear correlation between the mean W_0 value and the mean UV excess, showing that very few stars of high metallicity (i.e., δ < 0.1) can be expected at heights greater than ∿2 kpc.

To increase the sample, we have begun a radial velocity program for all stars in Figure 1 bluer than B-V = 1.0 (there are ∿1000 such stars) of all UV excess values so as to guard against selection effects of the kind that may be present in Figure 7 at small δ values. To date,

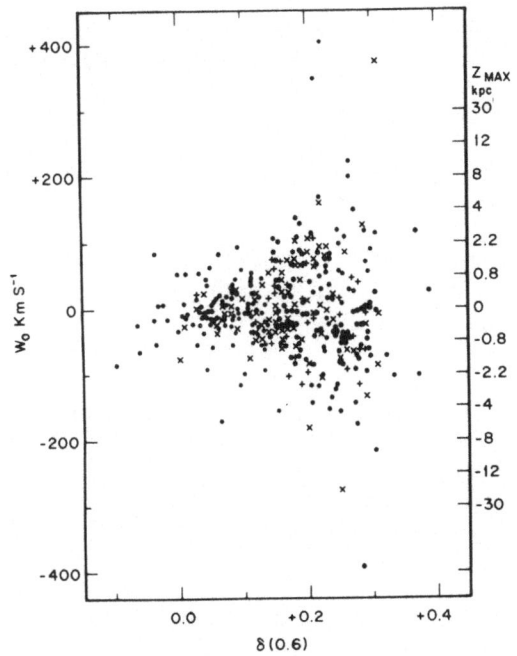

Figure 7. Correlation of W velocity in the plane (for nearby stars) with reduced ultraviolet excess. Height reached above the plane is on the right. Explanation of the data sample is given elsewhere (Sandage 1981).

about 300 new high velocity subdwarfs have been found from this radial velocity program, currently being done with the Mount Wilson Coudé spectrograph with a Reticon detector of Shectman's design.

b) To be obtained by distant stars in situ: The final aim of the program is to measure the metal abundance star by star, using individual $\delta(0.6)$ values for stars as faint as B = 19. Photometric parallaxes of these stars should then give the distribution of metallicity at each height. Clearly, this result obtained in situ must agree with that which can be derived by an anlaysis of Figure 7 which, with a given galactic acceleration K(Z), gives how high each of the stars then can climb out of the plane, and hence by summation, produces the in situ distribution. We expect to have UV excess data for ∿4000 F and G stars to Z ≃ 6 kpc within 3 years as part of this halo mapping program.

Part of this work has been supported by NSF grants AST-7800534 for the construction of the dual-channel automated Mount Wilson photometer, and AST-8015797 for the beginning of the actual halo mapping which has led to the data in Figures 5 and 7 and to the new subdwarf radial velocities for a complete sample of stars in Figure 1.

REFERENCES

Becker, W. and Fenkart, R.: 1976, Photometric Catalogue for Stars in SA51, 54, 57, 82, 107 (Basle).

Becker, W., Fenkart, R., Schaltenbrand, R., Wagner, R., and Yilmaz, F.: 1976, Photometric Catalogue for Stars in SA71, 94, 141, M3, M13, NGC 4147 (Basle).

Eggen, O. J., Lynden-Bell, D., and Sandage, A.: 1962, Astrophys. J. 136, 748.

Fenkart, R.: 1967, Zs. Astrophys. 66, 390.

Fenkart, R.: 1968, Zs. Astrophys 68, 87.

Hubble, E.: 1936, Astrophys. J. 84, 158.

Kinman, T. D.: 1972, Quart. Journ. R. A. S. 13, 258.

Kukarkin, B. B.: 1974, The Globular Star Clusters (Sternberg State Astron. Inst., Nauka, Moscow).

Perek, L.: 1951, Brno Cont. 1, No. 8.

Plaut, L.: 1966, Bull Astron. Inst. Neth. Suppl. 1, No. 3.

Plaut, L. and Souden, A.: 1963, Bull. Astron. Inst. Neth. 17, 70.

Roman, N.G.: 1954, Astron. J. 59, 307.

Sandage, A.: 1964, Astrophys. J. 139, 442.

Sandage, A.: 1969, Astrophys. J. 158, 1115.

Sandage, A.: 1970, Astrophys. J. 162, 841.

Sandage, A.: 1981, Astron. J. 86, 1643.

Sandage, A.: 1982, Astrophys. J. 252, 553.

Sandage, A. and Eggen, O. J.: 1959, Mon. Not. R. A. S. 119, 278.

Sandage, A. and Walker, M. F.: 1955, Astron. J. 60, 230.

Seares, F. H., Kapteyn, J. C., and van Rhijn, P. J.: 1930, Pub. Carnegie Institution of Washington, No. 402.

Searle, L. and Zinn, R.: 1978, Astrophys. J. 225, 357.

Smith, Horace: 1982 (Private Communication).

Stebbins, J., Whitford, A. E., and Johnson, H. L.: 1950, Astrophys. J. 112, 469.

Wildey, R. L., Burbidge, E. M., Sandage, A., and Burbidge, G. R.: 1962, Astrophys. J. 135, 94.

Zinn, R.: 1980, Astrophys. J. Suppl. 42, 19.

ORBITAL AND CHEMICAL PROPERTIES OF GLOBULAR CLUSTERS AND HALO STARS

K.C. Freeman
Mt Stromlo and Siding Spring Observatories
Research School of Physical Sciences
The Australian National University

When the Galaxy condensed, the stars and globular clusters of the galactic halo were among the first objects to form. To understand the processes that took place in the Galaxy at that early time, we need to learn as much as possible about the dynamics and chemical properties of these halo objects. I will discuss three topics: (i) the orbital and chemical properties of the galactic globular clusters, (ii) a program to study the halo giants in the outer parts of the galactic halo, and (iii) the orbital properties of the LMC globular clusters, which are very different from those for the clusters of the galactic halo, and which may give us some insight into the conditions necessary for globular clusters to form.

I. THE GALACTIC GLOBULAR CLUSTERS

Radial abundance gradients are common, both in the disk and spheroidal components of galaxies, including the globular cluster systems (at least in elliptical galaxies); see Pagel (1981) and Strom et al (1981). For the galactic globular cluster system, the metal-rich clusters are concentrated to the central regions of the galaxy; however, in the outer regions (8 < R < 40 kpc) there is a wide range of abundance among the clusters but no real evidence for a systematic radial abundance gradient (Zinn, 1980b). This is a surprising result. Seitzer and I have been looking at the properties of orbits for clusters now in this 8 to 40 kpc region, to see if there is any relation between the orbital and chemical properties that might help to understand this spread in abundance and the absence of radial abundance gradient among these clusters.

Two ways have been used to study the orbital properties of galactic globular clusters. The first is to use the cluster radial velocities: these are not a strong constraint in general on the orbital properties of an individual cluster but are very useful statistically. The velocities have been used to great advantage by Frenk and White (1980), to show that the cluster system is slowly rotating and kinematically

325

W. L. H. Shuter (ed.), Kinematics, Dynamics and Structure of the Milky Way, 325–331.

isotropic. The other procedure is to use the tidal radius for each
cluster. With a few assumptions, this gives the cluster's perigalactic
distance R_{min}; then, from the cluster's present distance R to the
galactic center, we can estimate statistically its orbital eccentricity.
Seitzer and I followed this procedure.

First we need to adopt a galactic potential field; we have taken
a spherical flat rotation curve potential, with rotational velocity
V = 220 km/s. We assume that the galactic tidal field at R_{min} defines
the cluster's tidal radius r_t, which is taken as the radius of the
inner Lagrangian point. These last two assumptions are fully supported
by Seitzer's (1982) detailed modelling of globular clusters in the
galactic tidal field. The relationship between R_{min} and r_t is then

$$R_{min} = [2V^2/GM]^{\frac{1}{2}} . r_t^{3/2} . g(e) \qquad (1)$$

Here M is the mass of the cluster (derived from its luminosity, with
M/L = 1.6) and g(e) is a function of the cluster's orbital eccentricity
$e = (R_{max} - R_{min})/(R_{max} + R_{min})$: g(e) increases monotonically from
$g(e) = 1$ at $e = 0$ to $g(e) = 1.6$ at $e = 0.8$. We can write equation (1)
as

$$R_{min} = R_{min,0} \, g(e) \qquad (2)$$

where $R_{min,0}$ is derived directly from the cluster luminosity and
tidal radius. For a cluster now at distance R from the galactic center,
the ratio $\alpha = R_{min,0}/R$ is a statistical estimator of the orbital
eccentricity. For a consistent choice of the parameters V, M/L (which
appear in equation (1)) and R_0 (we have adopted $R_0 = 8.5$ kpc), this
ratio α should not exceed unity except by observational error.

For the 48 clusters with well determined tidal radii, only three
have $\alpha > 1$ with the above values for V, M/L and R_0. The frequency
distribution of α gives an estimate of the frequency distribution of
the orbital eccentricity, f(e), for these clusters. [f(e)de is the
relative number of clusters with e between e and e + de]. Dynamical
simulations with different f(e) and including the obvious selection
effects were used to predict the corresponding f(α) distributions. It
turned out that f(e) $\propto$ e represents the observed α-distribution very
well. This f(e), inferred from the tidal radii of the clusters, is
consistent with Frenk and White's (1980) conclusion from the cluster
kinematics, that the cluster velocity distribution is close to isotropic.

Now we return to the problem of the orbital and chemical properties
for the clusters now in the 8 to 40 kpc region, which show no abundance gradient in the [Fe/H] - R plane. Figure 1 shows the abundance
[Fe/H] against the perigalactic distance R_{min} for all clusters in this
region with good r_t values and with absolute magnitudes $M_V < -6$ (ie it
excludes any of the low luminosity Palomar clusters). There is a
very obvious abundance gradient in this [Fe/H] - R_{min} plane. The
metal-richer clusters have small values of R_{min}, so these clusters are
in orbits of relatively high eccentricity, which plunge in close to

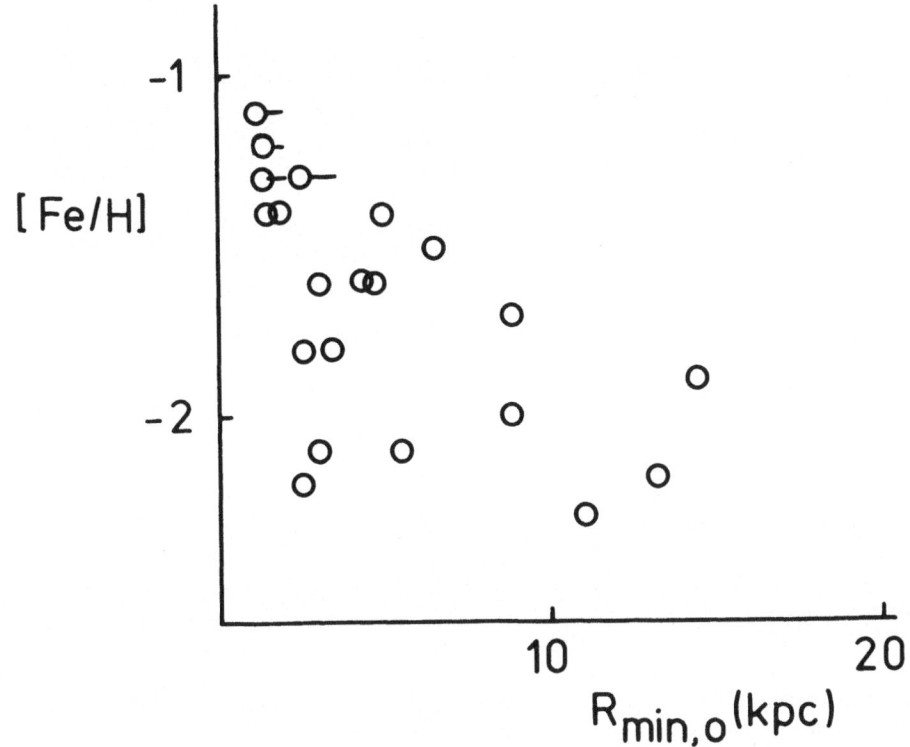

Figure 1. The abundance-perigalactic distance diagram for all clusters
with good values for the tidal radius, absolute magnitudes brighter
than M_V = -6 and galactocentric distances between 8 and 40 kpc. The
abundances are on the old scale (Zinn, 1980a). The short horizontal
lines on the four upper points indicate the correction g(e) to the
$R_{min,o}$ values for orbital eccentricity: see equation (2).

the galactic center. The metal weaker clusters have a wide range of
R_{min} values, and therefore a correspondingly wide range of orbital
eccentricity. In particular, there are metal weak clusters with low
eccentricity orbits in the outer parts of the galaxy.

    This is an unexpected result. It does not appear to fit in well
with the orbital properties of the nearby subdwarf stars, for which the
metal richer stars have the lower orbital eccentricities (Eggen et al,
1962). As a check, we should examine the cluster kinematics, to see
if they are consistent with our conclusion that the metal richer clusters
in the outer parts of the galaxy are in highly eccentric orbits. It
turns out from dynamical simulations that the shape of the frequency
distribution of cluster radial velocities, f(v), is sensitive to the
eccentricity distribution f(e) for the clusters in this outer region.
The observed distributions f(v) for the outer metal richer and metal
weaker clusters are significantly different, at the 90 percent level,

in the sense that the metal richer clusters are indeed in orbits of higher eccentricity. (Details of this study will be published later.)

To summarise this section: the metal richer clusters with enough energy to reach the outer parts of the galaxy have all fairly high orbital eccentricity and small perigalactic distances. This chemical association with the metal rich inner parts of the galaxy may suggest that these clusters formed near perigalacticon. The outer metal weaker clusters, on the other hand, have a wide range of e and R_{min}. Some of them have low orbital eccentricity. The presence of metal weak objects with low-e orbits in the outer parts of the galaxy leads us to the next section.

II. THE HALO FIELD STARS

We would like to know how the chemical properties and kinematics of halo objects change with distance from the galactic center. Two approaches have been used.

(i) Bright, easily identifiable halo objects, like globular clusters and RR Lyrae stars, can be studied in situ at large distances from the sun, and they show little evidence for an abundance gradient in the outer parts of the halo (see Zinn 1980b; Butler et al 1979; Butler et al 1982). However we should be cautious in adopting globular clusters and RR Lyrae stars as typical of the galactic halo. Although there are halo field stars with abundances down to at least [Fe/H] = -3, (Bond 1980; Norris, to be published), no clusters or RR Lyrae stars are known with [Fe/H] significantly less than about -2.2. For the RR Lyraes, this is probably a stellar evolution effect. The absence of really metal weak globular clusters is more difficult to understand. It may be a result of the conditions under which globular clusters were able to form early in the life of the galaxy.

(ii) From the work of Eggen et al (1962), we know that the metal weak halo subdwarfs in the solar neighborhood are in highly eccentric orbits. Although these stars were selected from their high proper motions, we know now that this high eccentricity is probably not an effect of kinematic selection. Norris (to be published) has recently studied a large sample of nearby spectroscopically selected metal weak stars. He has shown that the run of solar motion and velocity dispersion with abundance for these stars is similar to that for the kinematically selected subdwarfs. As the metal weak halo field stars are in highly eccentric orbits, they spend most of their orbital period far from the galactic center, and this is at least consistent with the presence of a radial abundance gradient in the halo. Sandage (1981) has used the distribution of W-velocities for subdwarfs in the solar neighborhhod to estimate the in situ distribution of abundance at different heights above the galactic plane; he has demonstrated the likely presence of a chemical gradient in the lower halo.

We would like to know the properties of the galactic halo at large distances from the galactic center (say 50 kpc). There are certainly halo stars in the solar neighborhood whose orbits reach out to such distances, but these stars are of course in highly eccentric orbits. Our work on the globular cluster orbits warns that there may be metal weak stars in orbits of low eccentricity far out in the halo, and these stars would never be seen in the solar neighborhood. We cannot then reliably derive the properties of the outer halo from those stars of the outer halo which happen to pass through the solar neighborhood. These stars in the solar neighborhood would give us a biased assessment of the orbital properties of outer halo stars, and maybe also of their chemical properties.

It seems that the only reliable way remaining to derive the properties of the outer halo of our galaxy is to observe field stars (not globular clusters or RR Lyrae stars) that are out there now, in the outer halo. Ratnatunga is engaged in such a program at Mt Stromlo, and I would like now to describe this program briefly.

The first part of the program is to find a sample of halo stars far out in the halo. The brightest population II stars are the giants, and the search is concentrated on giants with $0 > M_V > -2$ (0.9 < B-V < 1.4). A working limit of V = 17.5 corresponds to distances from the sun of 30 to 80 kpc. The observational problem is to identify the halo giants from the foreground disk dwarfs, disk giants and halo dwarfs in the same color range. Galactic models (eg Bahcall and Soneira 1981) show that the disk giants and halo dwarfs can be almost entirely excluded by restricting the search to stars with 13 < V < 17.5. The disk giants are mostly brighter and the halo dwarfs mostly fainter than this range of V. However the disk dwarfs are present in large numbers. For example, at the SGP we can expect about 200 disk dwarfs per square degree in this magnitude and color interval, and only about 10 halo giants per square degree more than 10 kpc from the sun. An efficient procedure is needed to separate the halo giants from the disk dwarfs.

Clark and McClure (1979) have shown quantitatively that an index which measures the strength of the Mg b and MgH feature near λ5100 A is an excellent discriminant between disk dwarfs and population II giants. It is ideal for a large survey, because it can be measured from objective prism spectra of relatively low resolution.

Before beginning the survey itself, a preliminary study was made, to check that the estimates of the numbers of halo giants were realistic. Bok and Basinski (1964) had made a fairly complete stellar photometry down to V = 16.2 in SA 141, near the SGP. There are 126 stars in the color interval 0.9 < B-V < 1.4 in this sample, and the Mg index was measured for all these stars, from slit spectra at the 74-inch telescope, using the Photon Counting Array. The dwarf-giant discrimination was unambiguous, and the estimates for the numbers of halo giants turned out to be fairly accurate.

The main survey is now under way. The stars are selected by color from direct Schmidt plates of 7 fields, and the giants are then identified from objective prism spectra. We are very much indebted to ESO, the UK Schmidt Telescope Unit, and the Tokyo Observatory at Kiso for taking these Schmidt plates for us. The whole measuring process, of microphotometry and reduction, is necessarily almost automatic.

The final part of this program will be to acquite higher resolution slit spectra of these distant halo giant, to measure their velocities and chemical abundances.

III. THE GLOBULAR CLUSTERS OF THE LMC

The LMC contains globular clusters of all ages. The oldest are similar in their age and stellar content to the halo clusters of the galaxy. The youngest are only about ten million years old. Although they are extreme population I objects, they are structurally similar to the globular clusters of the Milky Way, and have masses between 10^4 and 10^5 solar masses. We should recognise them as recently formed globular clusters, and seek to identify the reason why globular clusters can form now in the LMC (and SMC) but apparently not in the Milky Way at this time. See Freeman (1980) for a brief review of the properties of these young globular clusters in the LMC.

As a first step, we need to understand the dynamics of the LMC globular cluster systems of various ages. It is well known that the HI and HII components of the LMC lie in a flat rotating disk, so we would expect the very young globular clusters to lie in this disk also. The very old clusters could reasonably be expected to lie in a halo about the LMC, as in the Milky Way. At what age does the dynamical transition occur ? The way to tell if objects belong to a rotating disk is to measure their velocities: highly ordered rotational motion with a small velocity dispersion indicates a disk structure.

Illingworth Oemler and I have made a study of the kinematics of the LMC globular cluster system. Combining our data with data from Hartwick and Cowley and from Searle and Smith (which they have very generously allowed us to use before publication) gives velocities for about 60 globular clusters, covering the whole age range from about 10^7 to 10^{10} years. The results of the analysis were, in part, surprising.

The younger clusters (ages less than about 10^9 years) lie in a rotating disk similar to the disk defined by the HI and HII components. The apparent rotation amplitude for these young clusters is $V_{rot} = 37 \pm 5$ km/s, their galactocentric systemic velocity is $V_s = 40 \pm 3$ km/s, their kinematic major axis is in position angle $1^o \pm 5$, and the line of sight velocity dispersion σ about the rotation curve is about 13 km/s. The older clusters (ages greater than 1-2 billion years and including the halo-type clusters) are also flattened to a disklike system, with $\sigma = 14$ km/s and $V_{rot} = 41 \pm 4$ km/s. However this disk has $V_s = 26 \pm 2$

km/s and the kinematic major axis is in position angle $41° \pm 5$, very significantly different from the extreme population I disk. We do not fully understand this situation yet. However there seems no doubt that even the oldest clusters in the LMC lie in a flattened disklike system with a typical z-scaleheight of only about 600 pc. There is no evidence for a kinematic halo population among the globular clusters of the LMC, which is quite unlike the situation for the clusters of our galaxy. Globular cluster formation in the LMC appears then to have been associated at all times with a disk population. This is certainly worth knowing, but the important question remains: why have globular clusters continued to form throughout the life of the LMC, up to the present time ?

REFERENCES

Bahcall, J.N. and Soneira, R.M. 1981. Ap.J.Suppl. 47, pp 357-403.
Bok, B.J. and Basinski, J. 1964. Mt Stromlo Obs. Mem. 4, No. 16.
Bond, H.E. 1980. Ap.J.Suppl. 44, pp 517-533.
Butler, D., Kinman, T. and Kraft, R. 1979. A.J. 84, pp 993-1004.
Butler, D., Kemper, E., Kraft, R. and Suntzeff, N. 1982. A.J. 87, pp 353-359.
Clark, J.P.A. and McClure, R.D. 1979. P.A.S.P. 91, pp 507-518.
Eggen, O.J., Lynden-Bell, D. and Sandage, A. 1962. Ap.J. 136, pp 748-766.
Freeman, K.C. 1980. Star Clusters (IAU Symposium 85), ed J.E. Hesser, (Reidel, Dordrecht), pp 317-320.
Frenk, C.S. and White, S.D.M. 1980. M.N.R.A.S. 193, pp 295-311.
Pagel, B.E.J. 1981. The Structure and Evolution of Normal Galaxies, ed S.M. Fall, D. Lynden-Bell (Cambridge University Press, Cambridge), pp 211-235.
Sandage, A. 1981. A.J. 86, pp 1643-1657.
Seitzer, P. 1982. PhD thesis, University of Virginia, Charlottesville.
Strom, S.E., Forte, J.C., Harris, W.E., Strom, K.M., Wells, D.C. and Smith, M.G. 1981. Ap.J. 245, pp 416-453.
Zinn, R. 1980a. Ap.J.Suppl. 42, pp 19-40.
_____ 1980b. Ap.J. 241, pp 602-617.

COMMENTS ON THE GALACTIC HALO

K.A. Innanen
Physics Dept., York University,
Toronto, Canada

Summarised herein are the principal results of two papers to be
published in full elsewhere.   These are Paper I (Innanen, Harris and
Webbink, 1982) and Paper II (Valtonen, Innanen and Tahtinen, 1982).

Paper I (IHW): Globular Cluster Orbits and the Galactic
Mass Distribution

The shapes of typical globular cluster orbits and the distribution of
galactic mass with distance were investigated using the globular
cluster tidal radii as probes of the galactic tidal field at their
perigalactic points. From detailed consideration of the way in which
cluster energies and angular momenta interact with the galactic potential to determine the distribution of tidally-limited cluster densities,
and the observed properties of that distribution, we draw the following
conclusions: (1) Random observational errors (particularly of tidal
radii) are so large that individual estimates of cluster perigalactica
are largely meaningless. (2) If a cluster mass-to-light ratio $(m/\ell)_V$
= 1.7 is assumed, the orbital angular momenta implied by the tidal
radii are too large to be consistent with radial velocity analysis;
an m/ℓ ratio a factor of 2-3 larger is indicated. (3) The intrinsic
dispersion in cluster mean densities at a given galactocentric distance
is so small as to preclude any clusters having highly elongated orbits
($i.e.$, low dimensionless angular momenta). The globular clusters may
therefore have a toroidal, rather than spheroidal, distribution in
velocity-space, with the axis pointed toward the galactic center.
(4) The absence of radial variation in the dispersion of cluster densities indicates that radial trends in the kinematics of the cluster
system are negligible. (5) The observed variation of mean globular
cluster luminosity density with galactocentric distance, $\xi \sim R^{-1.73 \pm 0.18}$,
therefore reflects the true mass distribution in the Galaxy, $i.e.$, $M(R)$
$\sim R^{1.27 \pm 0.18}$. (6) This relationship holds to distances at least as
great as $R \sim 44$ kpc (the expected perigalactic distances of the farthest
globular clusters). at which point the galactic mass reaches (8.9 ± 2.6)
x 10^{11} $M_\odot$ and we estimate $M/L_B = (59 \pm 17)$ $M_\odot/L_B(\theta)$.   Finally, we
suggest that at distances typical of the dwarf spheroidal satellites of
the Galaxy, tidal stripping time scales become so great that these

W. L. H. Shuter (ed.), Kinematics, Dynamics and Structure of the Milky Way, 333–334.

systems have yet to reach tidal equilibrium.

Expressions were developed for (a) the theoretical tidal limit of a cluster, and its relationship to the observed tidal radius, in the elliptic restricted three-body problem for a power-law mass distribution; (b) orbit-averaged properties of bodies orbiting in such a gravitational potential, and (c) the statistical distributions of perigalactica and mean cluster densities for isothermal systems of clusters embedded in power-law mass distributions.

Paper II (VII): The Magellanic Clouds and the Galactic Halo

Extensive many-body numerical experiments are described which show that the Magellanic Clouds may cause considerable radial mobility of gravitating particles (e.g. stars, globular clusters) in the galactic halo. The results provide evidence for a new kind of (3-body) dynamical friction which in some cases is several times stronger than the classical (2-body) dynamical friction and which sets limits on the extent of possible dark halos.   Consequently, if some of the halo globular clusters have been ejected to their present orbits from smaller galacto-centric distances, the use of the radial velocities of these objects (as equilibrium test particles) to test for halo dark matter may not be justified.   The effect of radial mobility on the observed number density distribution of globular cluster-like objects is discussed.   Generally, an initial isothermal (R^{-2}) number density distribution ends up more nearly as an R^{-4} distribution together with a shell of excess density around 100 kpc radial distance from the Galaxy.   The fates of a test globular cluster initially in motion around the LMC and of the LMC-SMC binary are also briefly mentioned, and the tidal radius of the LMC cannot exceed 4 kpc.   Finally, we comment briefly on the relationship of this work to the dynamics of the Magellanic stream (Lin and Lynden-Bell, 1982).

References:

Innanen, K.A.,   Harris,W.E., and Webbink, R.F., 1982,
          submitted to the Astronomical Journal.

Valtonen,M.J.,   Innanen, K.A.and Tahtinen, L., 1982,
          submitted to the Astrophysical Journal.

Lin, D.N.C., and Lynden-Bell, D., 1982,
          M.N. Roy Astron Soc. 198, 707.

GLOBULAR CLUSTERS AND THE DISTANCE TO THE GALACTIC CENTRE

Carlos S. Frenk and Simon D.M. White
Astronomy Department, University of California
Berkeley, CA 94720, U.S.A.

An unbiased estimate of the centroid of the globular cluster population assuming the standard distance scale for metal-poor clusters gives a value of 6.8 ± 0.8 kpc for our distance from the Galactic Centre. A consistent centroid for the metal-rich clusters is obtained only if the absolute magnitude of their horizonal branch is taken to be 0.5 magnitudes fainter than that of metal-poor globular clusters.

1. INTRODUCTION

It is now more than six decades since Shapley (1919) added an important link to the Copernican chain of thought by identifying the centroid of the globular cluster population, rather than the position of the Sun, with the centre of the Milky Way. His original idea is still the basis for one of the few fundamental methods to determine R_0, the distance from the Sun to the Galactic Centre. While the technique is in principle straightforward, its application is hampered by two important complications; it is these complications which make it worthwhile to speak about this subject so many years after Shapley. The first is the occurrence of systematic and stochastic errors in the determination of distance moduli for individual clusters. The distance scale for these objects is based primarily on the value of $M_V(HB)$, the absolute magnitude of the horizontal branch near the RR Lyrae region. Some uncertainty in the calibration of this quantity remains, particularly concerning the issue of a metallicity dependence of $M_V(HB)$, first raised by Christy's (1966) theoretical models of pulsating stars. The prediction of these models that $M_V(HB)$ should be about 0.5 mag fainter for clusters with [Fe/H] ~ −1 than for clusters with [Fe/H] ~ −2 has proved difficult to test directly partly as a result of the uncertainties involved in fitting cluster main sequences as a function of metallicity. Recent work by Sandage and co-workers on the Oosterhoff period groups (Sandage 1982 and references therein) circumvent this problem and appear to support Christy's conclusions. In addition to possible systematic errors associated with the adoption of a distance scale, stochastic errors in the distance moduli can bias the determination of R_0 and their

335

W. L. H. Shuter (ed.), Kinematics, Dynamics and Structure of the Milky Way, 335–341.

effect must be modelled explicitly. These type of errors are likely to
be particularly large for the more metal-rich clusters which tend to
have few or no RR Lyrae stars, stubby and ill-defined horizontal
branches and large reddenings. The second complication in the method
arises from the incompleteness of the known sample of globular clusters
resulting from obscuration at low Galactic latitude. Since globular
clusters on the far side of the Galactic Centre are more likely to have
remained undiscovered than globular clusters on the near side, the
apparent centroid of the known sample is likely to be biased towards the
Sun. Again, it it important to model this effect explicitly. The
remainder of this paper will be mostly occupied with the treatment of
these two complications. A detailed discussion of their influence on
the apparent structure of the globular cluster system is given in our
recent paper (Frenk and White 1982).

2. METHOD

The first step in determining the distance from the centroid of the
cluster population is to adopt a value of $M_V(HB)$. There are three basic
ways to calibrate this quantity observationally: main-sequence fitting
of globular clusters, statistical parallax of field RR Lyrae stars, and
Cepheids in the Magellanic Clouds. All three methods give values of
$M_V(HB)$ close to 0.6 and inconclusive information on its metallicity
dependence (c.f. Harris 1976 and references therein). We shall
provisionally adopt a distance scale for the cluster system based on
this value independent of metallicity and inquire later whether this
assumption is consistent with the cluster data. We take the distance
information from the compilation of Harris and Racine (1979).

The second step is the construction of a suitable statistic to
determine the centroid of the globular cluster population. Let us
denote by X the direction towards the Galactic Centre. Traditionally,
the mean projection in this direction of the positions of the clusters
has been employed (Arp 1965, Harris 1976, 1980). This statistic,
however, is a rather poor choice since it is strongly affected by the
few objects at large galactocentric distance and is biased upwards in
the presence of distance errors. A better-behaved and more efficient
statistic is the median value, X_m, of the X coordinate of the clusters
in the sample (Frenk and White 1982).

The apparent value of X_m for a sample of globular clusters does not
measure the true centroid of the cluster distribution because the
apparent data values of X are biased as a result of incompleteness in
the sample and random errors in the distance moduli. The third and
final step of our method to determine R_0 consists of quantifying and
correcting these biases. We do this in the following manner.

2.1 Incompleteness

To eliminate incompleteness effects which are due primarily to obscuration we restrict our cluster sample both in galactic latitude and in galactocentric distance. Let b_{min} and r_{max} denote the limiting sample values of these coordinates. We can be fairly certain that in the region $r < r_{max} = 40$ kpc, $b > b_{min} = 10°$ the known sample is practically complete since no new clusters have been discovered in this region despite extensive searches while objects as sparse as Pal 3 and Pal 4 have been detected at distances considerably in excess of r_{max}. These values are in fact quite conservative (Woltjer 1975, for example, suggested $b_{min} = 5.7°$) and lead to a sample containing 65 globular clusters. To test for systematic effects associated with cluster abundance we consider two different subsamples separated at a value of [Fe/H] = -1.2. The metal-poor group contains 50 objects and the metal-rich one (adopting Woltjer's value of b_{min} to increase the size of this sample) 26 objects.

We can quantify the bias in the cluster distribution resulting from our selection criteria provided that we know its shape and radial density profile. These properties have been carefully modelled by Frenk and White (1982) and we summarize here the main results.

2.2 Models

The spatial distribution of the globular cluster population is remarkably well described by truncated power-law models of the form,

$$\rho \propto a^{-\nu} \qquad a < a_o$$

$$\rho = 0 \qquad a > a_o$$

(1)

where ρ is the number density of clusters and a the major axis of the equidensity spheroids. Such models are specified by a normalization factor and three dimensionless form parameters, ν, a_o/R_o and ε, the axial ratio of the equidensity surfaces. These parameters can be estimated using only the positions of the clusters on the sky. This is a particularly interesting feature since such estimates are independent of cluster distance and thus of either random or systematic distance errors. We obtain the following results:

Table 1.   Model parameters

Sample	ε	ν	a_o/R_o
Total	$0.84 \begin{smallmatrix} +0.16 \\ -0.14 \end{smallmatrix}$	3.0 ± 0.2	4.3
Metal-poor	$0.83 \begin{smallmatrix} +0.20 \\ -0.13 \end{smallmatrix}$	2.8 ± 0.3	4.3
Metal-rich	$0.75 \begin{smallmatrix} +0.25 \\ -0.18 \end{smallmatrix}$	3.6 ± 0.4	1.4

2.3   Stochastic distance errors

Observational errors in the distance moduli of an axially symmetric population concentrated towards the Galactic Centre will cause it to appear both less centrally concentrated and elongated towards the Sun. Such deviations from axial symmetry can be used to infer the size of the errors in the distance data. The distribution of our metal-poor and total samples exhibit only slight distortions consistent with a mean assumed error of 0.3 mag in their distance moduli; the metal-rich clusters, on the other hand, appear to form a bar-like structure pointing at and extending approximately as far as the Sun. Errors of about 1 mag in their distance moduli are most probably responsible for this asymmetry.

Distance errors even as large as those inferred from the metal-rich clusters will have only a small effect on the estimate of the median X coordinate of the cluster population. (This is one of the reasons why we have chosen this statistic.) Its sampling variance, however, is somewhat sensitive to the size of the errors since the dispersion of X_m around its expected value is clearly related to the central concentration of the apparent X-distribution.

3.   THE DISTANCE TO THE GALACTIC CENTRE

To translate the observed values of X_m for our samples into an estimate of R_0 we need to know the expected X-distribution of a cluster population subject to latitude and galactocentric cut-offs and distorted by stochastic distance errors. This distribution can be readily obtained from the number density laws of Table 1 by means of Monte Carlo simulations. For different assumed values of R_0, a large number of points is generated according to these laws; the resulting distribution is then subjected to the selection criteria of Section 2.1 and the "observed" distance moduli perturbed with Gaussian random errors of the size discussed in Section 2.3. Comparison of the resulting sampling distribution of X_m as a function of R_0 with the observed values of X_m gives the estimate of R_0 and of its variance from each of our samples.

For the 50 metal-poor clusters X_m = 5.63 kpc implying R_0 = 6.2 ± 0.9 kpc. The distribution of the X-coordinate for this sample is shown as a histogram in Fig. 1(a). The solid curves in this figure are the expected distributions predicted by models with R_0 = 6.2 kpc and R_0 = 9.0 kpc respectively, normalized to the sample size. It is clear that the first model is an excellent fit to the observed distribution while values of R_0 similar to those recommended by the IAU are inconsistent with these data. The 26 metal-rich clusters, on the other hand, have X_m = 8.56 kpc implying R_0 = 9.1 ± 1.4 kpc, almost 2σ larger than the previous estimate. The relative shift in the X-distributions of the two metallicity groups is shown quite dramatically in Figure 1(b); since the value of R_0 must be unique, this strongly suggests that the cluster distance scale is indeed a function

of metallicity. It it reassuring that the difference in the appropriate
distance scales implied by this figure is in the same sense and of about
the size predicted by Christy's (1966) theoretical pulsation models for
RR Lyrae variables. These models, however, as well as Sandage's (1982)
supporting data apply to clusters with [Fe/H] $\lesssim$ -1 and the possibility
remains that the shift seen in Figure 1(b) reflects a systematic error
associated with those processes responsible for the large random errors
in the metal-rich distance data discussed in Section 2.3 rather than any
intrinsic property of RR Lyrae stars. The data, however, are very
suggestive of the predictions of pulsation models.

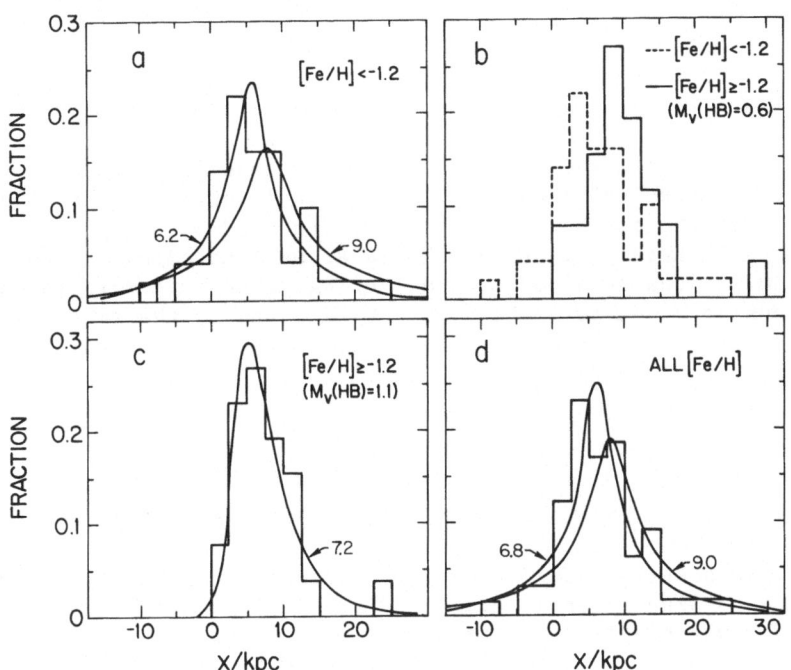

Figure 1. Distribution of the X coordinate for our
globular cluster samples. Histograms represent the
observed data and solid curves represent model
predictions for the values of R_0 shown. (a) Metal-
poor sample. (b) Metal-poor and metal-rich samples
assuming a metallicity independent distance scale.
(c) Metal-rich sample assuming an absolute magnitude
of the horizontal branch of 1.1 mag. (d) Total
sample based on the distance scale of (a) and (c).

To reconcile the apparent centroids of the metal-poor and
metal-rich groups a difference of about 0.5 mag in the relative distance
scales is required. The calibration $M_V(HB)$ = 0.6 seems fairly secure
for the metal-poor objects so we take $M_V(HB)$ = 1.1 for our metal-rich

sample. (A simple step function of this type represents a first-order approximation to a putative smooth dependence of $M_V(HB)$ on metallicity.) This correction reduces the value of X_m for this sample to 6.8 kpc and the implied distance from the centre to $R_0 = 7.2 \pm 1.1$ kpc, thus achieving reasonable consistency with the determination from metal-poor clusters. The resulting X-distribution for the high metallicity clusters is shown in Figure 1(c) together with the predictions of the "best-fit" model. Again the agreement between the observed and expected distributions is very good. Note that the latter is less centrally peaked and considerably more asymmetrical than the corresponding curves in Figure 1(a). This difference is mainly due to the large distance errors assumed in the models for the metal-rich population.

The final panel in Figure 1 displays the X-distribution of the 65 clusters in our total sample. For these objects $X_m = 6.12$ kpc leading to our best estimate, $R_0 = 6.8 \pm 0.8$ kpc from globular cluster data. The curve corresponding to the model with an assumed centre at 9 kpc shown in the figure reinforces our earlier conclusion that the large values of R_0 frequently quoted are not consistent with the globular cluster distribution and the currently accepted magnitude calibration of metal-poor RR Lyrae variables. It is as well to emphasize that Monte Carlo simulations show our estimate of R_0 to be extremely insensitive to variations in model parameters $(\epsilon, \nu, a_0/R_0)$ and to the magnitude of the distance errors; its sampling variance, on the other hand, is found to be slightly model dependent.

4. CONCLUSIONS

Our analysis of the globular cluster system yields two main results:

i) The distribution of metal-rich clusters appears to be centred at a significantly greater distance from the Sun than that of the metal-poor clusters unless a metallicity dependence of the cluster distance scale is assumed. The shift of +0.5 mag in the level of the horizontal branch of metal-rich clusters required to remove the discrepancy is in the sense and of about the size predicted by Christy's (1966) theoretical models of pulsating RR Lyrae stars and Sandage's (1982) supporting data, but it may also reflect other systematic calibration problems.

ii) If the currently accepted value of the absolute magnitude of the horizontal branch in metal-poor clusters is adopted ($M_V(HB) = 0.6$), the distance to the Galactic Centre inferred from the globular cluster population is $R_0 = 6.8 \pm 0.8$ kpc. This value scales as $dex\{0.2[0.6 -M_V(HB)]\}$ where $M_V(HB)$ refers to metal-poor systems. Apart from the use of globular clusters, the only direct method to determine R_0 consists of finding the location of the maximum in the density distribution of RR Lyrae stars in the Centre direction. The value derived by Oort and Plaut (1975) using this method is consistent with the low values we obtain provided that RR Lyrae stars near the Galactic

Centre are metal-rich and have an absolute magnitude similar to that
inferred for such stars from the globular cluster data.

A final remark concerns the compatibility of our proposed value of
R_0 with data on the rotation properties of the Galaxy. The value of
Oort's constant, A, can be obtained from the kinematics of nearby stars
and recent determinations give numbers close to 16 km s^{-1} kpc^{-1}. The
product AR_0 is also well determined by 21 cm data and a recent
reanalysis by Gunn, Knapp and Tremaine (1979) gives AR_0 = 110±5 km s^{-1}.
Data bearing on the constant B, on the other hand, are difficult to
interpret and its value is still uncertain; additional information on B
comes from the relation A − B = V_c/R_0, where V_c is the circular velocity
at the position of the Sun. From a dynamical analysis of the globular
cluster system we have found the latter quantity to be close to
220 km s^{-1} (Frenk and White 1980; Lynden-Bell and Frenk 1981; White and
Frenk, this volume), in agreement with the result of Knapp, Tremaine and
Gunn (1978) from a study of the HI distribution outside the Solar
circle. All these numbers fit together rather nicely. Our estimate of
R_0 together with the value of AR_0 of Gunn et al. gives
A = 15.9 ± 2 km s^{-1} kpc^{-1}. Adopting R_0 = $\overline{7 \text{ kpc}}$ and V_c = 220 km s^{-1} we
obtain a globally (but not necessarily locally) consistent set of
numbers by taking −B = A = 15.9 km s^{-1} kpc^{-1}.

REFERENCES

Arp, H.C.: 1965, in "Galactic Structure, Stars and Stellar Systems,"
        Vol. V, eds. Blaauw & Schmidt, University of Chicago Press.
Christy, R.F.: 1966, Astrophys. J. 144, p. 108.
Frenk, C.S. and White, S.D.M.: 1980, M.N.R.A.S. 193, p. 295.
Frenk, C.S. and White, S.D.M.: 1982, M.N.R.A.S. 198, p. 173.
Gunn, J.E., Knapp, G.R. and Tremaine, S.D.: 1979, Astron. J. 84, p. 1181.
Harris, W.E.: 1976, Astron. J. 81, p. 1095.
Harris, W.E.: 1980, in "Star Clusters," I.A.U. Symposium 85, ed. Hesser,
        D. Reidel Publishing Co., Dordrecht, Holland, p. 81.
Harris, W.E. and Racine, R.: 1979, Ann. Rev. Astron. Astrophys. 17,
        p. 241.
Knapp, G.R., Tremaine, S.D. and Gunn, J.E.: 1978, Astron. J. 83, p. 1585.
Lynden-Bell, D. and Frenk, C.S.: 1981, Observatory 101, p. 200.
Oort, J.H. and Plaut, L.: 1975, Astron. Astrophys. 41, p. 71.
Sandage, A.: 1982, Astrophys. J. 252, p. 553.
Shapley, H.: 1919, Astrophys. J. 49, p. 311.
Woltjer, L.: 1975, Astron. Astrophys. 42, p. 109.

GLOBULAR CLUSTERS AND THE POTENTIAL WELL OF THE GALAXY

Simon D.M. White & Carlos S. Frenk
Astronomy Department, University of California
Berkeley, CA 94720, U.S.A.

1. INTRODUCTION

The globular cluster system affords a unique opportunity to investigate the gravitational field of our Galaxy. Except very close to the Galactic plane the known sample of clusters is almost certainly complete. Both distances and velocities for clusters can be obtained with relative ease to an accuracy which is quite sufficient for the study of Galactic structure, at least outside the central few kiloparsecs. In addition clusters are seen at large radii where they provide information about the potential field that no other test particle population can at present supply. The fundamental limitation of the cluster system in this context is, of course, the fact that less than 100 clusters are known in regions of the Galaxy where we can hope to use them to study Galactic structure. This "small" sample will never be significantly increased and so we are forced to deal with the statistics of small numbers. Further statistical uncertainty is introduced into dynamical analysis of cluster data by the fact that only one of the three components of the space velocity is, in general, observable. Such uncertainties make it impossible to use the clusters to answer detailed questions about the mass distribution of the Galaxy, but their kinematics can give useful information about the large-scale distribution and pose quite severe constraints on any proposed mass model.

Unless there have been very violent events in the recent past, the globular cluster system should at present be in equilibrium in the potential well of the Galaxy, and so its space and velocity distribution should satisfy the equation of hydrostatic equilibrium:

$$\frac{\partial}{\partial x_j} \left(\rho_c \langle v_i v_j \rangle \right) = \rho_c \frac{\partial \Phi_g}{\partial x_i} , \qquad (1)$$

where ρ_c is the space density of clusters, Φ_g is the overall gravitational potential of the Galaxy and angle brackets denote an average over the velocity distribution at a fixed position. This

343

W. L. H. Shuter (ed.), Kinematics, Dynamics and Structure of the Milky Way, 343–348.
Copyright © 1983 by D. Reidel Publishing Company.

expression simplifies considerably if we assume that the spatial
distribution of the test particles may be taken to be spherically
symmetric; this appears to be a good approximation for the cluster
system. Averaging the radial component of (1) over a spherical shell
then leads to the equation:

$$\langle v_r^2 \rangle \frac{d \ln \rho_c}{d \ln r} + \frac{\partial \overline{\langle v_r^2 \rangle}}{\partial \ln r} + 2(\overline{\langle v_r^2 \rangle} - \overline{\langle v_t^2 \rangle}) = \frac{\overline{\partial \Phi_g}}{\partial \ln r} \equiv -\overline{v^2}_{circ} , \qquad (2)$$

where $\langle v_r^2 \rangle$ and $2\langle v_t^2 \rangle$ are the rms velocities at each point in the
directions along and normal to the radius vector, and an overbar denotes
an average over a spherical shell. The dynamical effects of any
systematic rotation of the cluster system are included in the term
$2\langle v_t^2 \rangle$, and are negligible in comparison with the effect of random
motions (Frenk & White 1980). We neglect such systematic motions
throughout the remainder of this paper. Since we know the
three-dimensional positions of globular clusters we can estimate ρ_c
directly, and for each cluster we know the relative extent to which the
measured line-of-sight velocity reflects v_r and v_t. From our single
vantage point within the Galaxy we cannot, however, derive unbiassed
estimates of the averaged velocity dispersions in equation (2). To
proceed further we are forced to make the additional simplifying
assumption that the phase-space distribution of the cluster system is
spherically symmetric so that (2) applies separately at each point as
well as over a spherical shell as a whole. We have no direct
observational justification for this assumption, and indeed, since it
implies (through (1)) that the potential well of the Galaxy is
spherically symmetric, it is almost certainly false. Nevertheless it
seems reasonable to hope that the inaccuracies incurred as a result of
treating a moderately flattened potential as though it were spherical
will not be too great.

We have recently completed a study of the structure and the
dynamical properties of the Galactic globular cluster system based on
new methods both for elucidating the spatial structure and for treating
the dynamics as embodied in equation (2) (Frenk & White 1980, 1982).
This talk presents an update and a few new results from our dynamical
analysis. A rediscussion of our findings for the spatial distribution
of clusters is given elsewhere in this volume.

II.   THE DATA

Our analysis of the distribution of globular clusters in space
confirmed earlier work by Woltjer (1975) and Harris (1976) in concluding
that outside of the central few kiloparsecs the globular cluster system
is, to a good approximation, spherical. We also showed that if the
standard distance scale for clusters is used the Galactic Centre should
be taken to be at a distance of only 7 kpc, and that in the range
2 kpc < r < 30 kpc the distribution of clusters about this centre is

well described by a power law:

$$\rho_c \propto r^{-3 \pm 0.2}. \tag{3}$$

These findings differ slightly from the assumptions we adopted from earlier work when carrying out our initial dynamical analysis (Frenk & White 1980). Since that paper a few more velocities for clusters have become available; there are now 68 objects in the apparent Galactocentric distance range 2 kpc < r < 40 kpc which have measured velocities and "good" distances quoted in Harris & Racine (1979) or in later work. The rest of this talk is based upon this sample of clusters.

In Table 1 we show the rms dispersions obtained if the velocities of these 68 clusters are corrected for a Solar motion around the Galactic Centre of 220 km/sec and then averaged in four equal groups according to apparent distance from the centre (assumed to be at a distance of 7 kpc).

Table 1:	r (kpc)	(km/sec)
	2.0 – 4.2	106 ± 18
	4.2 – 6.9	99 ± 17
	6.9 – 12.6	125 ± 22
	12.6 – 33.0	138 ± 24

A striking feature of these numbers is their small size; comparison with equations (2) and (3) shows that they will be difficult to reconcile with a typical circular velocity in the Galaxy much greater than 220 km/sec. Another interesting feature of the table is the fact that the dispersion increases with radius whereas the opposite would be expected if the material responsible for the Galactic potential were distributed in the same way as the clusters themselves. Notice, however, that while the observed dispersion is approximately $(\langle v_r^2 + 2v_t^2 \rangle / 3)^{1/2}$ in the central part of the Galaxy it is closer to $\langle v_r^2 \rangle^{1/2}$ in the outer regions. Because $\langle v_r^2 \rangle$ and $\langle v_t^2 \rangle$ are not necessarily equal, and because they enter the equilibrium equation (2) in different ways, the interpretation of Table 1 and of the velocity data in general is model-dependent.

III. MODELS FOR THE CLUSTER SYSTEM

(i) Constant Eccentricity Models

A simple assumption for the kinematic structure of the cluster system is that the shape of the velocity ellipsoid is the same every-where and that the velocity dispersions vary as a power law of galacto-centric distance. This leads to three-parameter models specified by:

$$\langle v_r^2 \rangle = \sigma_0^2 \, (r/r_\oplus)^{-2\alpha}$$

$$\langle v_t^2 \rangle \propto \langle v_r^2 \rangle; \quad e \equiv \langle v_r^2 \rangle / (\langle v_r^2 \rangle + 2\langle v_t^2 \rangle); \tag{4}$$

$$\rho_c \propto r^{-3}.$$

Substituting these relations in (2) shows that they imply a Galactic rotation curve:

$$V_{circ} = \sigma_0 \, (2\alpha + e^{-1})^{1/2} \, (r/r_\odot)^{-\alpha} \tag{5}$$

The shape parameters α and e can be constrained by comparing the observed dependence of cluster velocity on position with model predictions. For given values of these parameters the normalisation σ_0 is set by the overall rms dispersion. Figure 1 shows the confidence limits that can be placed on α and e using the statistical techniques described in Frenk & White (1980). The parameter α is required to be negative, showing that the rise in dispersion seen in Table 1 is inconsistent with any constant eccentricity model with a falling rotation curve. The eccentricity parameter e is required to be less than 0.6; strongly radial orbits are not consistent with the fact that large velocities are measured for clusters in the inner region of the Galaxy for which the line of sight is almost normal to the vector joining them to the Galactic centre. Dashed contours in Figure 1 show

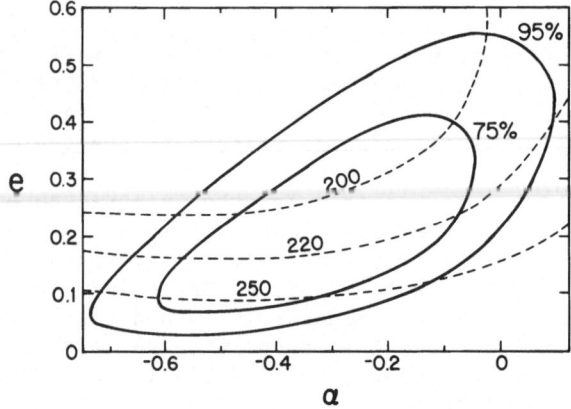

Fig. 1: Constraints placed on the shape parameters of constant eccentricity models for the cluster system by the observed radial velocities of 68 globular clusters. Closed contours join parameter values that can be rejected with 75% and 95% confidence. Dashed lines join models which predict certain values for the Galactic rotation velocity at the Sun when normalised using the observed dispersion of the cluster system.

parameter pairs for which the data imply (through (5)) particular values
for the circular velocity in the Galaxy at the Solar radius. The
statistical uncertainty in these normalisations, arising from the finite
number of velocities and from uncertainties in the cluster density
profile, is ± 9%. Values of the circular velocity in excess of 220
km/sec are only comfortably accomodated by models in which the cluster
velocity distribution is biassed in favour of circular orbits (e < 1/3).
Such models predict rotation curves which rise unrealistically quickly
beyond the Solar radius (cf equation 5).

(ii) Eddington-like Models

The above constant eccentricity models seem unrealistic in that
they allow cluster distributions which are dominated by near-circular
orbits and they do not allow the average shape of orbits to change as a
function of radius. Simple three-parameter models which avoid these
difficulties can be constructed by replacing the second of equations (4)
by:

$$\langle v_t^2 \rangle = \langle v_r^2 \rangle / (1 + r^2/r_a^2).$$ (6)

The dependence of the shape of the velocity ellipsoid on radius in these
models is similar to that in Eddington's (1914) modification of the
isothermal sphere. In the central regions the velocity distribution is
isotropic but beyond the anisotropy radius, r_a, it becomes more and more
radially biassed. This behaviour would seem quite natural for a
population which formed during violent dynamical collapse of the
protogalaxy. The rotation curve implied by such Eddington-like models
is:

$$V_{circ} = \sigma_o \, [2\alpha + (r^2 + 3r_a^2)/(r^2 + r_a^2)]^{1/2} \, (r/r_\oplus)^{-\alpha}.$$ (7)

In Figure 2 we display the constraints the data place on the two shape
parameters α and r_a of these models. Once more α is required to be
negative, this now implying that the radial velocity dispersion must
increase with radius to be consistent with Table 1. The anisotropy
radius is required to exceed the Solar distance from the centre, thus
avoiding strongly radially biassed orbits in the inner parts of the
Galaxy. The contours of rotation velocity at the Solar distance show
that even a Solar velocity of 220 km/sec is rather large in the context
of these models. This is clearly because they allow only isotropic or
radially biassed orbits, whereas Figure 1 shows that circularly biassed
orbits are necessary for the low observed velocity dispersions of the
cluster system to be consistent with a large circular velocity for the
Sun.

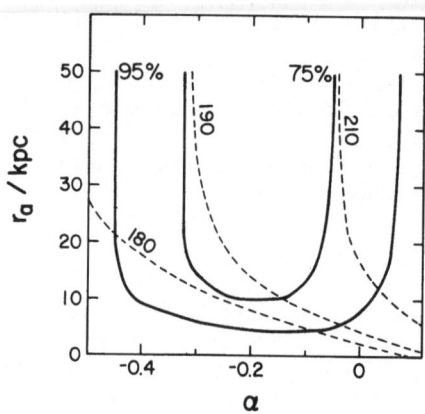

Fig. 2: Constraints on the shape parameters of
Eddington-like models for the kinematics of the
globular cluster system. Curves correspond to those
in Fig. 1.

IV.   CONCLUSIONS

       From our previous work and recent updatings of it reported here we
conclude that:
(1) The velocities of globular clusters are not strongly biassed in the
radial direction, at least in the part of the Galaxy inside the Solar
circle. An isotropic velocity distribution fits the data well.
(2) The velocity dispersion of the cluster system is low. Unless
cluster orbits are predominantly circular and the Galactic rotation
curve rises unusually fast, this dispersion requires a low
characteristic rotation velocity for the Galaxy at the Solar radius
($<$ 220 km/sec).
(3) The apparent rise in velocity dispersion with galactocentric radius
seen in the cluster population requires the "effective" Galactic
rotation curve to be flat or rising over the range 2 kpc $<$ r $<$ 30 kpc.

REFERENCES

Eddington, A.S.: 1914, M.N.R.A.S. 75, p.366.
Harris, W.E.: 1976, Astron. J. 81, p.1095.
Harris, W.E. and Racine, R.: 1979, Ann. Rev. Astron. Astrophys. 17,
       p.241.
Frenk, C.S. and White, S.D.M.: 1980, M.N.R.A.S. 193, p.295.
Frenk, C.S. and White, S.D.M.: 1982, M.N.R.A.S. 198, p.173.
Woltjer, L.: 1975, Astron. Astrophys. 42, p.109.

MASSES AND MYSTERY IN THE LOCAL GROUP

D. Lynden-Bell
Institute of Astronomy, Cambridge, U.K.

1. THE ANDROMEDA AND MILKY WAY SUBGROUPS

Any assessment of the dynamics of the Local Group must locate the main
mass concentrations. The galaxies of the Local Group show two major
concentrations: one of ten galaxies centred in Andromeda (see Fig. 1)
and one of nine centred on our Galaxy. Each of these concentrations is
about 200 kpc in radius. While the concentration around Andromeda
contains the five major satellites M32, NGC 205, NGC 185, NGC 147 and
M33, our Galaxy has only two major satellites, the Magellanic Clouds.
The large concentration of faint satellites about the Galaxy must, in
part, reflect our enhanced ability to detect objects that are close to
us. From the above data alone, it already seems likely that the
Andromeda subsystem is the more massive of the two. From the Fisher-
Tully relations it is then to be expected that the maximum circular
velocity to be found in our Galaxy is less than the 250 km/s found by
smoothing Newton & Emerson's curves for Andromeda. This is borne out
by the velocities of our Galaxy's globular clusters. Figure 2 shows
$\text{Log}(3v_\ell^2)$ plotted against log r. v_ℓ is the line of sight velocity of a
globular cluster, corrected to the galactic centre, and r is the
distance of that cluster from the galactic centre. The averages of $3v_\ell^2$
over the columns indicated by the blackenings in the margin give the
large black points. We show that these give an indication of the
Galaxy's circular velocity out to 100 kpc. Newton's law reads

$$\underline{\ddot{r}} = \underline{\nabla}\psi$$

dotting with $\underline{r}$ we find

$$\frac{d^2}{dt^2}\,|\underline{r}|^2 - \underline{v}^2 = \underline{r}.\underline{\nabla}\psi = -V_c^2$$

averaging in time for one orbit or in space for a statistically station-
ary system of many orbits, we have

$$\langle\underline{v}^2\rangle = \langle V_c^2\rangle$$

349

W. L. H. Shuter (ed.), Kinematics, Dynamics and Structure of the Milky Way, 349–363.
Copyright © 1983 by D. Reidel Publishing Company.

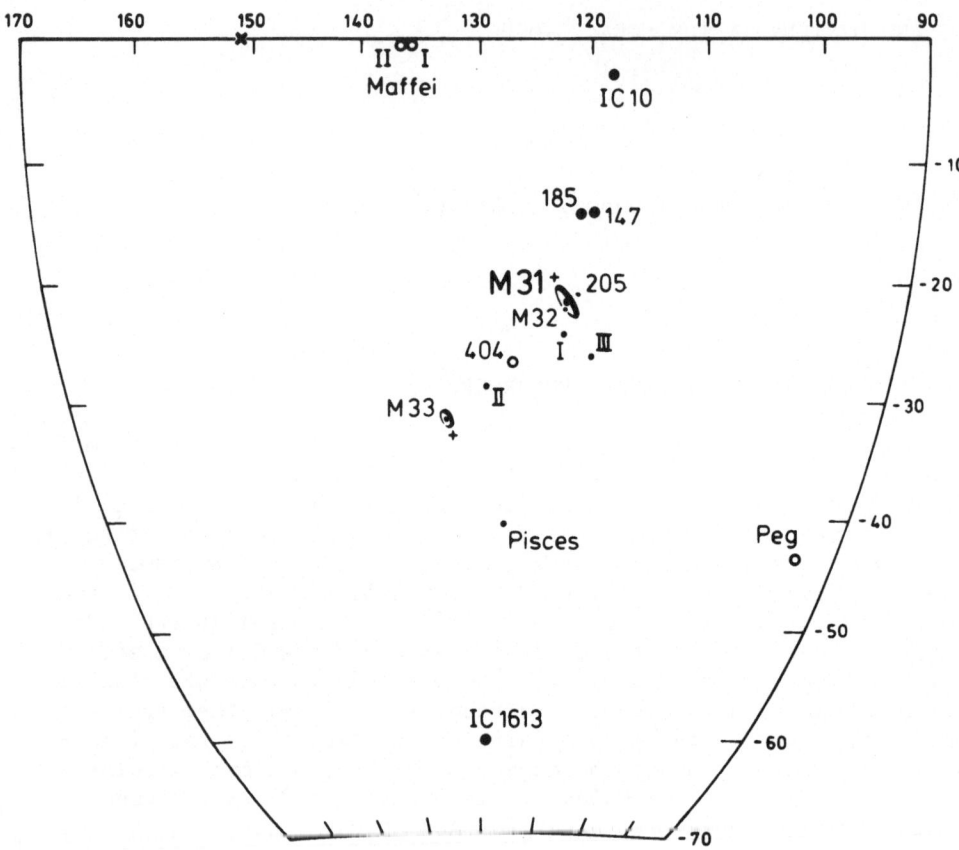

Figure 1: The sky in the direction of Andromeda.   The objects re-
          presented by hollow points are nearby but beyond the
          Local Group (e.g. Maffei I & II).   For the spirals,
          + indicates the end rotating away from us and the hard
          lines indicates the side towards us.   The numbers are
          NGC numbers, I, II & III are dwarf spheroidal companions
          in the Andromeda subgroup.

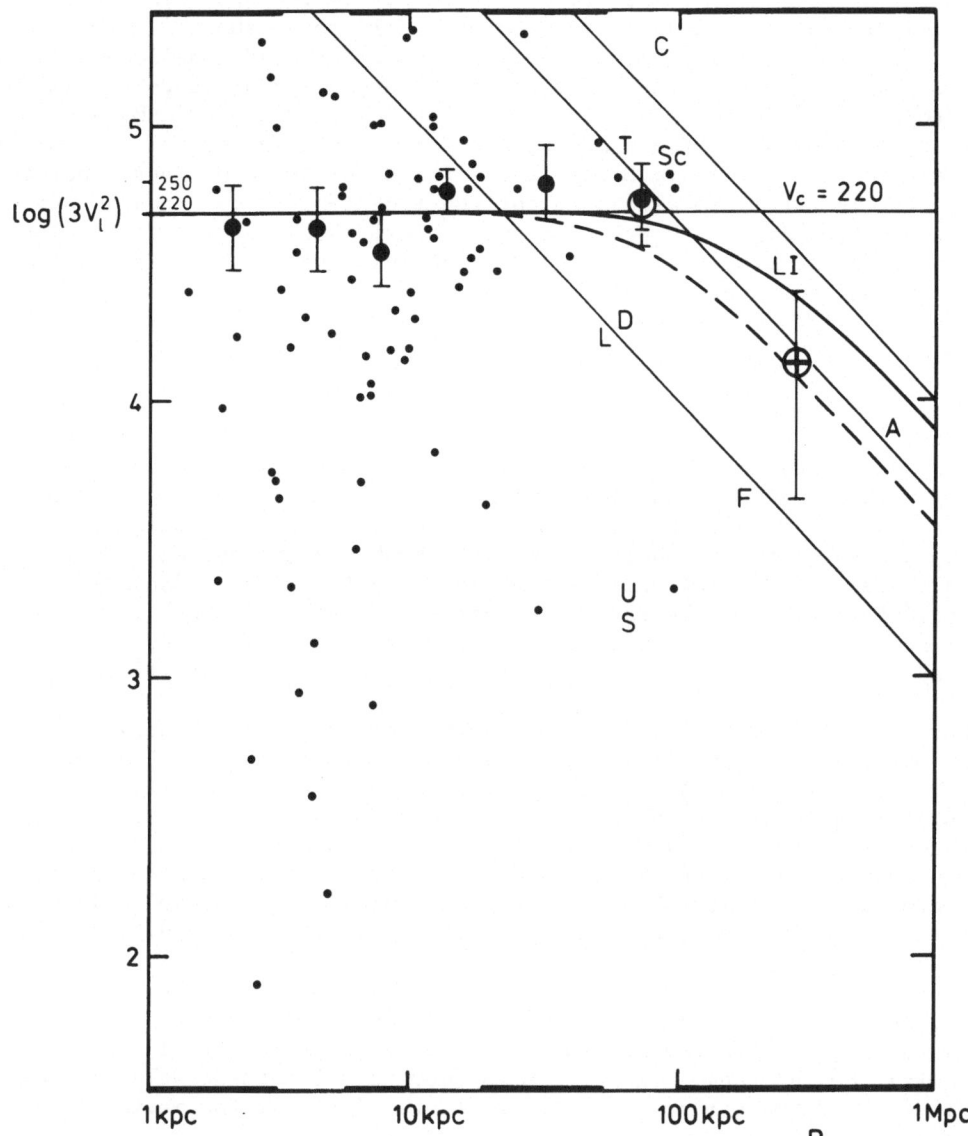

Figure 2: Velocities and distances of globular clusters and satellites
          of the Milky Way. T is the tip of the Magellanic Stream. A =
          Andromeda after suitable allowance for mass ratio. Carina,
          Sculptor, LMC, Draco, Ursa Minor, SMC, Leo I & Fornax are also
          plotted.

If the velocities are not too anisotropic, it will be sufficient to replace $\langle v^2 \rangle$ by $3\langle v_\ell^2 \rangle$. The circular velocity at the sun need not be assumed but has been deduced self-consistently to be about 220 km/s from such data (Lynden-Bell & Frenk 1981, for more detailed modelling see Frenk & White 1981). Added to the diagram are diagonal lines corresponding to total masses of 2.3×10^{11} $M_\odot$, 10^{12} $M_\odot$ and 2.3×10^{12} $M_\odot$ and the letters correspond to the names of the satellite galaxies. T stands for the tip of the Magellanic Stream and A for an estimate of where Andromeda ought to lie after due allowance for the fact that it is a heavy body. The two dotted lines are possible extrapolations of the circular velocity curve using the simplest interpolation formula between V_c = constant and the constant mass case $V_c \propto r^{-\frac{1}{2}}$. We use

$$V_c^2 = \frac{V^2}{(1 + r^2/r_h^2)^{\frac{1}{2}}}$$

The total mass of such a system is the same as that for a system which has constant velocity V out to r_h viz $V^2 r_h/G$. The two curves correspond to r_h = 70 and 140 kpc and total masses of 8×10^{11} and 1.6×10^{12} $M_\odot$ respectively. Such graphs must be treated with caution. Distance estimates to globular clusters sometimes have changed by factors of 2 or 3. The most distant ones may have been sent to these distances by error. Furthermore, the radial velocities of faint distant globulars will have greater errors than those of bright ones. The observed points probably fit such falling extrapolations better than a pure flat curve.

Other estimates of the Mass of the Galaxy out to great distances have been obtained from modelling the Magellanic Stream (Murai & Fujimoto 1980, Lin & Lynden-Bell 1982). However, such detailed modelling procedures inevitably involve detailed assumptions and so lack some of the directness of the determinations cited above. Recently it has been shown that the dwarf spheroidal galaxies Ursa Minor and Draco must be associated with the break-up of the Magellanic Clouds (Lynden-Bell 1981, Hunter & Tremaine 1977, Lynden-Bell 1976, Kunkel & Demers 1977) and it is not unlikely that Carina is also associated. These objects are all further away than the Magellanic Clouds. This makes it likely that the Magellanic Orbit is a large one with the Clouds just past pericentre in their orbit about the Galaxy and the Magellanic Stream trailing along the orbit behind them. Taking the maximum galactocentric velocity seen at the tip of the stream to be 190 km/s, we find that the Newtonian formula

$$v_r = \sqrt{\frac{GM}{\ell}} \, e \sin \phi \qquad\qquad (1)$$

fits the radial velocity along the stream provided

$$\frac{GM}{\ell} e^2 = (190 \text{ km/s})^2$$

For ℓ = 60 kpc we then find M = $\frac{0.49}{e^2} 10^{12}$ $M_\odot$

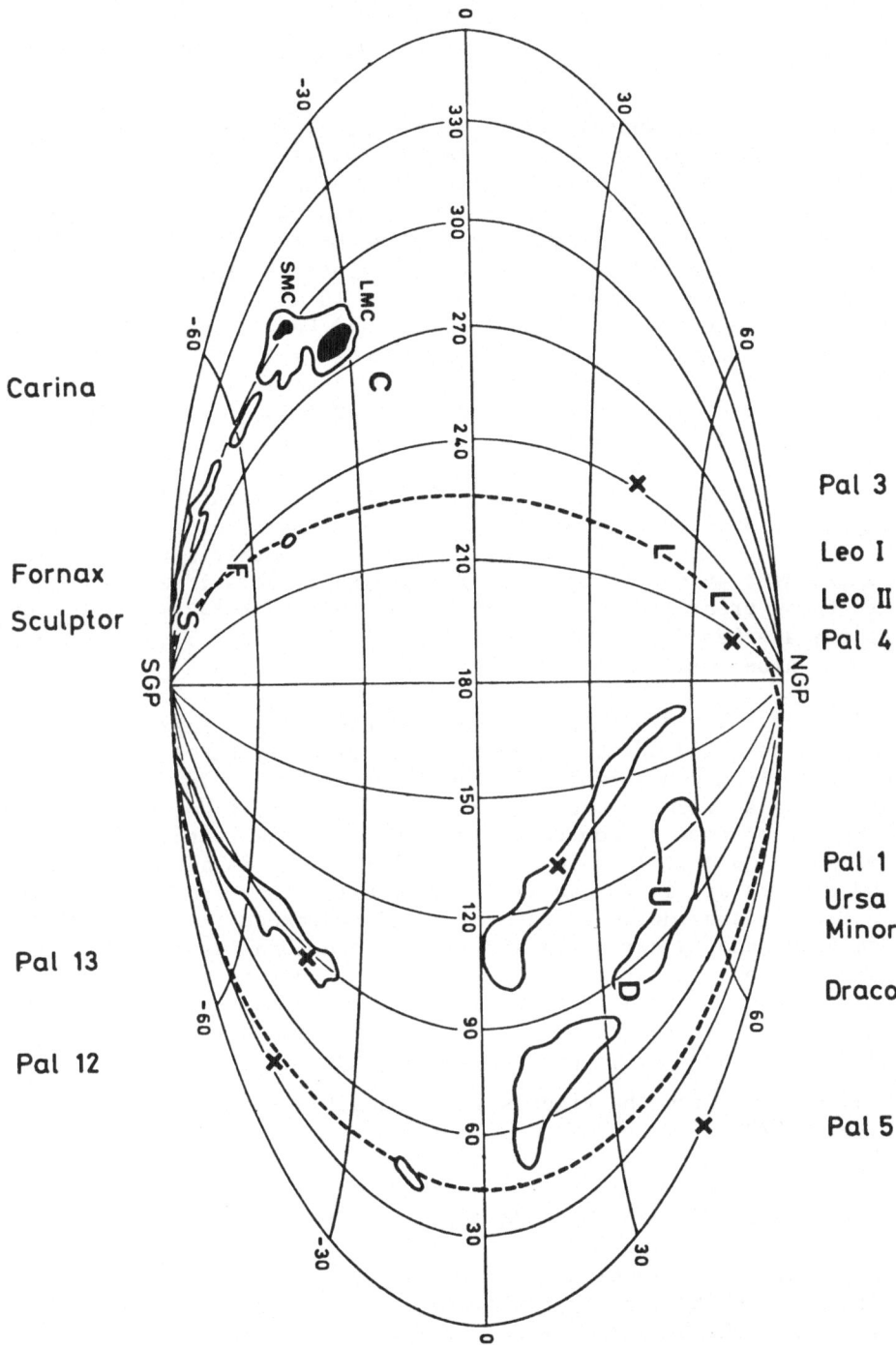

Figure 3: In the Galactocentric sky, Fornax, Leo I, Leo II & Sculptor
lie only 0.5°, 1.4°, 0.6° and 0.6° from a great circle. Sculptor's major axis lies along this circle to 3°±3°. Possibly all
these objects were torn from a proto-Fornax by our Galaxy's tide.

An e of .7 corresponds to an apogalacticon of 200 kpc, so the orbit is unlikely to be of greater eccentricity which at once leads to a mass of 10^{12} $M_\odot$ for the Galaxy.

More recently I found what seems to be the debris formed by the break up of another dwarf galaxy: Fornax, Leo I, Leo II and Sculptor lie almost exactly in a great circle as seen from the Galactic centre (see Fig. 3). Furthermore, Sculptor, the object closest to its tidal limit is oriented along this orbit just as Ursa Minor and Draco are oriented along the Magellanic Stream. Fitting an elliptical orbit with a pericentre at 80 kpc and an apocentre at 230 kpc (Sculptor at 85 kpc has a galactocentric velocity of -15 while Leo I and Leo II at 220 kpc may well be near apocentre) one finds e = .48 and ℓ = 120 kpc. For Fornax at 188 kpc one then finds ϕ = 130° and using formula (1) and the galactocentric radial velocity of Fornax of 40 ± 15 km/sec, we get $M = \ell v_r^2/(Ge^2 \sin^2 \phi) = (.5 \overset{\times}{\div} 1.9) \times 10^{12}$. Clearly more accurate parameters are needed to get a good value for the mass, but these objects do give us the hope that the total mass out to 120 kpc can eventually be measured. At present any such refinement as an orbit in a halo potential is not warranted by the accuracy of the data. If the hypothesis that both Leo systems are near apocentre is correct, then their velocities should be close to those predicted by pure reflections of the solar motion. Such predictions are Leo I v = 96 ± 10 and Leo II v = 48 ± 5. Any change from these values will imply a larger apocentric distance.

Figure 4 shows a log 3v² - log r graph for the satellite galaxies about Andromeda. The solid curve is the circular velocity taken from Newton and Emerson. From the position in the diagrams of NGC 185 it is clear that Andromeda's circular velocity remains fairly flat out to it. Even an extrapolation onward from these downward at 45° corresponding to a constant mass would still make M33 a satellite. It is not clear that IC 10 or Pegasus are in the Local Group at all - they are too far away to be seriously considered as satellites of Andromeda. IC 1613 at 400 kpc from Andromeda is certainly a member of the Local Group but probably an independent one rather than a member of the Andromeda subgroup. Comparing Figures 2 and 4 it is by no means clear that Andromeda is more massive than the Galaxy; however, our earlier arguments indicate that it must be. The Fisher-Tully relation would suggest a mass ratio of $\left(\frac{250}{220}\right)^4$ = 1.66 so, taking the Figures into account we shall use a ratio of M_A/M_G = 3/2.

2. THE TRANSVERSE VELOCITY OF ANDROMEDA

By looking for the reflection of our velocity in the stray members of the Local Group unattached to the two major subgroups, one may get some idea of the way the Galaxy moves. Since the two subgroups emit more than 99% of the light, it is reasonable to assume that they contain most of the mass so that their momenta must be equal and opposite with respect to the Local Group's barycentre. Thus Andromeda's velocity can be determined from the Galaxy's velocity and our estimate of the mass ratio.

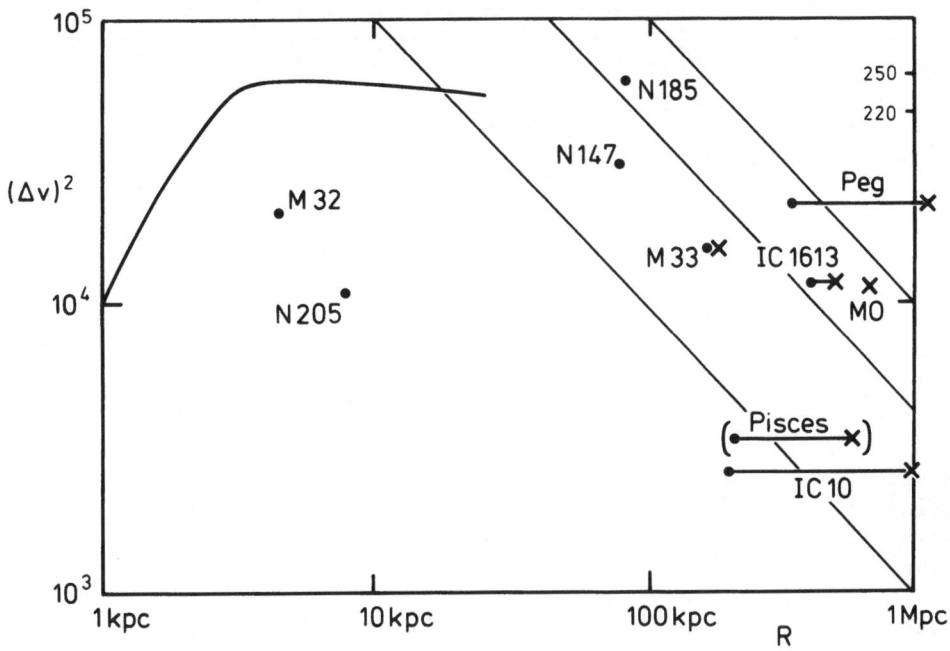

Figure 4: Possible satellites of Andromeda are plotted with its velocity
curve in a log $(\Delta v)^2$ versus log R plot. Δv is the best estimate of the velocity with respect to the centre of M31. The
satellites have only one component of velocity measured so
their $(\Delta v)^2$ have been multiplied by 3 to compensate. The
points ● are plotted with R, the projected distance from
Andromeda as seen on the sky. The points X are plotted at the
estimated true separations from Andromeda. For the Galaxy,
MO, the total velocity determined in this paper has been used
reduced by a factor $(3/5)^2$ to correct for a mass ratio that
is 2:3 rather than infinitesimal. Notice the fall in $(\Delta v)^2$
beyond 100 kpc. The slanting lines are $(\Delta v)^2 R/G$ equal to
$2.3 \ 10^{11} \ M_\odot$, $1.10^{12} \ M_\odot$ and $2.3 \ 10^{12} \ M_\odot$.

In practice there are too few stray members of the Local Group to make
an accurate determination of all three velocity components so we solve
for the Galaxy's velocity using the constraint that the momenta balance
along the line of sight to Andromeda and that the mass ratio is 3/2.
The observed heliocentric velocity of Andromeda is -301 km/s. The
correction to the LSR is +3 and the correction for a circular velocity
of 220 is +175 in that direction. Thus, in the Galaxy's system of rest,
Andromeda is approaching at 123 km/sec. Dividing this in the ratio of
3 to 2 we find that in the Local Group's system of rest the Galaxy moves
at 74 km/s towards Andromeda which moves at 49 km/s towards us.

We now look for the reflection of the transverse velocity of the Galaxy
in the stray members of the Local Group. These are:

	ℓ	b	v	d/Mpc	v_{LG}	Δv	δv
Wolf-Lundmark-Melotte	76	-74	-116	1.3	-19	3	7
IC 5152	344	-50	78		- 5	-2	5
NGC 6822	25	-18		.62		+4	0
Sagitarius	21	-16	- 69		+ 3	+3	0
Aquarius	34	-31	-131		- 1	+5	3
IC 1613	130	-61	-156	.77	-99	+4	+10
Sextans A	246	40	+324	1.3	+129	-7	- 8
IC 10	119	- 3	-343		- 75	+9	12
Pegasus	95	-44	-181	4.3,1.7	+ 44	+7	11
Leo A	197	52	26	3.0	- 26	-2	- 2

Δv is the change in the velocity of the object, v_{LG}, in the Local Group's
system of rest, should we change our estimate of the circular velocity
by + 10 km/sec.

δv is the change in v_{LG} should we change our estimate of the ratio
$M_A/(M_A + M_G)$ by +0.1. Such a change would give $M_A/M_G = 7/3$

Figure 5 is a map of the sky in Galactic Coordinates. A is the direction
to M31, X is the original line to M31 according to the tidal torque of
Thuan & Gott. The contour is the 1σ error ellipse for the direction of
our Galaxy's velocity in the Local Group using the primary set of 10
objects.

The direction of the Galaxy's motion is quite consistent with a straight
line orbit towards M31. This suggests that we should model Local Group
dynamics with the gravity field of two heavy masses emerging from the
big bang but gravitationally retarding each other's motion so that they
reapproach one another. The stray independent members of the Local
Group can be considered as moving in the resultant gravity field.

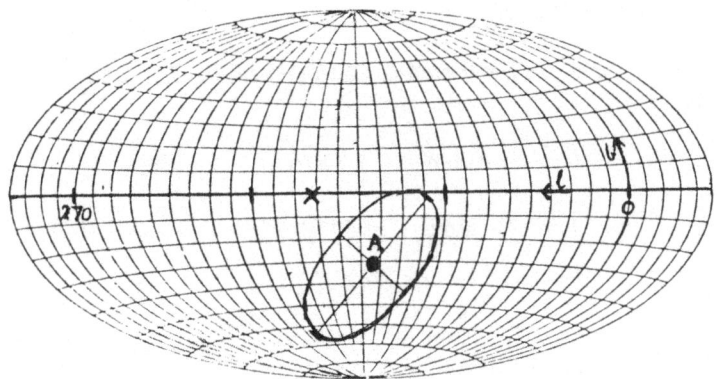

Figure 5: The direction of motion of the Galaxy is rather poorly
determined from the stray members of the Local Group,
but the most probable value lies close to the direction
of Andromeda A. X marks the original line to Andromeda
according to tidal torque theory (Gott and Thuan, Astro-
phisical Journal 223, 426, 1978).

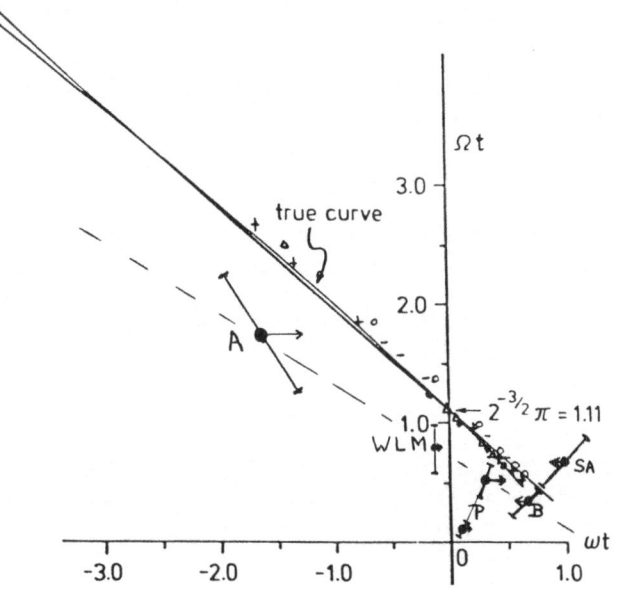

Figure 6: The true curve of $(\frac{GMt^2}{r^3})^{\frac{1}{2}} = \Omega t$ plotted against $vt/r = \omega t$ as
given by equations (4) and (5). Also plotted is the approxim-
ating straight line (equation 6) and the results of computing
5 orbits, illustrated in Figure 7, are shown by the symbols
+Δo- and •. Also plotted are observed points with errors and
the best dotted line through them. A = M31, WLM = Wolf Lund-
mark Melotte, P = Pegasus (plotted twice), SA = Sextans A,
B = Sextans B.

3. LOCAL GROUP DYNAMICS

For the straight line orbit of the Galaxy and Andromeda the energy equation reads

$$\tfrac{1}{2}\,\dot{r}^2 - GM/r = -\tfrac{1}{2}\,GM/a \tag{1}$$

where $M = M_A + M_G$ and $2a = r_{max}$. This equation is integrated by setting

$$r = a(1 - \cos 2\eta) \tag{2}$$

in the form

$$2\eta - \sin 2\eta = (GM/a^3)^{\tfrac{1}{2}}t \tag{3}$$

Eliminating a in favour of the observable r we find,

$$\Omega t \equiv \left(\frac{GMt^2}{r^3}\right)^{\tfrac{1}{2}} = 2^{-\tfrac{1}{2}}(\eta - \sin\eta\,\cos\eta)/\sin^3\eta \tag{4}$$

we also have $\omega \equiv \dot{r}/r = [GM/r^3\,(2 - r/a)]^{\tfrac{1}{2}} = (2GM/r^3)^{\tfrac{1}{2}}\cos\eta$
So using equation (4)

$$\omega t = \frac{\dot{r}t}{r} = (\eta - \sin\eta\,\cos\eta)\,\cos\eta/\sin^3\eta \tag{5}$$

For each value of η (5) yields a value of Ωt and (4) a value of ωt so these may be plotted against one another as η is varied. The result is very close to the straight line (see Fig. 6)

$$\left(\frac{GMt^2}{r^3}\right) + 0.85\,\frac{\dot{r}t}{r} = 2^{-3/2}\pi \cong 1.11 \tag{6}$$

This equation connects M, the total mass, with t the time since the big bang, since r and $\dot{r}$ are known observationally to be 690 kpc and -123 km/s.

Although equation (6) gives us an interesting relationship between the unknowns, greater interest is attached to any new method of solving for them. Evidently we need a second equation relating them. This may be obtained by studying the dynamics of one of the lightweight independent members of the Local Group. To start with consider a distant member and make the approximation that it was also significantly further than the separation of the two heavyweights. Then its motion may be approximated as that of a test particle in the gravity field of a single point mass $M = M_A + M_G$ at the Local Group's barycentre. It is interesting that in this approximation the radius vector r to the test body obeys the same equation (1) but with new values r_1 for r and a_1 for a. This assumes that the third body came from the big bang like everything else! Following the earlier line of reasoning that led to equation (6) we have

$$\left(\frac{GMt^2}{r_1^{\,3}}\right)^{\tfrac{1}{2}} + 0.85\,\frac{\dot{r}_1}{r_1}\,t = 2^{-3/2}\pi \tag{7}$$

In (7) the unknowns M and t have the same interpretation as the mass of the Local Group and the time since the big bang, but r_1 is now the distance of the small independent galaxy from the Local Group's barycentre. Equations (6) and (7) may now be solved for the unknowns Mt^2 and t so the "age of the universe" and the mass of the Local Group may be deduced (Lynden-Bell 1981). It is this path that we shall follow in the next section but we must first secure our foundations by showing that the point mass approximation is valid. To this end a number of 3-body orbits have been computed (see Fig. 7) and the results are plotted on Figure 6. For bodies that are still expanding or only just coming back towards the barycentre of the Local Group agreement with the point mass approximation is excellent (Lynden-Bell 1982).

It is of some interest to discuss these orbital computations in greater detail. The governing equations are

$$\ddot{\underline{r}}_1 = + GM_A \frac{(\underline{r}_A - \underline{r}_1)}{|\underline{r}_A - \underline{r}_1|^3} + GM_G \frac{(\underline{r}_G - \underline{r}_1)}{|\underline{r}_G - \underline{r}_1|^3} \tag{8}$$

where

$$\frac{\underline{r}_A}{M_G} = - \frac{\underline{r}_G}{M_A} = \frac{\underline{r}_A - \underline{r}_G}{M_A + M_G} = \frac{r(t)}{M} \underline{j}$$

and $\underline{j}$ is the unit vector along the y axis and $\underline{r}_A - \underline{r}_G = r\underline{j}$. The separation of Andromeda and the Galaxy r satisfies

$$\ddot{r} = - GM/r^2 \tag{9}$$

and we now rescale all our distances using r as a unit. Thus we define $r_1' = r_1/r$.

Similarly we choose a non-uniform time satisfying

$$d\tau = f(t)dt . \tag{10}$$

Then

$$\ddot{\underline{r}}_1 = rf^2 \{ \frac{d^2 \underline{r}_1'}{d\tau^2} + \frac{d}{d\tau} [\log (fr^2)] \frac{d\underline{r}_1'}{d\tau} + \frac{\ddot{r}}{rf^2} \underline{r}_1'\}$$

and the right hand side of equation (8) is proportional to r^{-2}. The main dependence on r cancels if we choose $f^2 r^3 = GM$ for then $\ddot{r}/(rf^2) = -1$ and we obtain

$$\frac{d^2 \underline{r}_1'}{d\tau^2} + \frac{1}{2} \frac{d}{d\tau} (\log r) \frac{d\underline{r}_1'}{d\tau} = \underline{r}_1' + \mu_A \frac{\underline{r}_A' - \underline{r}_1'}{|\underline{r}_A' - \underline{r}_1'|^3} + \mu_G \frac{\underline{r}_G' - \underline{r}_1'}{|\underline{r}_G - \underline{r}_1'|^3} \tag{11}$$

where $\mu_A = 1 - \mu_G = \frac{M_A}{M_A + M_G}$.

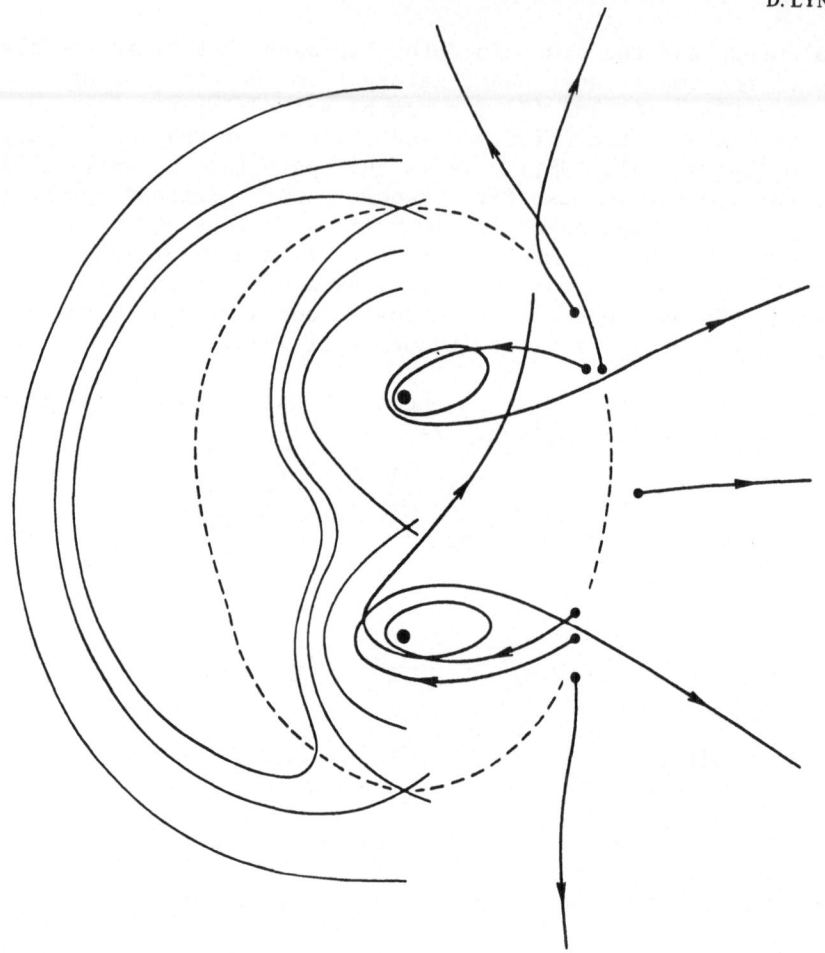

Figure 7: Equipotentials of equation 12 are plotted on the left.
The upper heavy dot represents Andromeda, the lower one
the Milky Way. The dotted line shows the ridge surrounding the "double crater". Orbits calculated from eqs.
(11) and (13) are plotted on the right. They start
from rest at $\tau = -2$.

The right hand side of (11) is derivable from the potential

$$\psi(\underline{r}_1') = \tfrac{1}{2}(\underline{r}_1')^2 + \frac{\mu_A}{|\underline{r}_A' - \underline{r}_1'|} + \frac{\mu_G}{|\underline{r}_G - \underline{r}_1'|} \tag{12}$$

which is large, both near the bodies and at large r_1'. For $\mu_A : \mu_G = 3:2$
the equipotentials are plotted in Figure 7; they are familiar equipotentials of the rotating restricted 3-body problem but the "damping"

term on the l.h.s. of equation (11) is not the coriolis term familiar from that problem. Notice that it damps when Andromeda is separating from the Galaxy but changes sign to "excitation" when these bodies start approaching one another.

One may integrate (9) in the form $\frac{1}{2}\dot{r}^2 + GM/r = \frac{1}{2}GM/a$ where 2a is the maximum separation if we now change variables to τ we find

$$(d \log r/d\tau)^2 = 2(1-r/2a)$$

If we choose the moment of maximum separation as the zero of τ, then we may integrate this in the form $\tau = \sqrt{2} \log[\tan(\eta/2)]$ where η is the substitution variable given by $r = 2a \sin^2 \eta$. Using these formulae we evaluate the damping term as a function of τ as follows:

$$\tfrac{1}{2}d \log r/d\tau = \frac{1}{\sqrt{2}} \cos \eta = \frac{1}{\sqrt{2}} \left[\frac{1-\tan^2 \eta/2}{1+\tan^2 \eta/2}\right] = \frac{1}{\sqrt{2}} \left[1 - \frac{2\exp(\sqrt{2}\ \tau)}{1+\exp(\sqrt{2}\ \tau)}\right] \quad (13)$$

We have integrated equation (11) with the friction coefficient (13) starting from comoving rest close to the big bang (normally $\tau = -2$). We find that orbits starting within the dotted ridge line rapidly fall into one of the major concentrations passing close to their centres – collisions and real dynamical friction will probably ensure that such bodies combine into these major galaxies. Orbits starting outside the ridge suffer the "apparent cosmic repulsion" and expand with the universe. Thus, only objects that start close to the ridge line remain in the outer parts of the Local Group. A number of orbits starting close to the ridge line have been computed. Ωt and ωt have been worked out for a number of points and these are plotted in Figure 6.

4. THE MASS AND DYNAMICAL AGE OF THE LOCAL GROUP

Each independent small member of the Local Group which has a known distance and radial velocity can, in principle, yield a determination of the mass and age. However, the distances even within the Local Group are unreliable, not all velocities are known with the necessary accuracy and even those that are, must be corrected to give the radial velocities that would be seen from the Local Group's barycentre rather than the sun. The objects concerned, all "ought" to lie on a line given by equation (7) when M and t are suitably chosen. Looking at equation (7) one sees that the gradient of the line (for given observations of r_1 and $\dot{r}_1$) is a measure of $(GM)^{\frac{1}{2}}$. Hence if we plot our points with any chosen value of M, 2.3 x 10^{12} $M_\odot$ say, then they should lie on a line but it will normally have the wrong gradient. We can however adjust the gradient by choosing another value of M. Indeed, the correct value of M will be 2.3 x 10^{12} g^2 $M_\odot$, where g is the ratio of the gradient of the correct line, $-1/0.85$, to the gradient of the line formed through the observed points. To plot the observed points we also need a value of t. If any value is chosen, 10^{10} years say, then the true value is found by scaling the whole graph so that it has the right extrapolated intercept on the x axis. In summary, for each object, Andromeda included, we plot $1/r_1^{3/2}$ with r_1 measured in Mpc against $\dot{r}_1/r_1$ with $\dot{r}_1$ measured in hundreds of km/s. In these units $G \times 2.3 \times 10^{12} M_\odot$ is unity and the unit of time is 10^{10} yrs. We then measure M from the gradient ratio g. Finally we find t from the intercept of the extrapolated curve onto $\Omega = 0$. The scaling is done on both axes so does not alter the gradient.

As pointed out earlier, the observational points are subject to both distance errors and velocity errors. Distance errors affect both co-ordinates, velocity errors only move the points horizontally. The error bars drawn correspond to 20% errors each way in the distances and the velocity errors that would be generated by changing the assumed circular velocity at the sun, from 220 to 250 km/sec. Notice that errors in velocity due to that cause are correlated in that all the points move to the points indicated by the heads of the arrows.

An awful warning is given by the small dwarf Pegasus which has been plotted twice - once with the distance of 4.3 Mpc given by Sandage & Tammann - that is so far away as to make the current theory quite invalid because masses outside the Local Group would be involved - and again with the distance of 1.7 Mpc given by Hoessel & Mould. Sandage & Tammann's new distance to Sextans A also raises problems, as it suggests the existence of random velocities that have not been generated by the gravity field of the two Local Group heavyweights. It is quite possible that such random motions of unknown origin will cause such a scatter of the points that these methods of mass and age determination will be of little use. However, at present, it is still possible to maintain that the scatter is really due to that fundamental bugbear of astronomers, errors in the estimated distances.

The gradient of the observed line gives us a total mass of the Local Group of $(5.1 \overset{\times}{\div} 1.7) \times 10^{12} M_\odot$. The extrapolated intercept at $\Omega = 0$ gives the dynamical age of the Local Group since the big bang as

$$t = (1.0 \overset{\times}{\div} 1.5) \times 10^{10} \text{ yrs.}$$

It should be emphasised that this is a real age and not the reciprocal of a Hubble constant.

The large mass found should make us stop and think. Even the heavy haloes of the Galaxy and Andromeda can hardly account for so much mass unless there is significant mass beyond 100 kpc. However, the extra mass is needed to reverse the supposed expansion generated by the big bang. A super-halo that was always outside the heavyweights of the Local Group would do nothing to slow or reverse their supposed expansion, so the mass must lie within about 600 kpc of the Local Group's mass centre. We cannot eliminate the need for such mass by neglecting our data on the small independent members of the Local Group. Even using any of the ages of the Universe proposed by the conflicting schools of cosmography, equation (6), together with the data for Andromeda, give a large mass (Kahn & Woltjer 1959, Einasto & Lynden-Bell 1982). However, there is

still an avenue of escape and one that may open wider as non-gravitation-
ally generated random velocities of galaxies become more widely accepted.
S. van den Bergh suggested that it was by no means impossible that the
Local Group is not a coherent system but rather that the Galaxy and M31
are passing each other as ships pass in the night. Against this we can
only offer the weak argument that the transverse velocity as deduced
from Local Group kinematics is small and this would not be expected on
such a picture. The weakness of this defence is further exposed when it
is realised that the Local Group kinematics themselves rest on the idea
that the Local Group is coherent, for we tacitly assumed that the mean
velocity of the independent small members of the Group was approximately the velocity of the Group's barycentre. Without such an assumption
we know nothing of the transverse velocity. Our argument is thus
reduced to the statement that a radial motion for the two heavy bodies
is predicted from the big bang and the coherent Local Group picture and
the observed motions, interpreted on that picture, do indeed show an
almost radial motion. This evidence is too weak for us to reject the
ships passing in the night picture which remains a viable alternative to
that given here. It should probably be explored in more detail.

What then of mystery? There seem to be heavy haloes for the Galaxy and
M31 but their nature is mysterious. It is likely that a larger halo
enveloping both has a similar total mass or even a slightly greater one.
It is natural to suppose that this is made of the same stuff but its
very existence is dependent on our accepting that the Local Group is
coherent. If we turn to the ships passing in the night alternative, we
find a different problem - where did the momenta that currently carry
M31 and the Galaxy towards each other come from? It is not clear that
any gravitational forces of the required strength are available, so this
raises a cosmogonic mystery itself.

REFERENCES

Einasto, J. & Lynden-Bell, D.: 1982, Mon.Not.R.astr.Soc.199,67.
Frenk, C.S. & White, S.D.M.: 1980, Mon.Not.R.astr.Soc.193,295.
Hoessel, J.G. & Mould, J.: 1982,Astrophys.J.254,38.
Hunter,C. & Tremaine, S.: 1977,Astron.J.82,262.
Kahn, F.D. & Woltjer, L.: 1959,Astrophys.J.130,705.
Kunkel, W.E. & Demers, S.: 1977,Astrophys.J.214,21.
Lin, D.N.C. & Lynden-Bell, D.: 1982,Mon.Not.R.astr.Soc.198,707.
Lynden-Bell, D.: 1976, Mon.Not.R.astr.Soc.174,695.
Lynden-Bell, D.: 1981, Observatory, 101,111.
Lynden-Bell, D.: 1982b,Vatican Conference 1981.
Lynden-Bell, D.: 1982a,Observatory 102,7.
Lynden-Bell, D. & Frenk, C.S.: 1981, Observatory 101,200.
Lynden-Bell, D. & Lin, D.N.C.: 1977, Mon.Not.R.astr.Soc.181,37.
Newton, K. & Emerson, D.T.: 1977, Mon.Not.R.astr.Soc.181,573.
Richer, H.B. & Westerlund, B.E.: 1982, preprint.
Tammann, G.A. & Sandage, A.R.: 1982, Vatican Conference 1981.
van den Bergh, S.: 1971,Astron.Astrophys.11,154.
Yahil, A., Sandage, A.R. & Tammann, G.A.: 1977, Astrophys.J.217,903.

CARBON STARS AS A CONSTRAINT ON MODELS OF THE MAGELLANIC STREAM
AND AS A KINEMATICAL PROBE OF THE LARGE MAGELLANIC CLOUD

Harvey B. Richer, Garry Joslin
Department of Geophysics and Astronomy
University of British Columbia
Vancouver B.C. Canada V6T 1W5

Nils Olander, Bengt E. Westerlund
Astronomical Observatory
Box 515 S751 20
Uppsala Sweden

ABSTRACT

The high latitude carbon star which appears superimposed on the
northern part of the Magellanic Stream (Sanduleak 1980) has been
found to be spectroscopically similar to RGO 55 in ω Centauri.
Assuming identical luminosities for these two objects, a distance
of 38 kpc can be derived to the Magellanic Stream carbon star.
The heliocentric radial velocity of the star is - 149 km sec^{-1};
rather different from the Stream velocity in this direction which
is near - 300 km sec $^{-1}$. Current models of the Stream (Lin and
Lynden-Bell 1982) can account for the radial velocity of the star
even though it differs considerably from that of the HI seen in
the same region of the sky only if it is at a distance of about
100 kpc. This large distance is very unlikely and questions
either the association of the carbon star with the Stream or current
models of it.

A grism survey of a number of Local Group Galaxies has recently been
completed (Richer and Westerlund 1983). Three carbon stars were
found in the Sculptor dwarf spheroidal and slit spectra of them
resulted in a radial velocity of +20 km sec^{-1} for the galaxy. This
is very different from a published value (Hartwick and Sargent 1978)
of +198 km sec^{-1} and makes it much more likely that Sculptor is
related to the Magellanic Stream.

Radial velocities were determined for a sample of carbon stars spread
throughout the Large Magellanic Cloud. The stars were selected from
the catalogues of Westerlund, Olander, Richer and Crabtree (1978),
Blanco, McCarthy and Blanco (1980) and Sanduleak and Philip (1977).
Rotation curves based on the derived radial volocities indicates that
the LMC rotates like a solid body. The velocity gradient of the carbon
stars along the major axis was found to be 10.4 km sec^{-1} deg^{-1}. For

W. L. H. Shuter (ed.), Kinematics, Dynamics and Structure of the Milky Way, 365–366.

circular Keplerian orbits, this places a lower limit of 14.0×10^9 M⊙ on the mass of the LMC. When carbon stars were compared with other stellar populations in the LMC an increase of velocity gradient along the major axis was found to correlate with a decrease in object age. This was interpreted as an increase in orbital eccentricity with age. Comparison with other stellar populations in the LMC further implied that the mean eccentricity of the carbon star orbits was 0.6 which place a lower limit of 34.0×10^9 M⊙ on the mass of the LMC. It was also found that the carbon stars in the catalogue of Sanduleak and Philip (1977) do not share in the general rotation of the LMC. Earlier observations of these stars (Richer, Olander and Westerlund [1979]) indicated that they are analogous to Galactic Population II objects and hence it seems clear that the halo of the LMC is kinematically different from its disk.

More detailed discussions of the results mentioned in this abstract can be found in Richer and Westerlund (1983) and Joslin et al (1983).

REFERENCES

Blanco, V.M., McCarthy, M.F., and Blanco, B. 1980, Ap.J. 242, 938.

Hartwick, F.D.A. and Sargent, W.L.W. 1978, Ap.J. 221, 512.

Joslin, G., Richer, H.B., Olander, N., and Westerlund, B.E. 1983, in preparation.

Lin, D.N.C., and Lynden-Bell, D. 1982, M.N.R.A.S. 198, 707.

Richer, H.B., Olander, N., and Westerlund, B.E. 1979, Ap.J. 230, 724.

Richer, H.B., and Westerlund, B.E. 1983, in press, Jan. 1, Ap.J.

Sanduleak, N. 1980, P.A.S.P. 92, 246.

Sanduleak, N., and Philip, A.G.D. 1977, Publications, of the Warner and Swasey Observatory, Vol. 2, No. 5, p. 105.

Westerlund, B.E., Olander, N., Richer, H.B., and Crabtree, D.R. 1978, Astr. Ap. Suppl. 31, 61.

MOLECULAR CLOUDS IN SPIRAL GALAXIES

Judith S. Young and Nick Z. Scoville
Five College Radio Astronomy Observatory, U. of Massachusetts

ABSTRACT

We are conducting a large observational program investigating the 2.6 mm CO line in spiral galaxies using the 14 m telescope of the Five College Radio Astronomy Observatory (HPBW = 50"). Thus far we have observed 46 galaxies of types Sa, Sb, Sc and Irr, detected 31, and mapped 16. The major findings from our CO observations are:

(1) In several late type spiral galaxies (IC 342, NGC 6946 and M51) the radial distribution of molecular gas out to 10 kpc follows the expo-nential blue luminosity profile of the disk within each galaxy (Young and Scoville 1982a, Scoville and Young 1982).

(2) From a comparison of the CO and blue luminosities of the central 5 kpc in a larger sample of Sc galaxies, we find that the blue lumino-sity is proportional to the first power of the CO content in both the nuclei and disks of Sc galaxies (Young and Scoville 1982b). We interpret this to mean that the star formation rate per H_2 in Sc galaxies (indicated by the blue luminosity) is constant.

(3) No molecular rings like the one in the Milky Way at radii 4 to 8 kpc were seen in the Sc galaxies.

(4) We have found molecular rings in two Sb galaxies, NGC 7331 and NGC 2841, with peaks at radii of 4-5 kpc (Young and Scoville 1982c). The central holes in the CO distributions are possibly related to the presence of large nuclear bulges in these galaxies.

1. INTRODUCTION

The structure and evolution of galaxies depend on both the distri-bution and abundance of dense interstellar matter. The giant molecular clouds are the potential sites of future star formation, and the massive stars which are formed there produce a major part of the galactic lumi-nosity. Surveys of CO emission from the Milky Way indicate that this gas shows strong maxima within 400 pc of the galactic center and in a ring at radii 4 to 8 kpc (Scoville and Solomon 1975; Burton et al.

W. L. H. Shuter (ed.), Kinematics, Dynamics and Structure of the Milky Way, 367–377.

1975). It is of great interest to determine the radial distributions of molecular gas in external galaxies of all types to increase our understanding of our own galaxy, of star formation, and of the structure and evolution of galaxies.

In order to investigate the dense molecular clouds and their relation to star forming activity in other galaxies, we have been observing the 2.6 mm CO line using the 14 m telescope of the Five College Radio Astronomy Observatory (HPBW = 50"). The aims of these observations are to determine (1) the radial distributions of molecular gas in spiral galaxies, (2) the dependence of the CO distributions on morphological type and galaxy luminosity, (3) the relative amounts of star-forming material in the nuclei of normal and active galaxies, and (4) the relative confinement of molecular clouds to spiral arms.

2. SC RADIAL DISTRIBUTIONS

In IC 342, NGC 6946 and M51 (types Sc and Scd) we have mapped the CO radial distributions out to 10 kpc at 1-2 kpc resolution (Young and Scoville 1982a, Paper I; Scoville and Young 1982, Paper II). Figure 1 shows the observed CO emission along the north-south strip in M51. In each galaxy the spectra exhibit velocity and intensity changes which correlate well with galactic location. The highest intensities are observed in the nuclei, with smooth falloffs at larger galactic radii. The mean velocities vary regularly across each galaxy in the sense expected from galactic rotation. Figure 2 shows the spatial-velocity map along the N-S axis in IC 342.

The CO radial distribution in IC 342 is shown in Figure 3; along with the integrated intensities expected for our own Galaxy. Although our effective resolution is clearly adequate, the Sc galaxies do not have "holes" in the CO distributions like that at R = 1-4 kpc between the nucleus and molecular cloud annulus in our Galaxy.

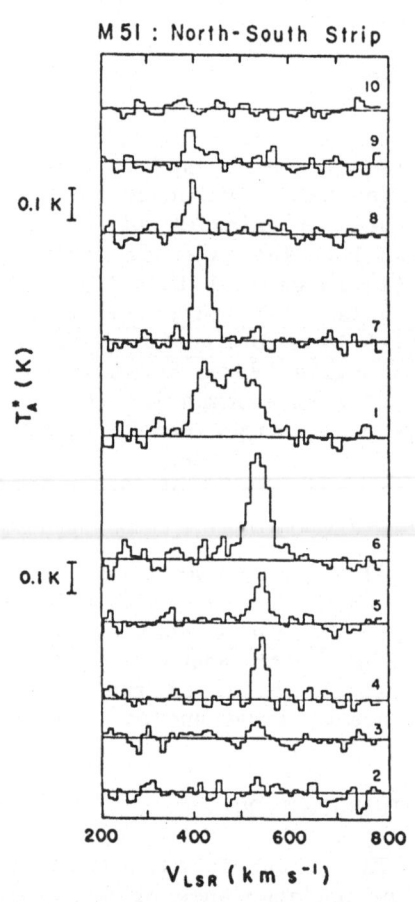

M5I : North-South Strip

Figure 1. Spectra of CO emission along the N-S strip in M51 smoothed to 10 km s^{-1}. This strip is 10° from the major axis.

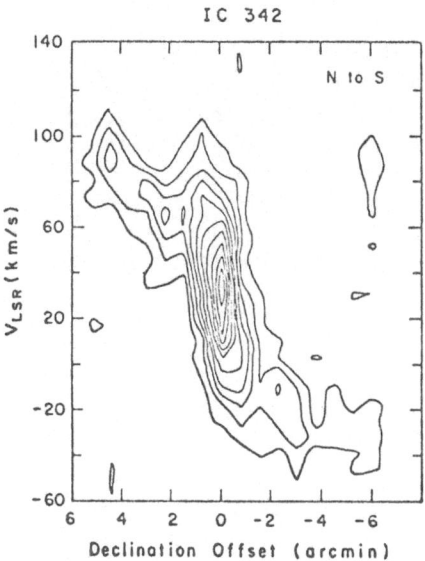

Figure 2.   Observed CO rotational velocities in IC 342 along the N-S strip.   Contours are in intervals of 0.05 K.

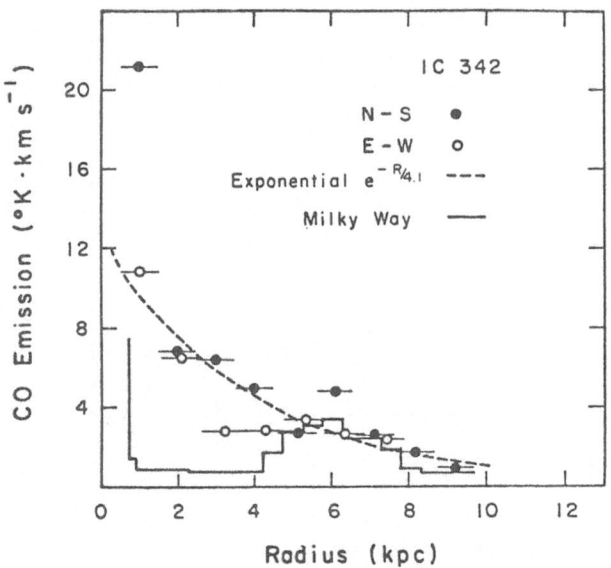

Figure 3.   CO radial distribution in IC 342.   Points plotted represent the observed integrated intensities, with horizontal bars showing the beam size on the galaxy.   The solid line represents the CO distribution for the Milky Way.

In IC 342, NGC 6946 and M51 we have compared the CO distributions with the optical light profiles and find that the radial distribution of molecular gas out to 10 kpc follows the exponential blue luminosity profile of the disk within each galaxy, as shown in Figure 4 for NGC 6946. Since the blue starlight probably originates from relatively young stars ($\leq 2 \times 10^9$ years old, cf. Tinsley 1980), and assuming the abundance of molecular gas is proportional to the CO emission (see Paper I, Appendix), this close correspondence between the CO emission and blue light indicates that the star formation rate per nucleon is constant within a particular galaxy; in the particular galaxies studied the bulk of the ISM is molecular inside R ~ 10 kpc.

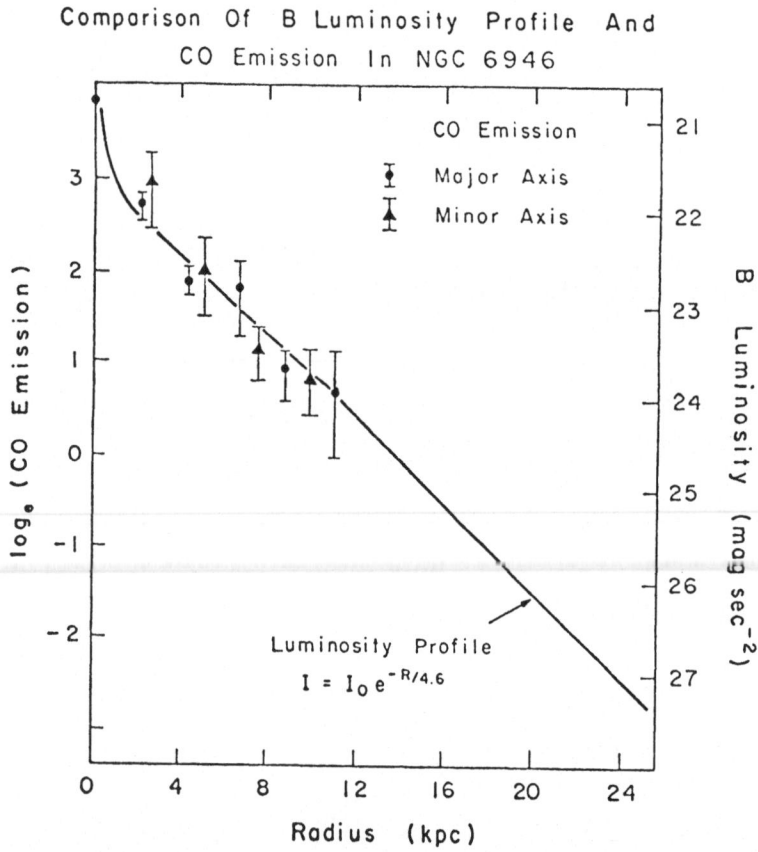

Figure 4. Comparison of the CO emission with the B luminosity profile of Ables (1971) for NGC 6946. The points plotted are the mean CO intensities at each radius, with bars indicating the spread in the observed intensities. The close agreement between the radial dependences of the CO emission and the luminosity profile is striking. The exponential nature of the luminosity profile is evident on this semi-log plot as a straight line.

We have derived an empirical relationship for determining H_2 column densities (N_{H_2}) from CO intensities (I_{CO}) based on visual extinction and CO observations in our own galaxy. In the Appendix of Paper I we showed that both dark cloud and giant cloud samples are consistent with $N_{H_2}/I_{CO} = 4 \pm 2 \times 10^{20}$ H_2 $cm^{-2}/(K$ km $s^{-1})$. Using this conversion in the external galaxies we find that H_2 dominates HI in the interiors of the Sc galaxies. The H_2 and HI mass distributions in NGC 6946 are shown in Figure 5. The H_2 and HI distributions diverge at the centers of these galaxies; the H_2 exhibits an exponential increase while the HI profiles are flat with central holes. In the Sc galaxies the H_2 masses within ~ 10 kpc are comparable to those of HI interior to 25 kpc. Global spiral structure is not evident in the molecular data at the present resolution.

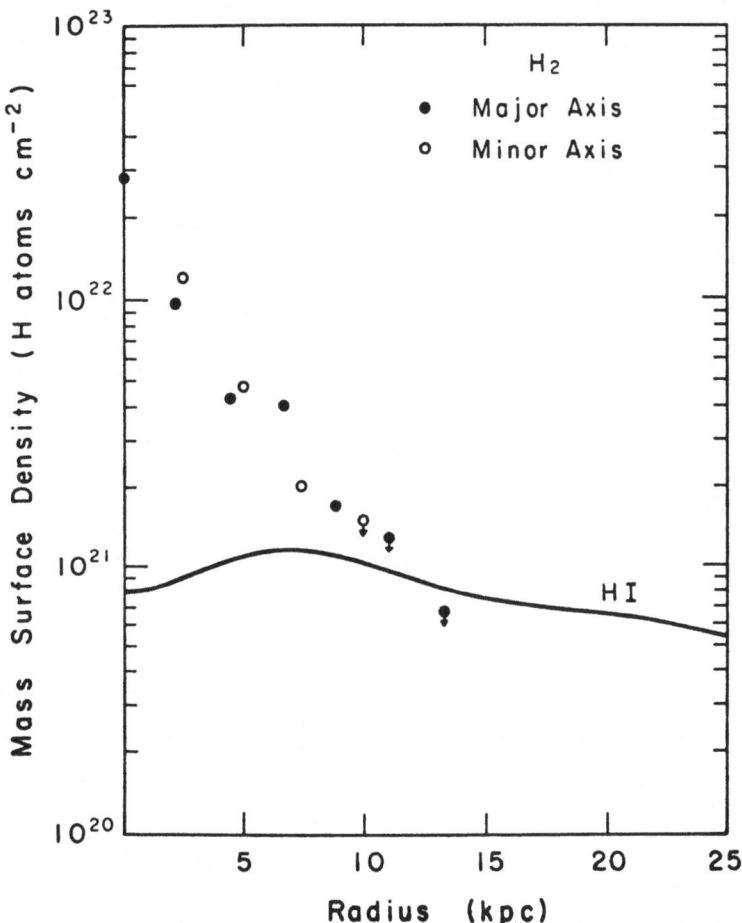

Figure 5. Mass surface densities in NGC 6946. H_2 values are derived assuming a linear relationship between CO flux and H_2 mass. The solid curve represents the HI mass surface density from Rogstad and Shostak (1972). H_2 dominates HI over the interior of the galaxy. The H_2 mass out to 12 kpc is 10^{10} M_Θ compared with 3×10^9 M_Θ for HI.

3. COMPARISON OF SC GALAXIES

The correlation of CO intensity with blue luminosity within a particular Sc galaxy led us to investigate a larger sample of Sc'c covering a wide range of size, mass, and total luminosity. The spectra observed in the central 50" of 8 galaxies are shown in Figure 6. We have compared the CO and blue luminosities of the central 50", the central 5 kpc, and 50" in the disks (Figure 7). An approximately <u>linear</u> correlation is revealed between the CO and blue luminosities in <u>both</u> the nuclei and the inner disks. Assuming the CO emission is proportional to the abundance of molecular gas, this correlation implies that low luminosity regions have little H_2 while high luminosity regions have large amounts. Within this sample the molecular masses out to a fixed radius, $R < 2.5$ kpc, range from $< 6 \times 10^7$ $M_\odot$ for M33 to $\sim 2 \times 10^9$ $M_\odot$ for NGC 6946. However, the H_2 mass to blue luminosity ratio is relatively constant, with $M_{H_2}/L_B = 0.17 \pm 0.08$ $M_\odot/L_\odot$ over two orders of magnitude in L_B. If the blue luminosity is an indicator of the recent star formation rate, these results suggest that <u>the star formation rate per nucleon in Sc galaxies is constant</u>. In contrast, these galaxies all have similar amounts of HI in the central $R < 2.5$ kpc, so that H_2/HI ratio varies from < 0.4 in M33 to ~ 32 in M51. These results are summarized in Table 1.

Table 1
Masses of H_2 and HI

Galaxy	Central R < 2.5 kpc[a]			$\dfrac{M(H_2)}{M(HI)}$	$\dfrac{M(H_2)}{L_B}$	$\dfrac{M(HI)}{L_B}$
	L_B $(L_\odot)$	$M(H_2)$ $(M_\odot)$	$M(HI)^a$ $(M_\odot)$		$(M_\odot/L_\odot)$	$(M_\odot/L_\odot)$
N1068	4.6×10^{10}	4.2×10^9	5×10^6	840	0.091	1×10^{-4}
N5236	1.7×10^{10}	1.9×10^9	9.5×10^7	20	0.11	5.6×10^{-3}
N6946	1.1×10^{10}	2.4×10^9	1.2×10^8	20	0.22	1.1×10^{-2}
N5194	1.0×10^{10}	2.0×10^9	6.3×10^7	32	0.20	6.3×10^{-3}
N4321	7.4×10^9	1.5×10^9			0.20	
I342	4.5×10^9	1.2×10^9	6.8×10^7	18	0.27	1.5×10^{-2}
N5457	2.8×10^9	5.6×10^8	9.1×10^7	6.2	0.20	3.2×10^{-2}
N2403	1.3×10^9	$<6.0\times10^7$	1.3×10^8	<0.46	<0.046	1.0×10^{-1}
N598	3.9×10^8	$<5.5\times10^7$	1.3×10^8	<0.42	<0.14	3.3×10^{-1}

[a]   B luminosities and HI surface densities are determined from observations reported in the literature, and H_2 masses are derived from our CO observations (see Young and Scoville 1982b).

4. MOLECULAR RINGS

No molecular rings like the one in the Milky Way at radii 4 to 8 kpc were seen in the Sc galaxies. In Paper I we suggested that the difference in the radial distributions is the absence of gas at $R \sim 1$ to 4 kpc in the Milky Way, and that this hole is due either to the exhaustion of gas in forming stars in the nuclear bulge of our galaxy, or to the

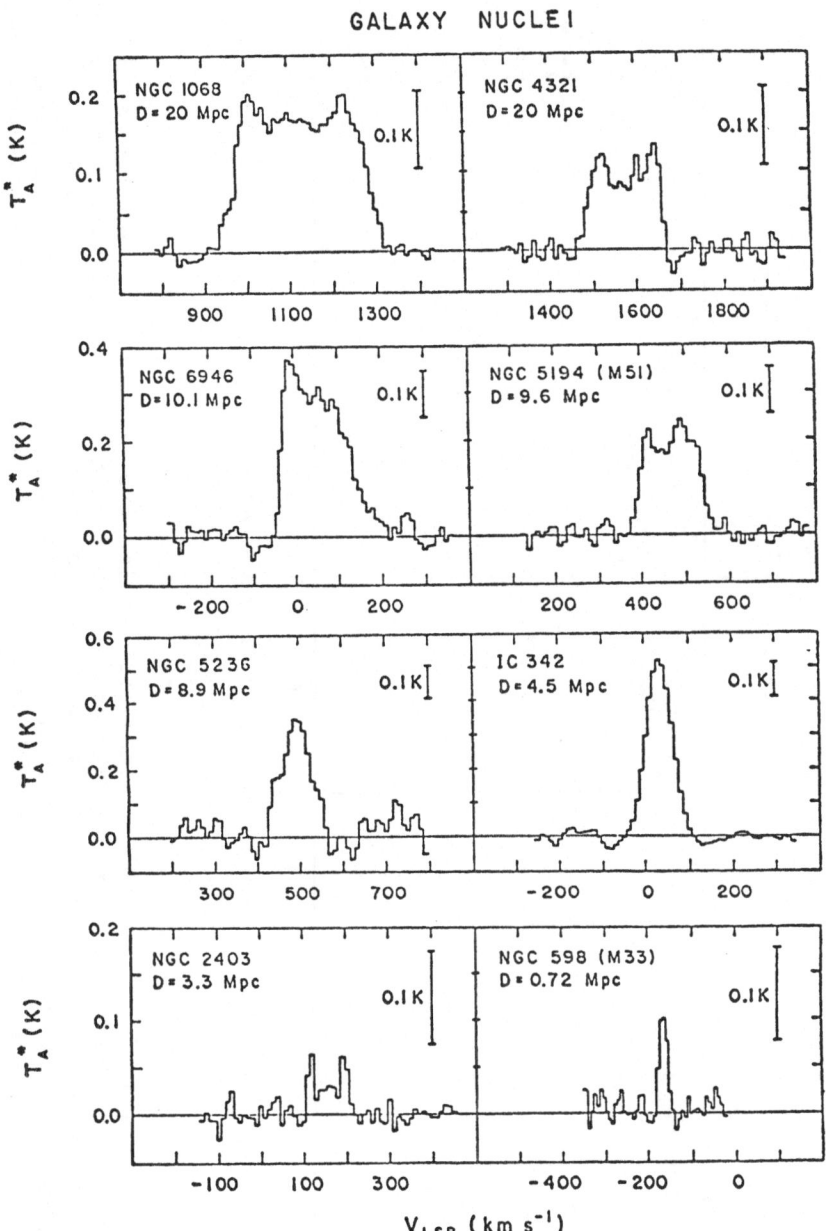

Figure 6. CO spectra in central 50" of the Sc galaxies and NGC 1068, smoothed to 15 km s^{-1} resolution.

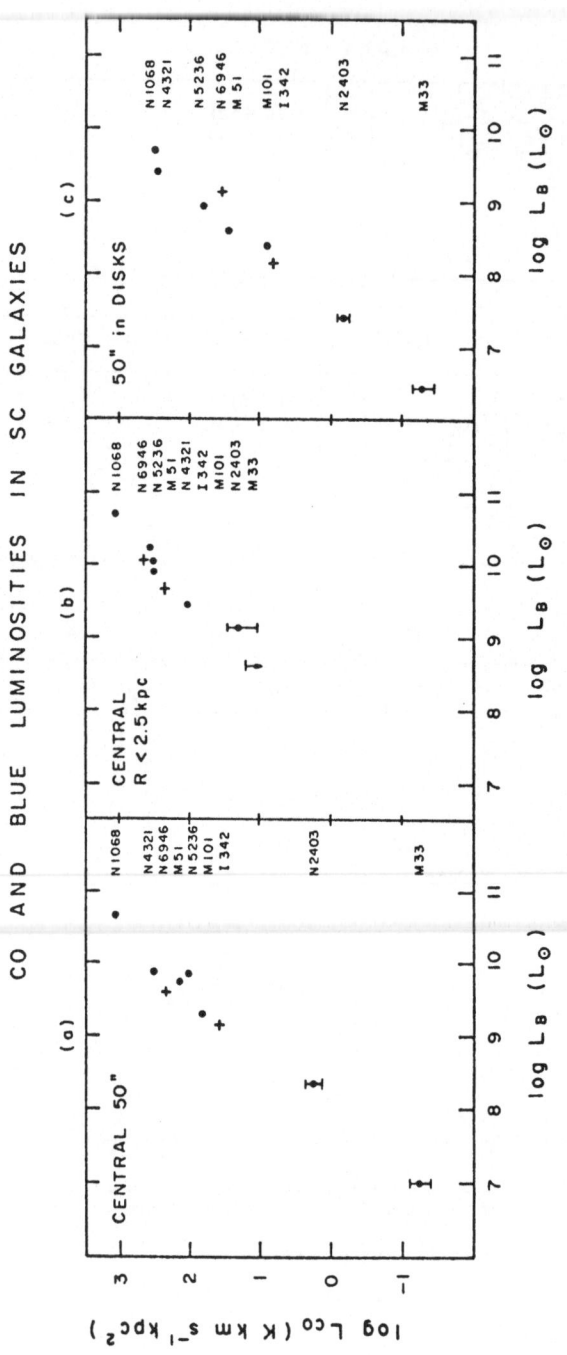

Figure 7. Comparison of CO and blue luminosities in (a) the central 50", (b) regions 5 kpc in diameter centered on the nuclei, and (c) regions 50" in diameter in the disks for nine galaxies ($L_{CO} \equiv I_{CO} \times$ Area Beam). Over four orders of magnitude in luminosity, a linear correlation is evident. The open circle represents the Seyfert galaxy NGC 1068 which has the highest luminosity. The uncertainties in L_{CO} are given for M33 and NGC 2043, and for the other galaxies are smaller than the dots plotted. Distance uncertainties do not alter the correlation since both luminosities depend on D^2. The best fit to the data in (b) is $L_B = 4 \times 10^7 L_{CO}^{1.0}$.

evacuating action of the inner Lindblad resonance (ILR). In IC 342 and
NGC 6946 it is not possible to distinguish between these two possibilities because the smooth rotation curves suggest the presence of neither
nuclear bulges nor ILR's (cf. Kormendy and Norman 1979).

     Recently, however, we discovered molecular rings in two Sb galaxies,
NGC 7331 and NGC 2841, with peaks at radii of ~ 4 kpc (Young and
Scoville 1982c). Figure 8 shows the CO radial distributions in these
galaxies, where the integrated intensities toward the nuclei clearly are
low relative to the disk. The presence of the central CO holes in these
Sb galaxies provides clues to the origin of such a distribution.

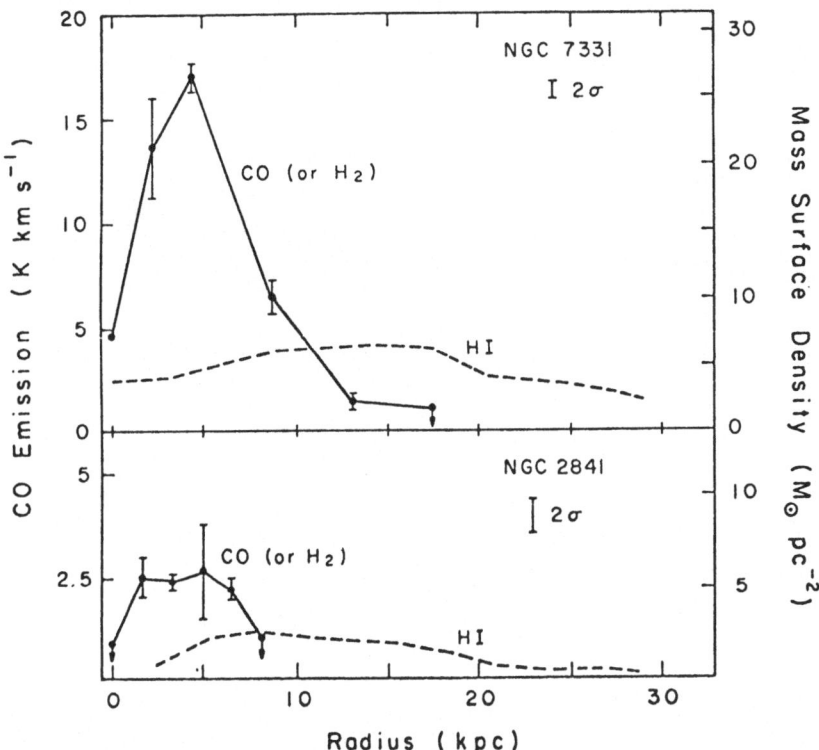

Figure 8. CO radial distributions in NGC 7331 and NGC 2841 (left-hand
axis) and mass surface densities (right-hand axis). The solid dots
represent the average CO emission at each radius, with bars indicating
the dispersion in intensities observed. The magnitude of the
2σ uncertainty on each individual measurement is also indicated. In
both galaxies the CO radial distributions exhibit molecular rings which
peak around 4-5 kpc. Also plotted are the HI distributions (Bosma
1978). NGC 7331 has higher surface densities in both H_2 and HI than
NGC 2841.

The observed CO rotational velocities in NGC 7331 are shown in Figure 9. The previously measured Hα and HI rotation curves for these galaxies (Rubin et al. 1965; Bosma 1978) reach their maximum velocities at the same radii as the peaks in the molecular rings, so that we can rule out the possibility that the hole in the distribution arises from dynamical action at the ILR. Instead, the CO distributions in NGC 7331, NGC 2841 and possibly the Milky Way are related to the nuclear bulges. In the external galaxies, the central CO holes are coincident with the size of the bulge components measured by Boroson (1981). Rather than being in molecular clouds the gas which was present during the formation of the galaxy may have been depleted in forming stars in the nuclear bulge.

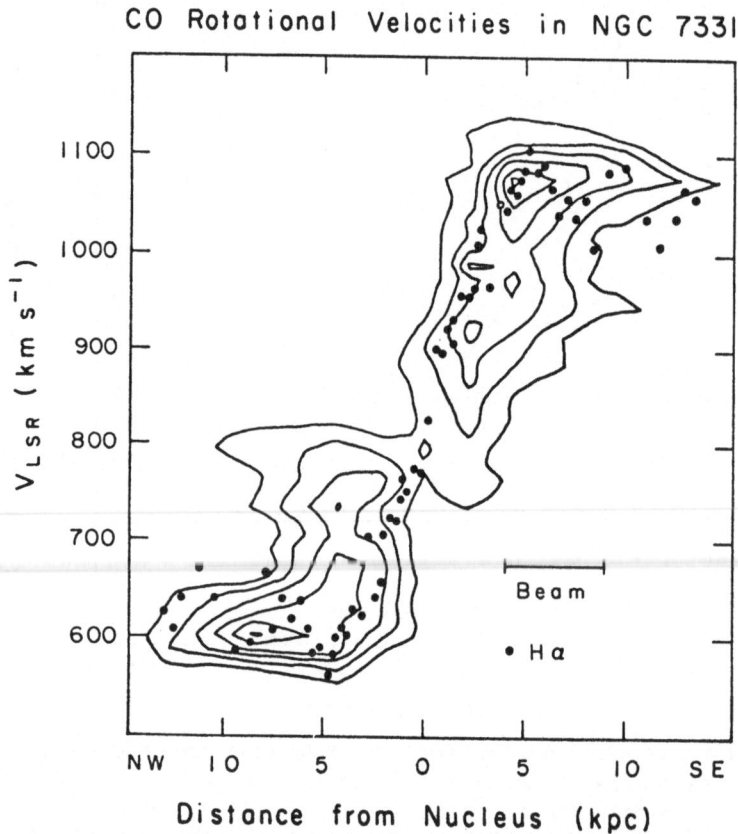

Figure 9.   Observed CO rotational velocities along the major axis of NGC 7331; the Hα emission (Rubin et al. 1965) shows similar velocities (solid dots). The peak in the rotation curve coincides in radius with the peak in the molecular emission. The hole in the gas distribution is probably related to the nuclear bulge.

We are presently expanding our observations of CO radial distributions to earlier Hubble types, as well as observing galaxies of all morphological types and luminosities in the Virgo cluster.

REFERENCES

Ables, H.D.: 1971, Publ. U.S. Naval Obs. Sec. Ser., Vol XX, Part IV, Washington, D.C.
Boroson, T.: 1981, Ap.J. Supp., 46, 177.
Bosma, A.: 1978, Ph.D. dissertation, University of Groningen.
Kormendy, J. and Norman, C.A.: 1979, Astrophys.J., 233, 539.
Rogstad, D.H. and Shostak, G.S.: 1972, Astrophys.J., 176, 315.
Rubin, V.C., Burbidge, E.M., Burbidge, G.R., Crampin, D.J., and Prendergast, K.H.; 1965, Astrophys.J., 141, 759.
Scoville, N.Z. and Solomon, P.M.: 1975, Astrophys.J.(Letters), 199, L105.
Scoville, N.Z. and Young, J.S.: 1982, Astrophys.J., in press.
Tinsley, B.M.: 1980, in Fundamentals of Cosmic Physics (London: Gordon Breach), Vol. 5, 287.
Young, J.S. and Scoville, N.Z.: 1982a, Ap.J., July 15 (Paper I).
_____ 1982b, Astrophys.J.(Letters), Sept. 1.
_____ 1982c, Astrophys.J.(Letters), Sept. 15.

NGC 3200: IS THIS WHAT OUR GALAXY IS LIKE?

Vera C. Rubin
Department of Terrestrial Magnetism
Carnegie Institution of Washington

ABSTRACT. Systematic dynamical properties of Sa, Sb, and Sc field galaxies are presented. The rotational properties of the Sb(r)I galaxy NGC 3200 suggest that it resembles our Galaxy. It is concluded that the general dynamical properties of our Galaxy are not yet sufficiently well known to place it accurately in the sequence of external galaxies.

1. SYSTEMATIC DYNAMICAL PROPERTIES OF FIELD SPIRAL GALAXIES

For the past several years, my colleagues W. Kent Ford, Jr. and Norbert Thonnard and I have been engaged in a study of the dynamical properties of field spiral galaxies. One aim is to learn the systematics of galaxy dynamics along the Hubble sequence and within a Hubble type, so as to relate the dynamical properties to other galaxy parameters.

Long slit optical spectra of about 60 Sa, Sb, and Sc galaxies have been obtained with the 4-m telescopes at Kitt Peak and Cerro Tololo Observatories at high dispersion (25A/mm) and high spatial scale (25"/mm). A few spectra were also obtained with the 100-in Las Campanas telescope. Within each Hubble type, we have observed galaxies with as large a range of luminosity as we could identify. Radio observations of the 21-cm line of neutral hydrogen for approximately 3/4 of the program galaxies come from the 300-ft and 140-ft NRAO telescopes. Details and results of the study, and a comparsion of the Sa, Sb, and Sc dynamics are available elsewhere. (Rubin, Ford, and Thonnard 1980; Burstein et al. 1982; Rubin et al. 1982; Rubin, Thonnard, and Ford 1982). Throughout the discussion, we use $H = 50$ km s^{-1}Mpc^{-1}.

Here I want to discuss some of the regularities in the galaxy dynamics, present data concerning NGC 3200, a galaxy which our Galaxy may resemble, and finally to discuss the relation between the properties of our Galaxy and those we observe externally.

Rotation curves for 23 Sb galaxies are shown in Fig. 1; rotational velocity is plotted against linear distance. The galaxies are arranged according to increasing luminosity. Low luminosity galaxies (i.e., small galaxies) have velocities which rise gradually from the nucleus, and reach a low maximum velocity only at the galaxy (isophotal) radius. For galaxies of higher luminosity (i.e., large galaxies), rotational

W. L. H. Shuter (ed.), Kinematics, Dynamics and Structure of the Milky Way, 379–384.

velocities rise more steeply from the nucleus, reach the "flat" portion in a smaller fraction of the galaxy radius, and rise to a higher maximum velocity.

From these data, we form "synthetic rotation curves" (see Thonnard and Rubin 1982 for a discussion of the method) showing rotational velocity as a function of the fractional isophotal radius. Fig. 2 plots such a set for the Sb galaxies. The systematic progression in the form of the rotation curves indicates that the form of the rotation curve is a clear luminosity indicator, within a Hubble type. Hence, a comparison of the rotation curve of a galaxy of known Hubble type with standard curves (such as those in Fig. 2), will permit its absolute magnitude to be estimated even if only a fraction of its rotation curve is available.

The velocity variation among Hubble types is equally systematic. At a fixed luminosity, Sa rotational velocities are higher than those of an Sb; Sc rotational velocities are lower. The increase of rotational velocity with increasing luminosity produces the correlation of V(max) with luminosity, i.e., the Tully-Fisher relation (Fig. 3). However, there is a significant displacement between the relations for galaxies of different Hubble types. At fixed luminosity, the maximum rotational velocity of an Sa is about 1.6 times higher than that of an Sc. Or stated another way, the blue luminosity of an Sc is

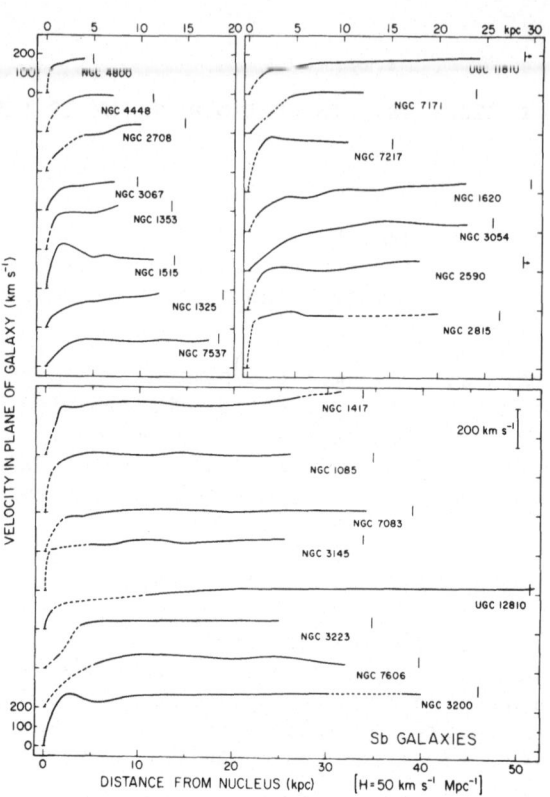

Fig. 1. Rotational velocities for 23 Sb galaxies arranged by increasing luminosity. Vertical bars: isophotal radius. Dashed lines: no measurements.

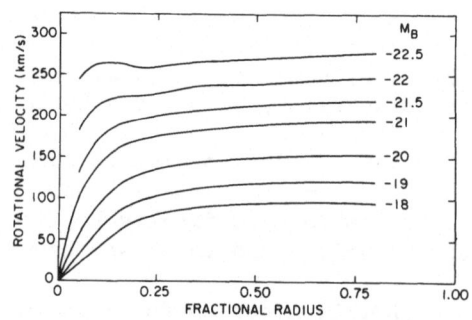

Fig. 2. Synthetic rotation curves for Sb galaxies of indicated luminosities as a function of fractional isophotal radius.

over two magnitudes brighter than an Sa of the same rotational velocity. The concordance of these correlations, based on blue magnitudes, with those based on red magnitudes (Aaronson, Huchra, and Mould 1979) is still under study.

We illustrate the correlations of blue luminosity with isophotal radius, with V(max), and with mass $\mathcal{M}$ within the isophotal radius $[\mathcal{M}(R) \propto RV^2(R)]$ on a single plot, Fig. 4. The correlations represented on this figure are those which we will employ below to relate the dynamical properties of our Galaxy to those derived for the program set.

2. NGC 3200, A HIGH LUMINOSITY SB GALAXY

NGC 3200 is an attractive multi-armed southern spiral [$\alpha = 10^h16^m12^s$; $\delta = -17°43'.9$ (1950); $\ell = 259°$, $b = +32°$] viewed at high inclination (i=72°). It is classified Sb(r)I in the RSA (Sandage and Tammann 1981). Its velocity with respect to the Local Group is V = 3265 $\pm$15 km s^{-1}, which corresponds to a distance of 65 Mpc. With an isophotal radius R_{25} of 2'.4, its linear radius is 46 kpc when corrected for external extinction and viewing angle. The apparent magnitude, $B_T = 12.29$, leads to an absolute magnitude $M_B = -22.87$ after correction for internal and external extinction. A central velocity dispersion of 130 $\pm$29 km s^{-1} has been measured over the inner 6" by Whitmore (unpublished) from Kitt Peak HGVS data.

I show in Fig. 5 a copy of a print of NGC 3200 taken with the Las

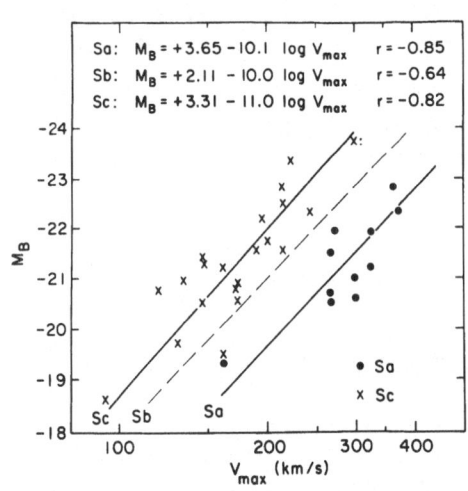

Fig. 3. Correlation of absolute magnitude with V(max), i.e., the Tully-Fisher relation, for program galaxies. Least squares mean lines for the Sa, Sb, and Sc samples are shown.

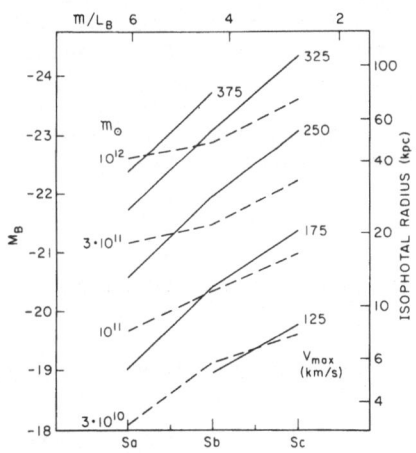

Fig. 4. Values of various parameters, as a function of Hubble type, H=50 km s^{-1}Mpc^{-1}. For H=100 km s^{-1}Mpc^{-1}, absolute magnitude is brighter by 1^m.5, radius decreases by 1/2, and $\mathcal{M}/L$ ratio increases x 2.

Fig. 5. (upper) Print of NGC 3200 taken by Sandage with the 100-in Las Campanas telescope. (mid) Spectrum at Hα taken with KPNO 4-m spectrograph plus image tube. (lower) Circular velocities in NGC 3200.

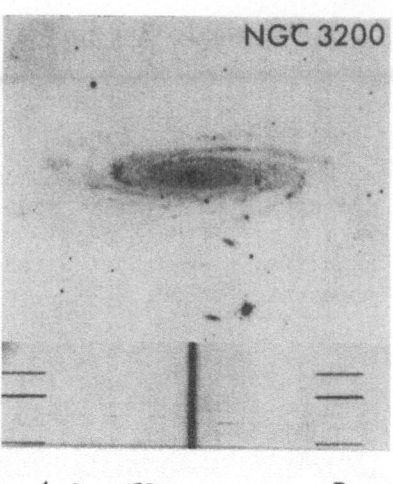

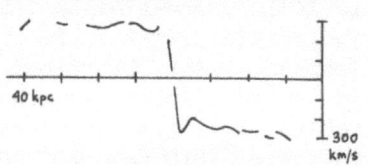

Campanas 100-in telescope, and kindly made available by A. Sandage, plus a copy of a spectrum and plot of rotational velocities. Well-defined spiral structure extends to 20 or 30 kpc, becoming patchy beyond. The nucleus exhibits a strong red stellar continuum, with [N II] emission stronger than Hα. These spectral features are characteristic of very high luminosity galaxies. The rotational velocities plotted in Fig. 6 have been smoothed to form the rotation curve for NGC 3200 shown as the last curve in Fig. 1.

There are several notable features of the rotational velocities in NGC 3200. Velocities from the two sides of the major axis are not identical, but are displaced about 25 km s^{-1} over the range of 5 to 17 kpc. At two locations (17 kpc NW; 29 kpc NW) velocity gradients amounting to about 40 km s^{-1} in 3 kpc regions are observed. Astronomers living at this radial distance in NGC 3200 will measure a local velocity which differs from the rotational velocity. In NGC 3200, the rotatioanl velocity at 9 kpc is 262 km s^{-1}; the maximum rotational velocity V(max)=282 km s^{-1} is reached only at the farthest measured radial distance, R_f=40 kpc (R_f/R_{25}=0.87).

For a spheroidal mass distribution, the mass within 9 kpc in NGC 3200 is $\mathcal{M}(R<10 \text{ kpc})$=1.5 x 10^{11} $\mathcal{M}_\odot$; $\mathcal{M}(R<40 \text{ kpc})$=7.5 x 10^{11} $\mathcal{M}_\odot$. Notable is the inner (R~2 kpc) velocity peak, a feature we observe only in galaxies of high luminosity. The combination of all of these characteristics makes the dynamical properties of NGC 3200, more than any other galaxy we have observed, resemble those of our Galaxy.

3. THE OVERALL DYNAMICAL CHARACTERISTICS OF OUR GALAXY

It is clear to all astronomers at this meeeting that we do not yet know the characteristics of our galaxy with an accuracy which will enable us to place it accurately among the galaxies we have observed. Although values of R(0)=8.5±1.5 kpc, V(local)=235±20 km s^{-1} seem realistic, we still are uncertain as to the value of the rotational velocity at R=8.5 kpc, and the maximum rotational velocity for the galaxy. Recent studies (Kulkarni, Blitz, and Heiles 1982) indicate that rotational velocity increases beyond the solar distance, as would be expected on the basis of the other rotation curves we have observed.

Hence it seems likely that V(max) ⩾ 250 km s⁻¹. This is an enormously high rotational velocity for an Sc galaxy, and suggests that our Galaxy is closer in type to a high luminosity Sb galaxy. Very likely, V(max) is not as high as that in NGC 3200, so the luminosity of our Galaxy is slightly smaller. However, as the type moves toward Sc, the luminosity must increase, for a fixed V(max).

There is increasing evidence that the effective diameter and mass of our Galaxy is large. Such evidence comes from the velocities of distant globular clusters (Hartwick and Sargent 1978; Frenk and White 1980), satellite galaxies (Einasto, Haud, and Kaasik 1977), and orbits of the Magellanic Clouds (Murai and Fujimoto 1980; Lin and Lynden-Bell 1982) when considered as test particles moving in the gravitational potential of our Galaxy. From the mean distances and velocities of these objects, the interior mass can be inferred, and is plotted on Fig. 6. Also shown is the mass of the Galaxy calculated for a simple spheroid, with the rotational velocity assumed constant at V=250 km s⁻¹ beyond the solar distance. It is clear that the satelite objects imply a large mass, which increases with radial distance. Moreover, a rotational velocity which remains high at enormous nuclear distances is consistent with these observations.

Finally, I point out that the calibration of rotational velocity with luminosity is dependent on the Hubble constant (Fig. 4). Thus, if we know the rotational properties, the isophotal radius, and the absolute magnitude of a galaxy, we can evaluate the Hubble constant by choosing that value for H which will cause the synthetic rotation curve (Fig. 2 for example) to match the known absolute magnitude. An attempt to do so for our Galaxy was unsuccessful due to the large uncertainty in the values. Therefore a second attempt was made to calibrate the system by using the rotation curve of M31 (Roberts, Whitehurst, and Cram 1978) and an adopted absolute magnitude. Adopting the external and internal extinction from Sandage and Tammann (1981), values of 57< H< 77 km s⁻¹ Mpc⁻¹ result. However, employing the extinction from de Vaucouleurs,

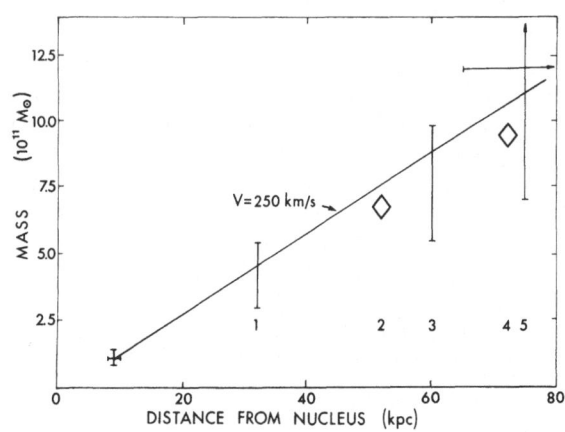

Fig. 6. Mass interior to R, as a function of R, for a galaxy with V = 250 km s⁻¹ for all R > 9 kpc. Also plotted are masses calulated by (1) Frenk and White 1980; (2) Murai and Fujimoto 1980; (3) Hartwick and Sargent 1978; (4) Lin and Lynden-Bell 1982; (5) Einasto et al. 1977.

de Vaucouleurs, and Corwin (1976), limits of $71 < H < 99$ km s^{-1} Mpc^{-1} are found. Thus the presently accepted range of extinction for M31 precludes determining the value of the Hubble constant to within a factor of two.

The dynamical properties of our Galaxy place it in the sequence of high luminosity Sb galaxies, perhaps toward Sbc. The motion of its satellite galaxies imply that its radius and mass are greater than 50 kpc and $7 \times 10^{11} \mathcal{M}_\odot$ respectively. Future work will permit us to place it even more accurately within the range observed for external galaxies. Ultimately, knowing the dynamical and luminosity properties of nearby galaxies will permit us to evaluate the local value for the Hubble constant. This is what the future is for.

ACKNOWLEDGEMENTS

I thank Dr. A. Sandage for the print of NGC 3200, and Drs. D. Burstein, R.J. Rubin, and B. Whitmore for comments on the manuscript. Figs. 1, 3, and 4 are copyright Astrophysical Journal and Astrophysical Journal Letters, and reproduced with permission.

REFERENCES

Aaronson, M., Huchra, J., and Mould, J. 1979, Astrophys. J. 229, p. 1.
Burstein, D., Rubin, V.C, Thonnard, N., and Ford, W. K. Jr. 1982, Astrophys. J. 253, p. 70.
de Vaucouleurs, G., de Vaucouleurs, A., and Corwin, H.G. 1976, Second Reference Catalogue of Bright Galaxies (Austin:University of Texas Press).
Einasto, J., Haud, U., and Kaasik, A. 1977, Tartu preprint A-5.
Frenk, C.S., and White, S.D.W. 1980, Mon. Not. Roy. Astron. Soc. 193, p. 308.
Hartwick, F.D.A., and Sargent, W.L.W. 1978, Astrophys. J. 221, p. 512.
Kulkarni, S.R., Blitz, L., and Heiles, C. 1982, preprint.
Lin, D.N.C., and Lynden-Bell, D. 1982, Mon. Not. Roy. Astron. Soc. 198, p. 707.
Murai, T., and Fujimoto, M. 1980, Publ. Ast. Soc. Japan 32, p. 581.
Roberts, M.S., Whitehurst, R.N., and Cram, T.R. 1978, in Structure and Properties of Nearby Galaxies, ed. E.M. Berkhuijsen and R. Wielebinski, (Dordrecht:Reidel), p. 169.
Rubin, V.C., Ford, W.K. Jr., and Thonnard, N. 1980, Astrophys.J. 238, p.471.
Rubin, V.C., Ford, W. K. Jr., Thonnard, N., and Burstein, D. 1982, Astrophys. J. 261 (in press).
Rubin, V.C., Thonnard, N., and Ford, W.K. Jr. 1982, Astrophys. J. (Letts), submitted.
Sandage, A., and Tammann, G.A. 1981, A Revised Shapley-Ames Catalog of Bright Galaxies, (Carnegie Institution of Washington: Washington, D.C.)
Thonnard, N., and Rubin, V.C. 1982, Carnegie Yrbk 80, p. 551.

AUTHOR INDEX

Bahcall, J.	209	Kerr, F.	91
Bash, F.	179	King, I.	53
Benson J.	109	Knapp, G.	233
van den Bergh, S.	43	Kulkarni, S.	97,109
van der Bij, M.	159		
Blitz, L.	97,143,151	Lin, C.	277
Bloemen, J.	31	Lindblad, P.	55
Bok, B.	1	Liszt, H.	135
Brand, J.	159	Lockman, F.	203
Brown, R.	197	Lynden-Bell, D.	349
Burton, W.	125		
Byl, J.	59	Manchester, R.	165
		Mark, J.	289
Caldwell, J.	249	Mayer-Hasselwander H.	223
Cheung, L.	183	McCutcheon, W.	165
Dickey, J.	109	Olander, N.	365
Dwek, E.	183	Ostriker, J.	249
		Ovenden, M.	59,67
Fich, M.	143,151		
Forbes, D.	217	Pryce, M.	67,73
Freeman, K.	325		
Frenk, C.	335,343	Richer, H.	365
Fuchs, B.	81	Roberts, W.	265
		Robinson, B.	165
Gezari, D.	183	Rubin, V.	379
van Gorkom, J.	109		
de Graauw, T.	159	Sandage, A.	315
		Sanders, D.	115
Haass, J.	283	Scoville, N.	367
Habing, H.	159	Seiden, P.	259
Hauser, M.	183	Shuter, W.	67,77
Heiles, C.	9,97,	Silverberg, R.	183
	105,109	Soneira, R.	209
		van de Stadt, H.	159
Innanen, K.	333	Stark, A.	127
Israel, F.	159	Stier, M.	183
Jenkins, E.	21	Troland, T.	9
Joslin, G.	365	Turek, G.	179
		Turner, B.	171
Kelsall, T.	183		

INDEX OF ASTRONOMICAL OBJECTS

SUBJECT INDEX

ASTROPHYSICS AND SPACE SCIENCE LIBRARY

Edited by

J. E. Blamont, R. L. F. Boyd, L. Goldberg, C. de Jager, Z. Kopal, G. H. Ludwig, R. Lüst,
B. M. McCormac, H. E. Newell, L. I. Sedov, Z. Švestka

1. C. de Jager (ed.), *The Solar Spectrum, Proceedings of the Symposium held at the University of Utrecht, 26–31 August, 1963.* 1965, XIV + 417 pp.
2. J. Orthner and H. Maseland (eds.), *Introduction to Solar Terrestrial Relations, Proceedings of the Summer School in Space Physics held in Alpbach, Austria, July 15–August 10, 1963 and Organized by the European Preparatory Commission for Space Research.* 1965, IX + 506 pp.
3. C. C. Chang and S. S. Huang (eds.), *Proceedings of the Plasma Space Science Symposium, held at the Catholic University of America, Washington, D.C., June 11–14, 1963.* 1965, IX + 377 pp.
4. Zdeněk Kopal, *An Introduction to the Study of the Moon.* 1966, XII + 464 pp.
5. B. M. McCormac (ed.), *Radiation Trapped in the Earth's Magnetic Field. Proceedings of the Advanced Study Institute, held at the Chr. Michelsen Institute, Bergen, Norway, August 16–September 3, 1965.* 1966, XII + 901 pp.
6. A. B. Underhill, *The Early Type Stars.* 1966, XII + 282 pp.
7. Jean Kovalevsky, *Introduction to Celestial Mechanics.* 1967, VIII + 427 pp.
8. Zdeněk Kopal and Constantine L. Goudas (eds.), *Measure of the Moon. Proceedings of the 2nd International Conference on Selenodesy and Lunar Topography, held in the University of Manchester, England, May 30–June 4, 1966.* 1967, XVIII + 479 pp.
9. J. G. Emming (ed.), *Electromagnetic Radiation in Space. Proceedings of the 3rd ESRO Summer School in Space Physics, held in Alpbach, Austria, from 19 July to 13 August, 1965.* 1968, VIII + 307 pp.
10. R. L. Carovillano, John F. McClay, and Henry R. Radoski (eds.), *Physics of the Magnetosphere, Based upon the Proceedings of the Conference held at Boston College, June 19–28, 1967.* 1968, X + 686 pp.
11. Syun-Ichi Akasofu, *Polar and Magnetospheric Substorms.* 1968, XVIII + 280 pp.
12. Peter M. Millman (ed.), *Meteorite Research. Proceedings of a Symposium on Meteorite Research, held in Vienna, Austria, 7–13 August, 1968.* 1969, XV + 941 pp.
13. Margherita Hack (ed.), *Mass Loss from Stars. Proceedings of the 2nd Trieste Colloquium on Astrophysics, 12–17 September, 1968.* 1969, XII + 345 pp.
14. N. D'Angelo (ed.), *Low-Frequency Waves and Irregularities in the Ionosphere. Proceedings of the 2nd ESRIN-ESLAB Symposium, held in Frascati, Italy, 23–27 September, 1968.* 1969, VII + 218 pp.
15. G. A. Partel (ed.), *Space Engineering. Proceedings of the 2nd International Conference on Space Engineering, held at the Fondazione Giorgio Cini, Isola di San Giorgio, Venice, Italy, May 7–10, 1969.* 1970, XI + 728 pp.
16. S. Fred Singer (ed.), *Manned Laboratories in Space. Second International Orbital Laboratory Symposium.* 1969, XIII + 133 pp.
17. B. M. McCormac (ed.), *Particles and Fields in the Magnetosphere. Symposium Organized by the Summer Advanced Study Institute, held at the University of California, Santa Barbara, Calif., August 4–15, 1969.* 1970, XI + 450 pp.
18. Jean-Claude Pecker, *Experimental Astronomy.* 1970, X + 105 pp.
19. V. Manno and D. E. Page (eds.), *Intercorrelated Satellite Observations related to Solar Events. Proceedings of the 3rd ESLAB/ESRIN Symposium held in Noordwijk, The Netherlands, September 16–19, 1969.* 1970, XVI + 627 pp.
20. L. Mansinha, D. E. Smylie, and A. E. Beck, *Earthquake Displacement Fields and the Rotation of the Earth, A NATO Advanced Study Institute Conference Organized by the Department of Geophysics, University of Western Ontario, London, Canada, June 22–28, 1969.* 1970, XI + 308 pp.
21. Jean-Claude Pecker, *Space Observatories.* 1970, XI + 120 pp.
22. L. N. Mavridis (ed.), *Structure and Evolution of the Galaxy. Proceedings of the NATO Advanced Study Institute, held in Athens, September 8–19, 1969.* 1971, VII + 312 pp.

23. A. Muller (ed.), *The Magellanic Clouds. A European Southern Observatory Presentation: Principal Prospects, Current Observational and Theoretical Approaches, and Prospects for Future Research, Based on the Symposium on the Magellanic Clouds, held in Santiago de Chile, March 1969, on the Occasion of the Dedication of the European Southern Observatory*. 1971, XII + 189 pp.

24. B. M. McCormac (ed.), *The Radiating Atmosphere. Proceedings of a Symposium Organized by the Summer Advanced Study Institute, held at Queen's University, Kingston, Ontario, August 3–14, 1970*. 1971, XI + 455 pp.

25. G. Fiocco (ed.), *Mesospheric Models and Related Experiments. Proceedings of the 4th ESRIN-ESLAB Symposium, held at Frascati, Italy, July 6–10, 1970*. 1971, VIII + 298 pp.

26. I. Atanasijević, *Selected Exercises in Galactic Astronomy*. 1971, XII + 144 pp.

27. C. J. Macris (ed.), *Physics of the Solar Corona. Proceedings of the NATO Advanced Study Institute on Physics of the Solar Corona, held at Cavouri-Vouliagmeni, Athens, Greece, 6–17 September 1970*. 1971, XII + 345 pp.

28. F. Delobeau, *The Environment of the Earth*. 1971, IX + 113 pp.

29. E. R. Dyer (general ed.), *Solar-Terrestrial Physics/1970. Proceedings of the International Symposium on Solar-Terrestrial Physics, held in Leningrad, U.S.S.R., 12–19 May 1970*. 1972, VIII + 938 pp.

30. V. Manno and J. Ring (eds.), *Infrared Detection Techniques for Space Research. Proceedings of the 5th ESLAB-ESRIN Symposium, held in Noordwijk, The Netherlands, June 8–11, 1971*. 1972, XII + 344 pp.

31. M. Lecar (ed.), *Gravitational N-Body Problem. Proceedings of IAU Colloquium No. 10, held in Cambridge, England, August 12–15, 1970*. 1972, XI + 441 pp.

32. B. M. McCormac (ed.), *Earth's Magnetospheric Processes. Proceedings of a Symposium Organized by the Summer Advanced Study Institute and Ninth ESRO Summer School, held in Cortina, Italy, August 30–September 10, 1971*. 1972, VIII + 417 pp.

33. Antonin Rükl, *Maps of Lunar Hemispheres*. 1972, V + 24 pp.

34. V. Kourganoff, *Introduction to the Physics of Stellar Interiors*. 1973, XI + 115 pp.

35. B. M. McCormac (ed.), *Physics and Chemistry of Upper Atmospheres. Proceedings of a Symposium Organized by the Summer Advanced Study Institute, held at the University of Orléans, France, July 31–August 11, 1972*. 1973, VIII + 389 pp.

36. J. D. Fernie (ed.), *Variable Stars in Globular Clusters and in Related Systems. Proceedings of the IAU Colloquium No. 21, held at the University of Toronto, Toronto, Canada, August 29–31, 1972*. 1973, IX + 234 pp.

37. R. J. L. Grard (ed.), *Photon and Particle Interaction with Surfaces in Space. Proceedings of the 6th ESLAB Symposium, held at Noordwijk, The Netherlands, 26–29 September, 1972*. 1973, XV + 577 pp.

38. Werner Israel (ed.), *Relativity, Astrophysics and Cosmology. Proceedings of the Summer School, held 14–26 August 1972, at the BANFF Centre, BANFF, Alberta, Canada*. 1973, IX + 323 pp.

39. B. D. Tapley and V. Szebehely (eds.), *Recent Advances in Dynamical Astronomy. Proceedings of the NATO Advanced Study Institute in Dynamical Astronomy, held in Cortina d'Ampezzo, Italy, August 9–12, 1972*. 1973, XIII + 468 pp.

40. A. G. W. Cameron (ed.), *Cosmochemistry. Proceedings of the Symposium on Cosmochemistry, held at the Smithsonian Astrophysical Observatory, Cambridge, Mass., August 14–16, 1972*. 1973, X + 173 pp.

41. M. Golay, *Introduction to Astronomical Photometry*. 1974, IX + 364 pp.

42. D. E. Page (ed.), *Correlated Interplanetary and Magnetospheric Observations. Proceedings of the 7th ESLAB Symposium, held at Saulgau, W. Germany, 22–25 May, 1973*. 1974, XIV + 662 pp.

43. Riccardo Giacconi and Herbert Gursky (eds.), *X-Ray Astronomy*. 1974, X + 450 pp.

44. B. M. McCormac (ed.), *Magnetospheric Physics. Proceedings of the Advanced Summer Institute, held in Sheffield, U.K., August 1973*. 1974, VII + 399 pp.

45. C. B. Cosmovici (ed.), *Supernovae and Supernova Remnants. Proceedings of the International Conference on Supernovae, held in Lecce, Italy, May 7–11, 1973*. 1974, XVII + 387 pp.

46. A. P. Mitra, *Ionospheric Effects of Solar Flares*. 1974, XI + 294 pp.

47. S.-I. Akasofu, *Physics of Magnetospheric Substorms*. 1977, XVIII + 599 pp.

48. H. Gursky and R. Ruffini (eds.), *Neutron Stars, Black Holes and Binary X-Ray Sources*. 1975, XII + 441 pp.

49. Z. Švestka and P. Simon (eds.), *Catalog of Solar Particle Events 1955–1969. Prepared under the Auspices of Working Group 2 of the Inter-Union Commission on Solar-Terrestrial Physics*. 1975, IX + 428 pp.

50. Zdenĕk Kopal and Robert W. Carder, *Mapping of the Moon*. 1974, VIII + 237 pp.

51. B. M. McCormac (ed.), *Atmospheres of Earth and the Planets. Proceedings of the Summer Advanced Study Institute, held at the University of Liège, Belgium, July 29–August 8, 1974*. 1975, VII + 454 pp.

52. V. Formisano (ed.), *The Magnetospheres of the Earth and Jupiter. Proceedings of the Neil Brice Memorial Symposium, held in Frascati, May 28–June 1, 1974*. 1975, XI + 485 pp.

53. R. Grant Athay, *The Solar Chromosphere and Corona: Quiet Sun*. 1976, XI + 504 pp.

54. C. de Jager and H. Nieuwenhuijzen (eds.), *Image Processing Techniques in Astronomy. Proceedings of a Conference, held in Utrecht on March 25–27, 1975*. 1976, XI + 418 pp.

55. N. C. Wickramasinghe and D. J. Morgan (eds.), *Solid State Astrophysics. Proceedings of a Symposium, held at the University College, Cardiff, Wales, 9–12 July, 1974*. 1976, XII + 314 pp.

56. John Meaburn, *Detection and Spectrometry of Faint Light*. 1976, IX + 270 pp.

57. K. Knott and B. Battrick (eds.), *The Scientific Satellite Programme during the International Magnetospheric Study. Proceedings of the 10th ESLAB Symposium, held at Vienna, Austria, 10–13 June 1975*. 1976, XV + 464 pp.

58. B. M. McCormac (ed.), *Magnetospheric Particles and Fields. Proceedings of the Summer Advanced Study School, held in Graz, Austria, August 4–15, 1975*. 1976, VII + 331 pp.

59. B. S. P. Shen and M. Merker (eds.), *Spallation Nuclear Reactions and Their Applications*. 1976, VIII + 235 pp.

60. Walter S. Fitch (ed.), *Multiple Periodic Variable Stars. Proceedings of the International Astronomical Union Colloquium No. 29, held at Budapest, Hungary, 1–5 September 1976*. 1976, XIV + 348 pp.

61. J. J. Burger, A. Pedersen, and B. Battrick (eds.), *Atmospheric Physics from Spacelab. Proceedings of the 11th ESLAB Symposium, Organized by the Space Science Department of the European Space Agency, held at Frascati, Italy, 11–14 May 1976*. 1976, XX + 409 pp.

62. J. Derral Mulholland (ed.), *Scientific Applications of Lunar Laser Ranging. Proceedings of a Symposium held in Austin, Tex., U.S.A., 8–10 June, 1976*. 1977, XVII + 302 pp.

63. Giovanni G. Fazio (ed.), *Infrared and Submillimeter Astronomy. Proceedings of a Symposium held in Philadelphia, Penn., U.S.A., 8–10 June, 1976*. 1977, X + 226 pp.

64. C. Jaschek and G. A. Wilkins (eds.), *Compilation, Critical Evaluation and Distribution of Stellar Data. Proceedings of the International Astronomical Union Colloquium No. 35, held at Strasbourg, France, 19–21 August, 1976*. 1977, XIV + 316 pp.

65. M. Friedjung (ed.), *Novae and Related Stars. Proceedings of an International Conference held by the Institut d'Astrophysique, Paris, France, 7–9 September, 1976*. 1977, XIV + 228 pp.

66. David N. Schramm (ed.), *Supernovae. Proceedings of a Special IAU-Session on Supernovae held in Grenoble, France, 1 September, 1976*. 1977, X + 192 pp.

67. Jean Audouze (ed.), *CNO Isotopes in Astrophysics. Proceedings of a Special IAU Session held in Grenoble, France, 30 August, 1976*. 1977, XIII + 195 pp.

68. Z. Kopal, *Dynamics of Close Binary Systems*. XIII + 510 pp.

69. A. Bruzek and C. J. Durrant (eds.), *Illustrated Glossary for Solar and Solar-Terrestrial Physics*. 1977, XVIII + 204 pp.

70. H. van Woerden (ed.), *Topics in Interstellar Matter*. 1977, VIII + 295 pp.

71. M. A. Shea, D. F. Smart, and T. S. Wu (eds.), *Study of Travelling Interplanetary Phenomena*. 1977, XII + 439 pp.

72. V. Szebehely (ed.), *Dynamics of Planets and Satellites and Theories of Their Motion. Proceedings of IAU Colloquium No. 41, held in Cambridge, England, 17–19 August 1976*. 1978, XII + 375 pp.

73. James R. Wertz (ed.), *Spacecraft Attitude Determination and Control*. 1978, XVI + 858 pp.

74. Peter J. Palmadesso and K. Papadopoulos (eds.), *Wave Instabilities in Space Plasmas. Proceedings of a Symposium Organized Within the XIX URSI General Assembly held in Helsinki, Finland, July 31–August 8, 1978.* 1979, VII + 309 pp.

75. Bengt E. Westerlund (ed.), *Stars and Star Systems. Proceedings of the Fourth European Regional Meeting in Astronomy held in Uppsala, Sweden, 7–12 August, 1978.* 1979, XVIII + 264 pp.

76. Cornelis van Schooneveld (ed.), *Image Formation from Coherence Functions in Astronomy. Proceedings of IAU Colloquium No. 49 on the Formation of Images from Spatial Coherence Functions in Astronomy, held at Groningen, The Netherlands, 10–12 August 1978.* 1979, XII + 338 pp.

77. Zdeněk Kopal, *Language of the Stars. A Discourse on the Theory of the Light Changes of Eclipsing Variables.* 1979, VIII + 280 pp.

78. S.-I. Akasofu (ed.), *Dynamics of the Magnetosphere. Proceedings of the A.G.U. Chapman Conference 'Magnetospheric Substorms and Related Plasma Processes' held at Los Alamos Scientific Laboratory, N.M., U.S.A., October 9–13, 1978.* 1980, XII + 658 pp.

79. Paul S. Wesson, *Gravity, Particles, and Astrophysics. A Review of Modern Theories of Gravity and G-variability, and their Relation to Elementary Particle Physics and Astrophysics.* 1980, VIII + 188 pp.

80. Peter A. Shaver (ed.), *Radio Recombination Lines. Proceedings of a Workshop held in Ottawa, Ontario, Canada, August 24–25, 1979.* 1980, X + 284 pp.

81. Pier Luigi Bernacca and Remo Ruffini (eds.), *Astrophysics from Spacelab.* 1980, XI + 664 pp.

82. Hannes Alfvén, *Cosmic Plasma,* 1981, X + 160 pp.

83. Michael D. Papagiannis (ed.), *Strategies for the Search for Life in the Universe,* 1980, XVI + 254 pp.

84. H. Kikuchi (ed.), *Relation between Laboratory and Space Plasmas,* 1981, XII + 386 pp.

85. Peter van der Kamp, *Stellar Paths,* 1981, XXII + 155 pp.

86. E. M. Gaposchkin and B. Kołaczek (eds.), *Reference Coordinate Systems for Earth Dynamics,* 1981, XIV + 396 pp.

87. R. Giacconi (ed.), *X-Ray Astronomy with the Einstein Satellite. Proceedings of the High Energy Astrophysics Division of the American Astronomical Society Meeting on X-Ray Astronomy held at the Harvard-Smithsonian Center for Astrophysics, Cambridge, Mass., U.S.A., January 28–30, 1980.* 1981, VII + 330 pp.

88. Icko Iben Jr. and Alvio Renzini (eds.), *Physical Processes in Red Giants. Proceedings of the Second Workshop, helt at the Ettore Majorana Centre for Scientific Culture, Advanced School of Agronomy, in Erice, Sicily, Italy, September 3–13, 1980.* 1981, XV + 488 pp.

89. C. Chiosi and R. Stalio (eds.), *Effect of Mass Loss on Stellar Evolution. IAU Colloquium No. 59 held in Miramare, Trieste, Italy, September 15–19, 1980.* 1981, XXII + 532 pp.

90. C. Goudis, *The Orion Complex. A Case Study of Interstellar Matter.* 1982, XIV + 306 pp.

91. F. D. Kahn (ed.), *Investigating the Universe. Papers Presented to Zdenek Kopal on the Occasion of his retirement, September 1981.* 1981, X + 458 pp.

92. C. M. Humphries (ed.), *Instrumentation for Astronomy with Large Optical Telescopes, Proceedings of IAU Colloquium No. 67.* 1981, XVII + 321 pp.

93. R. S. Roger and P. E. Dewdney (eds.), *Regions of Recent Star Formation, Proceedings of the Symposium on "Neutral Clouds Near HII Regions – Dynamics and Photochemistry", held in Penticton, B.C., June 24–26, 1981.* 1982, XVI + 496 pp.

94. O. Calame (ed.), *High-Precision Earth Rotation and Earth-Moon Dynamics. Lunar Distances and Related Observations.* 1982, XX + 354 pp.

95. M. Friedjung and R. Viotti (eds.), *The Nature of Symbiotic Stars,* 1982, XX + 310 pp.

96. W. Fricke and G. Teleki (eds.), *Sun and Planetary System,* 1982, XIV + 538 pp.

97. C. Jaschek and W. Heintz (eds.), *Automated Data Retrieval in Astronomy,* 1982, XX + 324 pp.

98. Z. Kopal and J. Rahe (eds.), *Binary and Multiple Stars as Tracers of Stellar Evolution,* 1982, XXX + 503 pp.

99. A. W. Wolfendale (ed.), *Progress in Cosmology,* 1982, VIII + 360 pp.

100. W. L. H. Shuter (ed.), *Kinematics, Dynamics and Structure of the Milky Way,* 1983, XII + 392 pp.